Development and Application of Optical Coherence Tomography (OCT)

Special Issue Editor
Michael Pircher

MDPI • Basel • Beijing • Wuhan • Barcelona • Belgrade

Special Issue Editor
Michael Pircher
Medical University Vienna
Austria

Editorial Office
MDPI AG
St. Alban-Anlage 66
Basel, Switzerland

This edition is a reprint of the Special Issue published online in the open access journal *Applied Sciences* (ISSN 2076-3417) from 2016–2017 (available at: http://www.mdpi.com/journal/applsci/special_issues/oct).

For citation purposes, cite each article independently as indicated on the article page online and as indicated below:

Lastname, F.M.; Lastname, F.M. Article title. *Journal Name.* **Year.** *Article number, page range.*

First Edition 2018

ISBN 978-3-03842-744-5 (Pbk)
ISBN 978-3-03842-743-8 (PDF)

Table of Contents

About the Special Issue Editor

Michael Pircher, Assoc. Prof, graduated from the Experimental Physics Department of the University of Technology of Graz, Austria in 2000. In 2003, he finished his PhD at the University of Technology of Vienna. From 2003 to 2006, he worked as a Postdoc at the Center for Biomedical Engineering and Physics of the Medical University of Vienna. From 2006 to 2012, he worked as an Assistant Professor at the Center for Medical Physics and Biomedical Engineering of the Medical University of Vienna where he gained his Habilitation in 2008. Currently, he is Associate Professor and faculty member of the same department. His scientific interests lie in retinal imaging, adaptive optics, new contrast techniques in optical coherence tomography, and coherence microscopy. He has published more than 100 research papers, 4 review papers and 4 book chapters in the field.

 *applied sciences*

Editorial

Special Feature Development and Application of Optical Coherence Tomography (OCT)

Michael Pircher *

Center for Medical Physics and Biomedical Engineering, Medical University of Vienna, Waehringerguertel 18-20, 1090 Vienna, Austria

Received: 8 October 2017; Accepted: 12 October 2017; Published: 13 October 2017

Introduction

To celebrate the 25th anniversary of the introduction of OCT, the special feature issue entitled "Development and Application of Optical Coherence Tomography (OCT)" had been initiated. OCT originated from low coherence interferometry [1] and was adapted for tomographic imaging in 1991 [2]. In OCT, broad bandwidth light is used in order to produce cross-sectional images of turbid and translucent samples with high axial resolution (in the order of a few μm). Thereby, the imaging speed of OCT can be as high as several millions of depth scans (A-scans) per second, which allows for volumetric investigations of dynamic processes [3]. Nowadays, OCT can be regarded as a well-established technique that offers various applications in different fields. Some of these applications and latest developments are covered in the special feature issue, consisting of three overview papers and nine research papers. These are outlined below in the order of their appearance in the feature issue.

The high imaging speed of OCT enables the contrasting of dynamic structures (e.g., blood vessels) from static tissue. Thereby, two or more cross-sectional OCT tomograms (B-scans) are recorded at the same sample location at high speed. Any motion introduced by flow will cause changes of the amplitude and phase between the recorded OCT B-scans. As a result, vessel maps of tissue are generated and the technique is referred to as OCT angiography (OCTA), mainly because of the similarity of the images to conventional angiographic imaging. However, in contrast to most angiographic techniques, OCTA is completely non-invasive and does not rely on the administration of contrast agents. J. Zhu et al. [4] present an overview on the various algorithms used for OCTA and provide a detailed analysis on the question of whether the technique can be extended from a purely contrasting method to a more quantitative assessment of the underlying flow.

By exploiting the polarization sensitive information carried by light, additional information on sample composition can be obtained. The corresponding technique is called polarization sensitive (PS-)OCT, which enables the contrasting of different tissue types based on their polarization sensitive properties. PS-OCT is not limited to the biomedical field, and can be very attractive for material sciences and non-destructive testing. Technological principles as well as the various applications of PS-OCT are summarized by B. Baumann [5].

An interesting and growing field of application for OCT is presented by H. Schneider et al. [6]. In their work, they provide an overview of OCT in dentistry and evaluate the performance of the technique for caries diagnosis and for the monitoring of composite restorations.

OCT-based image-guided surgery is another growing field of application. Thus, fast imaging processing will be essential for translating this technology to clinical routine. In order to address this issue, M. Zhou et al. [7] present new segmentation approaches for localizing surgical needles that can be inserted into the eye ball for microsurgical procedures.

The sintering process is an important step towards high quality fabrication of tooth prostheses. State of the art quality metrics of this process are either time-consuming or subjective.

C. Sinescu et al. [8] propose the use of OCT for a non-destructive investigation of temperature effects during this process.

For specific applications, the image contrast provided by OCT might not be sufficient. To increase the visibility of structures, contrast agents can be employed that are usually highly scattering. A very elegant method is the use of iron oxide particles in combination with magnetic-induced motion of the particles. The corresponding technique is referred to as magnetomotive-OCT (MM-OCT). In their paper, P. Cimalla et al. [9] present an improved method for MM-OCT and provide first imaging results in cells. Doppler OCT enables the non-invasive quantification of flow in tissue by measuring the axial velocity component. However, lateral motion may influence the corresponding quantitative data. With the aim of minimizing this influence, J. Walther et al. [10] extend the concept of resonant Doppler OCT by moving the OCT scanning unit lateral to the sample.

Protective coatings are very important for various objects. Corresponding thickness measurements are essential for determining the quality and degradation of coatings. The ability of OCT in comparison with state of the art methods for determining degradation effects are presented by M. Lenz et al. [11]. Coating thickness measurements for the automotive industry using a line field OCT approach are presented by S. Lawmann et al. [12].

As outlined above, dynamic processes can be investigated using OCT because of the fast imaging speeds provided by the technique. These dynamic processes may occur at different time scales and are investigated in two papers of the special issue. C. Schnabel et al. [13] use OCT to investigate the rather fast alveolar dynamics in an in vivo animal model. On the other hand, O. Thouvenin et al. [14] use full field OCT in combination with dynamic, mechanical, and molecular contrast with the aim of providing tissue- and disease-specific information on a microscopic level and improving the sensitivity/specificity of disease diagnosis. The depth range of spectral domain OCT instruments is generally limited and may not be sufficient for a specific application. For an extension of this depth range, T. Wu et al. [15] use two separate reference arms in combination with a phase modulation. The latter allows the reconstruction of the full complex signal and a doubling of the depth range, which is doubled again through the implementation of the second reference arm.

With all these contributions, this special feature issue underlines the very active and rapidly growing field of OCT. Meanwhile, various applications of the technology have been demonstrated and further improvements of instrumentation may be developed in the future. This may pave the way for additional applications of this powerful imaging technique.

Acknowledgments: I would like to take this opportunity to thank all the authors for their contribution to this special feature issue. In addition, I would like to thank the reviewers for their time and efforts in reviewing the manuscripts. Reviewing can very often be a thankless task, but I believe that it represents an essential part of the scientific publishing process as it significantly improves the overall quality of the papers that are published. Finally, I would like to acknowledge the hard work of the organizational and editorial team of Applied Sciences. Without their excellent support this special feature issue would not have been possible.

Conflicts of Interest: The author declares no conflict of interest.

References

1. Hitzenberger, C.K.; Drexler, W.; Leitgeb, R.A.; Findl, O.; Fercher, A.F. Key Developments for Partial Coherence Biometry and Optical Coherence Tomography in the Human Eye Made in Vienna. *Investig. Ophthalmol. Vis. Sci.* **2016**, *57*, 460–474. [CrossRef] [PubMed]
2. Huang, D.; Swanson, E.A.; Lin, C.P.; Schuman, J.S.; Stinson, W.G.; Chang, W.; Hee, M.R.; Flotte, T.; Gregory, K.; Puliafito, C.A.; et al. Optical Coherence Tomography. *Science* **1991**, *254*, 1178–1181. [CrossRef] [PubMed]
3. Wieser, W.; Draxinger, W.; Klein, T.; Karpf, S.; Pfeiffer, T.; Huber, R. High definition live 3D-OCT in vivo: design and evaluation of a 4D OCT engine with 1 GVoxel/s. *Biomed. Opt. Express* **2014**, *5*, 2963–2977. [CrossRef] [PubMed]
4. Zhu, J.; Merkle, C.W.; Bernucci, M.T.; Chong, S.P.; Srinivasan, V.J. Can OCT Angiography Be Made a Quantitative Blood Measurement Tool? *Appl. Sci.* **2017**, *7*, 687. [CrossRef]

5. Baumann, B. Polarization Sensitive Optical Coherence Tomography: A Review of Technology and Applications. *Appl. Sci.* **2017**, *7*, 474. [CrossRef]

6. Schneider, H.; Park, K.J.; Hafer, M.; Ruger, C.; Schmalz, G.; Krause, F.; Schmidt, J.; Ziebolz, D.; Haak, R. Dental Applications of Optical Coherence Tomography (OCT) in Cariology. *Appl. Sci.* **2017**, *7*, 472. [CrossRef]

7. Zhou, M.; Roodaki, H.; Eslami, A.; Chen, G.; Huang, K.; Maier, M.; Lohmann, C.P.; Knoll, A.; Nasseri, M.A. Needle Segmentation in Volumetric Optical Coherence Tomography Images for OphthalmicMicrosurgery. *Appl. Sci.* **2017**, *7*, 748. [CrossRef]

8. Sinescu, C.; Bradu, A.; Duma, V.F.; Topala, F.; Negrutiu, M.; Podoleanu, A.G. Effects of Temperature Variations during Sintering of Metal Ceramic Tooth Prostheses Investigated Non-Destructively with Optical Coherence Tomography. *Appl. Sci.* **2017**, *7*, 552. [CrossRef]

9. Cimalla, P.; Walther, J.; Mueller, C.; Almedawar, S.; Rellinghaus, B.; Wittig, D.; Ader, M.; Karl, M.O.; Funk, R.H.W.; Brand, M.; Koch, E. Improved Imaging of Magnetically Labeled Cells Using Rotational Magnetomotive Optical Coherence Tomography. *Appl. Sci.* **2017**, *7*, 444. [CrossRef]

10. Walther, J.; Koch, E. Flow Measurement by Lateral Resonant Doppler Optical Coherence Tomography in the Spectral Domain. *Appl. Sci.* **2017**, *7*, 382. [CrossRef]

11. Lenz, M.; Mazzon, C.; Dillmann, C.; Gerhardt, N.C.; Welp, H.; Prange, M.; Hofmann, M.R. Spectral Domain Optical Coherence Tomography for Non-Destructive Testing of Protection Coatings on Metal Substrates. *Appl. Sci.* **2017**, *7*, 364. [CrossRef]

12. Lawman, S.; Williams, B.M.; Zhang, J.; Shen, Y.C.; Zheng, Y. Scan-Less Line Field Optical Coherence Tomography, with Automatic Image Segmentation, as a Measurement Tool for Automotive Coatings. *Appl. Sci.* **2017**, *7*, 351. [CrossRef]

13. Schnabel, C.; Gaertner, M.; Koch, E. Optical Coherence Tomography (OCT) for Time-Resolved Imaging of Alveolar Dynamics in Mechanically Ventilated Rats. *Appl. Sci.* **2017**, *7*, 287. [CrossRef]

14. Thouvenin, O.; Apelian, C.; Nahas, A.; Fink, M.; Boccara, C. Full-Field Optical Coherence Tomography as a Diagnosis Tool: Recent Progress with Multimodal Imaging. *Appl. Sci.* **2017**, *7*, 236. [CrossRef]

15. Wu, T.; Wang, Q.Q.; Liu, Y.W.; Wang, J.M.; He, C.J.; Gu, X.R. Extending the Effective Ranging Depth of Spectral Domain Optical Coherence Tomography by Spatial Frequency Domain Multiplexing. *Appl. Sci.* **2016**, *6*, 360. [CrossRef]

 *applied*
sciences

Review

Polarization Sensitive Optical Coherence Tomography: A Review of Technology and Applications

Bernhard Baumann

Medical University of Vienna, Center for Medical Physics and Biomedical Engineering,
Waehringer Guertel 18-20, 4L, 1090 Vienna, Austria; bernhard.baumann@meduniwien.ac.at;
Tel.: +43-1-40400-73727

Academic Editor: Totaro Imasaka
Received: 10 February 2017; Accepted: 25 April 2017; Published: 4 May 2017

Abstract: Polarization sensitive optical coherence tomography (PS-OCT) is an imaging technique based on light scattering. PS-OCT performs rapid two- and three-dimensional imaging of transparent and translucent samples with micrometer scale resolution. PS-OCT provides image contrast based on the polarization state of backscattered light and has been applied in many biomedical fields as well as in non-medical fields. Thereby, the polarimetric approach enabled imaging with enhanced contrast compared to standard OCT and the quantitative assessment of sample polarization properties. In this article, the basic methodological principles, the state of the art of PS-OCT technologies, and important applications of the technique are reviewed in a concise yet comprehensive way.

Keywords: optical coherence tomography; polarization sensitive devices; biomedical imaging; birefringence; scattering; depolarization

1. Introduction

Optical coherence tomography (OCT) is an imaging modality providing 2D and 3D images with micrometer scale resolution [1–3]. Often considered an optical analog of ultrasound imaging, OCT detects light backscattered from sample structures. However, since the speed of light is much greater than that of sound, subtle differences in time delays corresponding to optical path lengths from different scatter locations within the sample cannot easily be measured in a direct way. For instance, the time delay corresponding to an optical path length of 10 μm is only on the order of ~30 fs. In order to assess such short delay times, OCT employs the interference of low coherent light [4,5]. Low-coherent light sources span a broad wavelength range, usually covering several tens of nanometers when used for OCT. Before the light interacts with the sample, it is split into two. One portion is directed onto the sample, while the other portion—the so-called reference beam—travels a defined path length before being recombined and interfered with the light beam scattered by the sample. Now, since low-coherent light consists of a continuum of wavelengths, the interference spectrum will be subject to modulations which depend on the path length difference between the sample beam and the reference beam. From the interference signal in the time or spectral domain, the axial position of scattering structures within the sample can be reconstructed [4,5]. The respective depth profile (backscatter intensity vs. depth) is called an axial scan (A-scan) and forms the basic unit of OCT images. By scanning the beam laterally across the sample, two- and three-dimensional images can be assembled from the acquired A-scans. OCT is a rapid imaging method providing 3D data comprising up to several millions of axial scans within few seconds [6]. The axial resolution of OCT is usually in the order of a few micrometers and, despite the high imaging speeds, the interferometric approach enables detection sensitivities of reflected light signals as low as 10^{-10} of the input.

During the past 25 years since its invention, OCT experienced a multitude of technological advances leading to higher imaging speeds, improved resolution, and novel contrast mechanisms. One so-called functional extension of OCT is polarization sensitive (PS) OCT. PS-OCT adds polarization contrast to the technique. While the standard OCT is based solely on the intensity of light backscattered or reflected by the sample, PS-OCT also detects its polarization state. Since the polarization state can be measured for every pixel in a depth scan, PS-OCT can enhance the image contrast and also enables quantitative measurements of a sample's polarization properties. These properties are often linked to the micro- or even ultrastructure of the sample which themselves are below the optical resolution limit of OCT [7]. Hence, the detection of changes of polarization properties—be they due to disordered microstructure in pathological tissue or due to stress and strain in a technical sample—may provide access to quantities and markers that are of interest for a broad variety of applications.

PS-OCT has been applied in many biomedical as well as non-medical fields. Figure 1 shows results of a search for scientific publications on PS-OCT using the free literature search engine PubMed (https://www.ncbi.nlm.nih.gov/pubmed/). In Figure 1a, the number of publications per year was plotted beginning with the first article on PS-OCT published in 1997 [8]. A rising trend of published documents per year can be observed, peaking with more than 30 publications per year for the last three years. The set of 360 publications was classified into seven medical fields and one non-medical field. The latter included articles primarily focusing on technological aspects or on measurements of technical (i.e., non-biological) samples. Figure 1b shows a pie chart with the break-up into the eight fields. The most prominent field was ophthalmology including roughly a third of all publications, followed by the 'technical' publications, reports on dental applications, PS-OCT in bones, cartilage, muscles and tendons, and skin imaging. Further fields of application were imaging of cardiac and vascular tissue, of cancerous tissue, and of neural tissue. The evolution of the eight groups over the past 20 years is shown in Figure 1c. Here, an increasing number of research papers on PS-OCT in the eye during the last decade can be observed. The relative share of publications grouped in the respective fields is shown in Figure 1d.

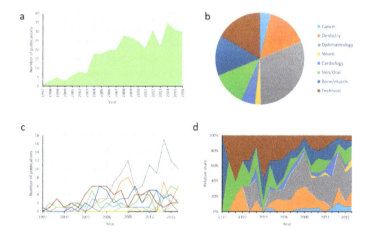

Figure 1. PS-OCT in the scientific literature. Results from a literature search for "polarization sensitive optical coherence tomography" and "polarization sensitive optical coherence tomography" on PubMed (https://www.ncbi.nlm.nih.gov/pubmed/, accessed on 5 January 2017). (**a**) Number of documents published per year since 1997; (**b**) Break-up of the publications into seven medical and one technical area. The latter includes PS-OCT measurements of technical samples and phantoms; (**c**) Absolute numbers of publications per year are shown for the eight fields of application; (**d**) Time course of the relative share of the eight subgroups. The color legend in (**b**) also applies to (**c**) and (**d**).

This review paper strives to provide an overview of the state of the art in PS-OCT. While other review papers provided a more in-depth discussion of the physical principles of the technique [9,10], this article is particularly focusing on applications of PS-OCT. In the following, we will first introduce basic concepts for describing polarization of light as well as commonly used technical approaches for realizing PS-OCT. Then, using the eight categories mentioned above, we will discuss important applications of PS-OCT in the biomedical field and for non-biomedical use.

2. Principles of Light Polarization and PS-OCT

2.1. Polarization of Light

Polarization of light describes the geometrical orientation of the oscillations of electromagnetic waves. When considering a light beam as a transverse wave propagating in z-direction, its electric field vector and magnetic field will not only be perpendicular to each other but also be perpendicular to z. At any spatial coordinate and time point, such a wave can be characterized by the complex-valued field components $e_{x,y}$ describing the oscillations in x- and y-direction, respectively. For a monochromatic plane wave travelling in z-direction, these two complex-valued components form the so-called Jones vector [11]. Depending on the amplitudes of e_x and e_y and on their respective phase delay δ, different states of polarization will be observed, as illustrated in Figure 2a. A Jones vector having a relative delay of 0° or 180° (i.e., half a wave) describes a linear polarization state. In case this linear state oscillates only in x-direction or only in y-direction, it is termed a horizontal or vertical linear state, respectively. When the x- and y-amplitudes are equal, the state is linear with an orientation of +45° for δ = 0° and linear with an orientation of −45° for δ = 180°. If the amplitudes in x- and y-direction are equal but δ is a quarter of a wave (i.e., δ = ±90°), the Jones vector will describe a right- or left-hand circular polarization state. In the general case of arbitrary δ and amplitudes, the wave will be in an elliptical polarization state.

As an alternative to describing polarization states by Jones vectors, a three-dimensional space spanned by horizontal/vertical linear state, +45°/−45° linear state, and right-/left-hand circular state can be used (Figure 2b). Four-component, real-valued vectors $[I\ Q\ U\ V]^T$, so-called Stokes vectors, describe polarization states in this space [12]. Here, I corresponds to the intensity of light, and Q, U, and V are the components along the three above-mentioned axes. Unlike Jones vectors, Stokes vectors enable the characterization of light depolarization. In case of fully polarized light, I corresponds to the length of the vector [Q U V], i.e., $I = \sqrt{Q^2 + U^2 + V^2}$. Then, the light's degree of polarization $DOP = \sqrt{Q^2 + U^2 + V^2}/I$ equals unity. In case of depolarization, DOP is less than unity, and equals zero for completely depolarized light. The unit sphere in Figure 2b is called a Poincaré sphere.

In order to describe the interaction of light with an optical element (or a sample investigated by a PS-OCT system), an operator acts on the polarization vector of the interrogating light beam. In Jones calculus, this operator is a complex-valued 2 × 2 matrix called Jones matrix (J) [11]. The evanescent light beam is represented by $e_{out} = J\ e_{in}$, where e_{in} is the input Jones vector (Figure 2c). If the light beam traverses several optical elements (or sample structures), the resulting Jones vector can be calculated by multiplying a cascade of Jones matrices to the input vector, $J_N J_{N-1} \cdots J_2 J_1 e_{in}$ where J_1 through J_N represent the polarization properties of N elements (or sample layers). Analogously, real-valued 4 × 4 matrices—so-called Müller matrices (M)—are used to describe the interaction of Stokes vectors with optical elements by the Stokes–Müller formalism: $S_{out} = M S_{in}$ (see Figure 2d) [12].

Different approaches of PS-OCT enable the pixelwise measurement of Jones vectors, Jones matrices, Stokes vectors, or Müller matrices. Since the display and interpretation of these multidimensional quantities is often not straight-forward (or even impossible), PS-OCT imagery usually displays physical polarization measures directly related to relevant sample properties.

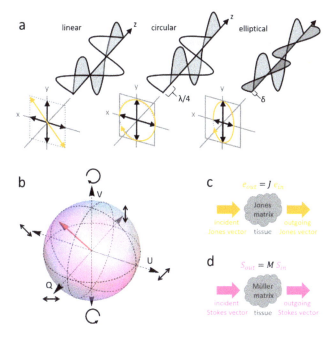

Figure 2. Polarization of light. (**a**) Jones vectors. Depending on the amplitudes $A_{x,y}$ and the relative phase retardation δ of the transverse light wave components oscillating in x and z direction, different polarization states such as linear for $\delta = 0°$ and $\delta = 180°$, circular for $\delta = \pm45°$ and equal amplitudes $A_x = A_y$, or elliptical for all other combinations can be described. The vector $e = [A_x, A_y \exp(i\delta)]^T$ is called Jones vector; (**b**) Stokes–Müller representation of polarization states. Vectors in the 3D space spanned by horizontal/vertical linear state, +45°/−45° linear state, and right-/left-hand circular state are called Stokes vectors. Stokes vectors (red) describe any elliptical polarization state and their length scales with intensity of polarized light. The unit sphere is called a Poincaré sphere; (**c**) The interaction of a Jones vector with a sample can be described by multiplication with a Jones matrix describing the tissue properties; (**d**) Likewise, the interaction of a Stokes vector with a sample can be calculated by multiplying with a so-called Müller matrix describing the sample polarization properties.

2.2. Polarization Effects

The polarization state of a light beam can be affected by interaction with optical components or sample structures [7]. In the following, four polarization effects will be described which are relevant for PS-OCT imaging.

- Preserved polarization. Many optical components and materials do not (or only negligibly) change the polarization state of light traversing them, i.e., $J = 1$ and $M = 1$. Their interaction can be described by $e_{out} = Je_{in} = e_{in}$ and $S_{out} = MS_{in} = S_{in}$.
- Birefringence. In birefringent media, differently oriented polarization states experience different speeds of light. When these basis polarizations are orthogonal linear polarizations (as in Figure 2a along the x- and y-axes), the effect is referred to as linear birefringence. Birefringence may also be circular or elliptical for respective different bases (eigenvectors); in this review, however, we restrict our discussion to linear birefringence. Linear birefringence occurs for instance in retarders such as wave plates, in crystals, and in many tissues with an oriented (e.g., fibrous) microstructure. Birefringence—i.e., the difference Δn of the refractive indices along the two axes–produces a phase retardation δ, which is proportional to the length L of the retarder, $\delta = \Delta n \cdot L$. The retardation δ of a

quarter wave plate (QWP) for example amounts to 90° ($\pi/2$ rad). If aligned with its slow and fast axes at 45° with respect to the x- and y-axes, the QWP would render a horizontal or vertical linear state into a circular polarization state (Figure 2a, center). The Jones matrix of a retarder with a retardation δ and aligned with an orientation $\vartheta = 0$ is represented by

$$J_{ret}(\delta, \vartheta = 0) = \begin{bmatrix} e^{i\delta/2} & 0 \\ 0 & e^{-i\delta/2} \end{bmatrix} \tag{1}$$

If an optical element such as a retarder is rotated by an angle ϑ, its Jones matrix $J = J(\vartheta = 0)$ is transformed into $T(\vartheta) \, J \, T(-\vartheta)$ where $T(\vartheta)$ is a rotation matrix. In Stokes-Müller formalism, the propagation of light through birefringent tissue is represented by a circular rotation of the Stokes vector tip on the Poincaré sphere. PS-OCT approaches based on Jones calculus or Stokes-Müller formalism enable depth-resolved measurements of phase retardation and of birefringent axis orientation [8,13–16].

- Diattenuation. Diattenuation (or dichroism) refers to a polarization dependent attenuation in an optical medium. When the axes of a diattenuating optical element or structure are aligned with the x- and y-direction, its Jones matrix is represented by

$$J_{diatt}(p_1, p_2, \vartheta = 0) = \begin{bmatrix} p_1 & 0 \\ 0 & p_2 \end{bmatrix} \tag{2}$$

where p_1 and p_2 correspond to the respective signal attenuation $p_{1,2} = \exp\left(-\mu_{a_{1,2}}L\right)$ with attenuation coefficients μ_a and the length L. Similar to the linear birefringence, the Jones matrix for a diattenuating element oriented at ϑ can be computed by sandwiching Equation (2) by rotation matrices. An extreme case of a diattenuating element with $p_1 = 1$ and $p_2 = 0$ is a linear polarizer which only transmits light along axis 1. It should be noted that diattenuation in biological tissue is usually very weak and thus is often assumed negligible for PS-OCT [16,17], although attempts have been made to quantify diattenuation using PS-OCT [15,17,18].

- Depolarization. Depolarization or polarization scrambling refers to a more or less random change of the incident polarization state at spatially adjacent sample locations. Using Stokes–Müller polarimetry, depolarization can be described by the DOP discussed in Section 2.1. However, owing to the coherent detection in OCT, DOP will always equal unity in any pixel of a PS-OCT dataset [9]. Therefore, in order to analyze polarization scrambling using PS-OCT, the randomization of polarization states among neighboring speckles is investigated [19,20]. The Stokes vectors of adjacent speckles will be more or less parallel in polarization preserving or weakly birefringent media, while they will point in different directions in depolarizing media (Figure 3a). For the purpose of depolarization assessment, the average Stokes vector can be calculated within a small kernel including several speckles (Figure 3b). Typical kernel sizes for this calculation are on the order of ~100 pixels spanning 2–3 times the axial resolution in depth and 2–3 times the transverse resolution laterally [20]. Note also that pixels with low reflectivity (i.e., less than several decibels above the noise floor) are usually excluded from the analysis. The length of the average normalized Stokes vector is referred to as the degree of polarization uniformity (DOPU) [20]:

$$DOPU = \sqrt{\overline{Q}^2 + \overline{U}^2 + \overline{V}^2}/\overline{I} \tag{3}$$

where the overbar indicates the ensemble average and $\overline{I}$ denotes the average Stokes vector length for normalization. As shown in Figure 3b, DOPU or the average Stokes vector length will be close to unity in polarization preserving tissue, while it will be lower in the case of polarization scrambling where the orientation of the Stokes vectors is more diverse. In DOPU images, DOPU values at every spatial coordinate are color-coded. Recently, advanced depolarization measures have been developed including DOPU with noise floor normalization [21], spectral DOPU [22], the depolarization index independent of the incident polarization state [23], and the differential depolarization index providing a larger dynamic range for depolarization mapping [24]. Different mechanisms can cause polarization scrambling, and the actual cause of depolarization observed in DOPU images has not always been completely clarified. In biological tissues, depolarization can be caused by multiple scattering or by scattering at non-spherical particles such as melanin granules [25]. Since the strength of depolarization is proportional to the concentration of polarization scrambling scatterers, depolarization measures such as DOPU may for instance enable a quantitative assessment of the melanin concentration in ocular tissues [26].

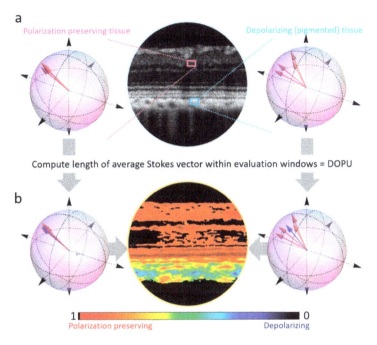

Figure 3. Depolarization measurement by PS-OCT. (**a**) Stokes vectors are plotted in Poincaré spheres for polarization preserving tissue (left, pink window) and for depolarizing tissue (right, blue window). Different Stokes vectors (red) correspond to the polarization states in adjacent image locations. In order to assess polarization scrambling within the evaluation kernels represented by the pink and blue rectangles overlaid on the OCT reflectivity image (center) of a rodent retina, the length of the average Stokes vectors is computed; (**b**) The length of the average Stokes vector (shown in blue in the Poincaré spheres) is decreased in case of depolarization. The DOPU image (center) assigns a color to the DOPU value of each pixel. The color map plots values from DOPU = 0 (blue, completely depolarized) to DOPU = 1 (red, uniform polarization). Pixels with low reflectivity (typically up to several decibels above the noise level) are masked in black. Note that the resolution in the DOPU image is slightly reduced by convolution with the evaluation kernel. In the DOPU image, most retinal structures appear to be polarization preserving, while pigmented structures such as the retinal pigment epithelium and the choroid scramble the polarization.

In general, polarization effects may be subject to dispersion, that is, their strength depends on wavelength. Since OCT is based on broadband light covering a wide range of wavelengths, efforts have been made to mitigate effects such as polarization mode dispersion in PS-OCT systems based on optical fibers [18,27–30].

2.3. Brief Basics of OCT

OCT is based on low coherence interferometry, i.e., the interference of broad band light [1,4]. A multitude of different interferometer designs have been used for OCT. A sketch of a basic Michelson interferometer is shown in Figure 4a. Here, light from a low coherent light source such as a superluminescent diode or a broadband laser is split into one beam that is directed on the sample and another beam that serves as a reference. After the beam splitter, the beam in the sample arm is directed onto the sample, whereas the reference beam is reflected by a mirror. Light backscattered and reflected by the sample e_S and light reflected by the reference mirror e_R is recombined at the beam splitter. The sample and reference beam interfere and their interference signal is detected at the interferometer exit. The interference signal in the time domain can be described by

$$I(z) = I_R + I_S + 2\sqrt{I_R I_S}|\gamma(z - z_0)|\cos[2k_0(z - z_0)]. \tag{4}$$

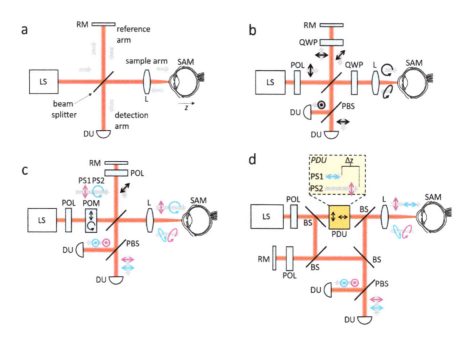

Figure 4. Basic schemes for PS-OCT. (**a**) Basic scheme of OCT with a Michelson interferometer. (**b**) PS-OCT with a single circular input state; (**c**) PS-OCT based on probing with several input states; (**d**) PS-OCT based on a Mach–Zehnder interferometer with two input states generated by a polarization delay unit. See text for descriptions of different approaches. The grey arrows indicate the directions of light beams. Black pictograms indicate linear, circular, and elliptical polarization states. LS—light source, POL—polarizer, QWP—quarter wave plate, RM—reference mirror, L—lens, SAM—sample, PBS—polarizing beam splitter, DU—detection unit, POM—polarization modulator, PS1/PS2—polarization state 1/2, BS—beam splitter, Δz—path delay between polarization states, PDU—polarization delay unit.

Here, $I_R \sim |e_R|^2$ is proportional to the intensity of the reference beam and $I_S \sim |e_S|^2$ is proportional to the intensity of light backscattered or reflected by the sample. The third term contains the interference information. Its first component $\sqrt{I_R I_S}$ indicates that the strength of the interference signal will scale with both the sample and the reference amplitude. The high sensitivity of OCT is based on the fact that even weak light scatter signals I_S from the sample can be amplified by a strong reference signal I_R. The second component $|\gamma(z - z_0)|$ contains the complex degree of coherence $\gamma(z)$ which is inversely related to the spectral bandwidth of the light source (via the Fourier transform of the spectral density). For broad bandwidth sources common in OCT, $|\gamma(z)|$ will only be greater than zero for a shallow depth around every scattering sample interface along the propagation direction z. Hence, the high resolution of OCT is contained in this term. The third component $\cos[2k_0(z - z_0)]$ describes a sinusoidal modulation of the interference signal along z. Here $k_0 = 2\pi/\lambda_0$ is the central wavenumber of the spectrum. In order to acquire depth scans in time domain OCT, the signal intensity $I(z)$ is recorded at the interferometer exit while the reference mirror is axially translated in beam direction z.

Most modern OCT systems rely on frequency (or Fourier) domain detection of the interference signal [5,31]. In such Fourier domain OCT systems, the reference mirror position is fixed and the interference spectrum is acquired. This can be achieved by using a spectrometer at the interferometer exit which disperses the interference signal into its spectral intensity components. Alternatively, a broadband wavelength-swept light source can be used to rapidly tune the spectrum with a narrow instantaneous bandwidth. In this case, the interference spectrum is recorded as a function of time by a detector at the interferometer exit. For each interface, the acquired spectral interference signal

$$S(k, \Delta z) = S_R(k) + S_S(k) + 2\sqrt{S_R(k)S_S(k)} \cos[2\Delta z \, k]. \tag{5}$$

contains three terms, similar to Equation (4). The first two terms contain the spectral densities returning from the reference arm (R) and the sample arm (S). Via $\sqrt{S_R(k)S_S(k)}$, the last term again is proportional to the spectral densities of the reference beam and the sample beam. Note that the last term is subject to a modulation $\cos[2\Delta z \, k]$ across wavenumber k, whose modulation frequency is proportional to the path length difference between the light path to sample interface and the light path to the reference mirror, $\Delta z = z_S - z_R$. The factor 2 accounts for the double pass through the interferometer arms. Since every path length difference Δz is encoded by a different spectral modulation frequency, the interference signals from multiple depth locations can be recorded simultaneously. A frequency analysis using the Fourier transform then provides the axial depth scan similar to Equation (4).

2.4. Technical Approaches to PS-OCT

During the past 25 years, a great variety of PS-OCT layouts has been devised. PS-OCT schemes differ in terms of optical technology (fiber optics vs. bulk optics), number of input states, number of detected variables, and reconstruction algorithm. The use of free-space beams in bulk optics permits defined polarization states at any location within the interferometer. Fiber optics provide easier system alignment, but the polarization of light will in general be influenced by birefringence and polarization mode dispersion in optical fibers. PS-OCT has been performed with as little as one input state and one detected intensity signal. Such settings correspond to regular OCT, however with altered reference polarization for cross-polarization imaging [32,33] or for imaging with variable reference polarization [34]. In contrast, the most comprehensive PS-OCT approaches detected up to 16 elements of the sample's Müller matrix—in every single image pixel [35–38]. In the following, we are going to describe two major categories of PS-OCT schemes: PS-OCT with a single circular input stage and PS-OCT based on sample illumination by multiple polarization states.

2.4.1. PS-OCT with a Single Circular Input State

PS-OCT with a single circular input state relies on a polarization sensitive low coherence interferometer design devised by Hee et al. in 1992 [13]. The basic scheme using a Michelson interferometer is shown in Figure 4b. Light from a low coherent light source is linearly polarized before being split up into a reference arm (top) and a sample arm (right). In the sample arm, the beam passes a QWP oriented at 45° which renders the original linear polarization into a circular polarization state and then illuminates the sample. Sample illumination by circular light offers sensitivity to any transverse orientation of birefringent media. If linearly polarized light were used for sample illumination, the sample's fast or slow birefringent axis could align with the interrogating linear polarization such that no birefringence would be observed. In the case of circular sample illumination shown in Figure 4b, a birefringent sample will in general produce an elliptical state. The reflected or backscattered light will transmit the QWP again and interfere with the reference beam at the beam splitter. Due to double passing a QWP oriented at 22.5° in the reference arm, the reference beam is a linearly polarized light beam oscillating at 45° which provides equal intensity components in the horizontal and vertical orientation, respectively. At the interferometer exit, the OCT light beam is split up into its horizontal (H) and vertical (V) component, which are detected by separate detection units. By the respective amplitudes $A_{H,V}$ and the relative phase difference $\Delta\Phi$, Jones vectors are detected for every image pixel. These Jones vectors enable the calculation of the sample's birefringent properties, namely of phase retardation δ [13,39] and fast birefringent axis orientation ϑ [14] as well as sample reflectivity R:

$$R \propto A_H^2 + A_V^2 \tag{6}$$

$$\delta = arctan\left(\frac{A_V}{A_H}\right) \tag{7}$$

$$\vartheta = \frac{\pi - \Delta\Phi}{2} \tag{8}$$

Since $\delta = \Delta n \cdot L$ accumulates as a function of the light path L travelled in a birefringent material, phase retardation measurements are cumulative. However, the measurement of δ is restricted to 0–90° due to the arctangent, which leads to cumulative retardation images with a banded structure caused by increasing and artificially decreasing δ in strongly birefringent samples (cf. Figures 9 and 10). From the detected amplitudes $A_{H,V}$ and the relative phase difference $\Delta\Phi$, the Stokes vector elements can also be calculated for every image pixel [19]. These may then serve as the input for depolarization images, for instance based on DOPU [20].

The beauty of the above scheme lies in its simplicity. Most implementations were done using free-space optics [14,40–44], however fiber optic prototypes have also been reported based on polarization maintaining (PM) fiber optics [45–52] and regular single mode fibers [53–55].

2.4.2. PS-OCT Based on Multiple Input States

PS-OCT systems using multiple polarization states as an input may provide access to additional polarization quantities. The scheme described in Section 2.4.1 is based on a single circular input state and relies on the assumptions that the sample is not diattenuating (which is a valid assumption for most biological tissues [17,56]) and that the axis orientation of the birefringent structure does not change along depth [43,44]. A method to overcome the latter limitation for retinal PS-OCT has been developed to remove the impact of corneal birefringence on birefringence measurements in the back of the eye [57]. Nevertheless, for applications such as PS-OCT in samples with strongly varying birefringent fiber orientations or for many approaches based on single-mode fiber optics, implementations based on multiple polarization states can provide access to Stokes vector quantification, Jones matrix characterization, and Müller matrix measurements [15–17,35–38,58–63].

In order to provide measurements of several polarization states, different approaches have been proposed, only a few of which are described here. By adding a polarization modulator (e.g., an electrooptic modulator) in the source arm, different input states can be produced in a sequential manner. In a commonly used scheme depicted in Figure 4c, a consecutive pair of polarization states corresponding to Stokes vectors perpendicular in a Poincaré sphere representation is generated at the input of the interferometer. From the polarization states detected at the output of the interferometer, depth-resolved Stokes vectors can be computed [64,65]. The retardation induced by birefringent tissue is related to the angle of rotation of Stokes vectors on the Poincaré sphere. By computing this angle between the Stokes vectors at the sample surface and those within the tissue, cumulative phase retardation can be computed at any sample position [66,67]. Furthermore, the direction of the optic axis can be determined from rotations of the pair of perpendicular Stokes vectors on the Poincaré sphere [68]. In such dual-input PS-OCT systems, polarization parameters such as retardation, axis orientation, and diattenuation can also be assessed using the Jones formalism. In this approach, the measured polarization states in the sample originating from the pair of input polarization states are used to reconstruct the Jones matrix in every image pixel. An eigenvalue analysis of these measured Jones matrices enables the calculation of phase retardation and diattenuation [15,16]. The Jones matrix approach to PS-OCT also enables the measurement of the optic axis orientation [15,16]. Depth-resolved measurements of the birefringent axis orientation have recently gained interest for mapping the orientation of birefringent fibers in PS-OCT based tractography of collagenous tissue [69,70].

Alternatively, using a Mach–Zehnder type interferometer, only the polarization state in the sample arm can be varied [71,72] from one scan to the next. By multiplexing two different states using a passive polarization delay unit as shown in Figure 4d, the four elements of a Jones matrix can be measured simultaneously [18,73,74]. In that case, the sample beam is split into two orthogonal input Jones vectors which travel different path lengths in the sample arm and therefore generate signals at different depths in the OCT image. These two input vectors provide an orthogonal system and their response—i.e., the two Jones vectors measured via multiplexing—readily provides a Jones matrix [15]. Jones matrix OCT relates the Jones matrix J_{meas} measured at each sample position to a reference matrix (e.g., $J_{surface}$ at the surface of the sample), thereby yielding a unitary transformation of the sample matrix J_{sample} [15,16,75]

$$\widetilde{J}_{sample} = J_{meas}\, J_{surface}^{-1}. \tag{9}$$

Here the tilde denotes the unitary transformation. From $\widetilde{J}_{sample}$, the polarization properties can be computed. As such, Jones matrix PS-OCT can not only measure phase retardation but also diattenuation [15], local birefringence [16], and local optic axis orientation [76]. Compared to cumulative retardation measurements, local measurements of birefringent properties provide a more intuitive approach to tissue architecture and composition. For instance, in collagenous tissue such as skin, local birefringence can be used for the depth-resolved assessment of the collagen content [77,78]. Different applications of birefringence imaging of collagen in healthy and diseased tissues will be discussed in Section 3.

2.5. Recent Advances in PS-OCT Technology

The development of PS-OCT has greatly advanced since Hee and coworkers first presented birefringence-sensitive ranging [13]. Not only have PS-OCT devices become faster and the detection schemes become more sophisticated, as briefly described in the previous section, but also the analysis of PS-OCT images has improved a lot.

The first PS-OCT prototypes provided axial scan rates on the order of several hertz [8,13,14]. Later rapid reference scanning schemes [79,80] and advanced beam scanning approaches such as transverse-scanning PS-OCT [40] sped up the technique to several frames (B-scans) per second. The advent of Fourier domain OCT (or: frequency domain OCT), which computes A-scan signals by a Fourier transform of the interference spectrum, provided a huge increase in detection sensitivity and the possibility to scan even faster since no more mechanical reference mirror movement was required to perform depth scanning [5,81–83]. First high-speed PS-OCT systems with spectrometer-based detection provided scan rates of several tens of A-scans per second [41,68]. These spectral domain (SD) PS-OCT prototypes employed two spectrometer cameras, one for each orthogonal polarization channel. In order to reduce system complexity, cost, and alignment efforts, SD PS-OCT approaches based on single camera detection were developed [42,49,84–89].

Fourier domain OCT can also be performed by using a frequency-swept laser and a high-speed detector, such that interference spectra are acquired as a function of time rather than in parallel with a spectrometer [90–92]. This variant of OCT is usually called swept-source (SS) OCT and sometimes also referred to as optical frequency domain imaging (OFDI) or time-encoded frequency domain OCT.

Providing the same sensitivity and speed advantages as spectrometer-based Fourier domain OCT, SS-OCT was soon expanded by polarization sensitivity [47,71,75,93–97]. In particular the advent of commercial laser technology providing longer imaging ranges led to the development of PS-OCT at ultrahigh imaging speeds of 100,000 axial scans per second [18,54,73,74,98]. Even higher imaging speeds were achieved by experimental swept lasers operating at several hundred kilohertz [50,99–102].

While PS-OCT technology has greatly advanced, there are still some limitations to the technique. Being an optical method, its applicability is limited to imaging of superficial locations in tissues and other objects. Further, PS-OCT has been used for qualitative imaging mostly; the exploitation of quantitative measurements however bears great potential for diagnostics and other applications, as will be demonstrated in the next sections. PS-OCT was also combined with other functional OCT extensions such as Doppler OCT or OCT angiography [30,68,80,103–105]. Such combinations may not only improve the contrast for vascular tissue components but also provide additional, complementary insight into disease patterns [30,103,105,106]. In order to perform PS-OCT beneath the body surface, endoscopic and needle-based PS-OCT was developed [28,29,107–112]. To further increase the contrast and image range, PS-OCT has been combined with other technologies such as ultrasound and fluorescence imaging [113,114].

In parallel to the impressive evolution of PS-OCT hardware, PS-OCT image processing also underwent massive improvements. Real-time display of PS-OCT data was enabled by parallel computing [115]. Computational methods were devised for removing polarization artifacts in order to produce clearer PS-OCT images [21,27–29,57,116,117]. As PS-OCT is an interferometric technique based on coherent light, images are subject to speckling which sometimes obscures structural details. The size of speckles can be kept small by using broadband light sources and optics providing high transverse resolution [46,118]. In image processing, speckle noise can be reduced by image averaging and dedicated algorithms [119,120]. PS-OCT also enables the segmentation of structures based on common polarization properties and the determination of interfaces between different tissue segments based on changing polarization properties. Segmentation and image feature assessment was developed based on depolarization [20,103,121–125] and birefringence [98,116,126–129]. Practical examples of PS-OCT applications will be shown in the following sections.

3. PS-OCT Applications

3.1. PS-OCT in the Eye

OCT is most established in ophthalmology, where it has become a standard diagnostic method in everyday clinical routine [130]. Also PS-OCT has been successfully applied for ophthalmic imaging using experimental prototypes [131]. The eye features a variety of tissues exhibiting birefringence or depolarization, which enable PS-OCT to provide additional contrast for discerning, segmenting, and quantifying ocular structures. Birefringence can be found in fibrous tissues such as the retinal nerve fiber layer (RNFL), the sclera (i.e., the white outer shell of the eye), the cornea, as well as in extraorbital muscles and tendons. Depolarization is pronounced in structures containing melanin pigments such as the retinal pigment epithelium (RPE), the choroid, and the pigment epithelium of the iris. Other structures such as the photoreceptor layer, conjunctive tissue, and the stroma of the iris are rather polarization preserving and do not markedly influence the polarization state of light.

The RNFL consists of the axons of the retinal ganglion cells. Since the RNFL is damaged in glaucoma—the second leading cause of blindness worldwide [132]—and since RNFL birefringence is connected to layer integrity [133,134], the polarization properties of the RNFL were investigated as potential diagnostic markers for glaucoma. PS-OCT based assessment of the RNFL's birefringent properties might be particularly interesting since it was shown that polarization changes in experimental glaucoma can be observed earlier than RNFL thickness changes [135]. Peripapillary RNFL thickness is currently a key OCT parameter for glaucoma diagnostics in state-of-the-art clinical routine [136]. In the vein of earlier scanning laser polarimetry approaches [137–139], PS-OCT was applied to investigate the RNFL using PS-OCT. After initial experiments in the primate retina [140], RNFL birefringence was measured in vivo in the human eye by performing circular scans around the optic nerve head [79,141]. Faster Fourier domain PS-OCT later enabled 2D mapping of RNFL birefringence and retardation as well as comparisons to scanning laser polarimetry [18,73,142–145]. Exemplary PS-OCT fundus images mapping reflectivity and RNFL retardation in a human eye are shown in Figure 5. Also in preclinical research, PS-OCT was used to investigate the birefringence properties of the RNFL and their relation to the intraocular pressure, which is an important parameter for glaucoma, in animals [135,146–148]. Aside from measuring their birefringence, PS-OCT was also demonstrated for tracing nerve fiber bundles in the RNFL [149].

PS-OCT images of the human retina exhibit strong depolarization in pigmented structures such as the RPE [150,151]. In the RPE, this depolarization is most pronounced around the fovea [152] and correlates with the pigmentation status, i.e., it is reduced or even absent in albino patients [25,153]. Comparative measurements of PS-OCT and histology in rat eyes have revealed a correlation between DOPU and the density of melanin pigments in the RPE and choroid (Figure 5e) [26]. Depolarization has proven a particularly useful contrast for the assessment of the RPE in clinical cases, where it is often hard to distinguish ocular structures in pathological eyes [20,103,154,155]. Based on DOPU images, algorithms were developed to assess areas and volumes of lesions quantitatively [121]. In age-related macular degeneration (AMD), PS-OCT was not only used to distinguish drusen characteristics but also to quantify the area and volume of drusen during disease progression (Figure 5f–h) [122,124,156]. In late stage non-exudative (dry) AMD, PS-OCT enables the assessment of atrophic areas lacking RPE (Figure 5a–d) [121,157,158]. In exudative diseases such as wet AMD, central serous chorioretinopathy, and diabetic macular edema, PS-OCT was demonstrated for imaging and identifying fibrotic scars, hard exudates, as well as pigment epithelial features [123,159–162]. Finally, PS-OCT also proved useful to enhance contrast for imaging pathologic structures in less common retinal diseases such as macular telangiectasia and Stargardt disease [163,164].

PS-OCT of the anterior eye markedly improves the contrast for birefringent, collagenous tissues such as the cornea, sclera, and tendons as well as for the trabecular meshwork [40,84,94,127]. The additional contrast has been exploited for automated, feature-based tissue discrimination [126]. Substantial changes in the birefringent appearance of the cornea can be observed in keratoconus as shown in Figure 6a–e, such that PS-OCT was proposed as a diagnostic method for this disease [128,165]. Since corneal birefringence depends on the microstructure, PS-OCT was also proposed for imaging changes during corneal crosslinking therapy [166]. After trabeculectomy, which is a surgical procedure for glaucoma treatment, the evolution of filtering blebs was monitored by PS-OCT (Figure 6f) [167–169]. In the sclera, PS-OCT was used to image necrotizing scleritis [170] and to study birefringence changes related to increased intraocular pressure [148,171].

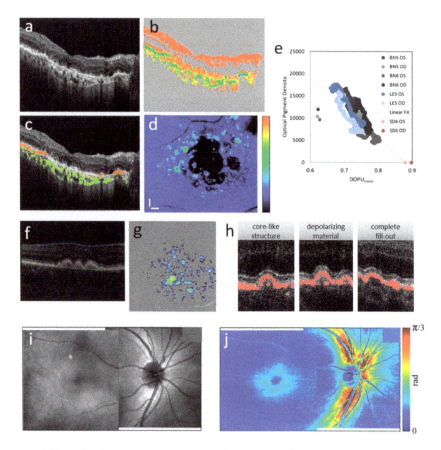

Figure 5. PS-OCT in the posterior eye. (**a–d**) PS-OCT of a patient with late-stage AMD (adapted with permission from [120], Optical Society of America, 2014). (**a**) Reflectivity B-scan; (**b**) DOPU B-scan (color map: 0–1); (**c**) Depolarization in the retinal pigment epithelium (RPE, red) and choroid (green) overlaid on reflectivity image; (**d**) Fundus map indicating thickness of depolarizing pixels at the level of the RPE; (**e**) Relation of DOPU and melanin density assessed by histology in RPE/choroid of rat eyes (adapted with permission from [26], ARVO, 2015). (**f,g**) PS-OCT based layer segmentation in an AMD patient with drusen (adapted with permission from [121], SPIE, 2010). (**f**) Reflectivity B-scan with segmented inner limiting membrane (blue), RPE (red), and Bruch's membrane (green). Drusen can be observed as bumpy elevations of the retina at the RPE level; (**g**) Drusen thickness map; (**h**) Drusen characteristics assessed by PS-OCT (courtesy by Dr. F. G. Schlanitz, Medical University of Vienna, Austria). Depolarizing pixels are marked in red. (**i,j**) Birefringence in the retinal nerve fiber layer (RNFL) imaged by PS-OCT (adapted with permission from [18], Optical Society of America, 2014). (**i**) Fundus reflectivity map; (**j**) Fundus map showing increased phase retardation caused by the RNFL around the optic nerve head and by Henle's fibers around the fovea.

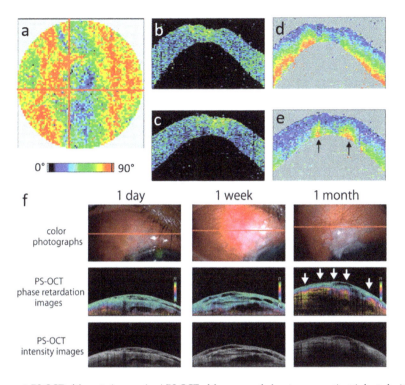

Figure 6. PS-OCT of the anterior eye. (**a–e**) PS-OCT of the cornea of a keratoconus patient (adapted with permission from [165], ARVO, 2007). (**a**) PS-OCT en-face image showing retardation at the posterior surface of the cornea. The red lines indicate the locations of the B-scans shown in (**b–e**); (**b,c**) Horizontal and vertical reflectivity B-scan images. Decreased corneal thickness can be observed in the center; (**d,e**) Corresponding retardation B-scans exhibiting irregular pattern. (**f**) PS-OCT of blebs in the anterior eye after glaucoma surgery (adapted with permission from [167], ARVO, 2014). Photographs (red lines indicate locations of OCT scans), phase retardation and intensity images with PS-OCT at one day, one week, and one month after surgery. Cases of partial increase of phase retardation after surgery. Arrows indicate irregular and abnormal phase retardation.

3.2. PS-OCT in Skin and Oropharyngeal Tissue

Since the imaging regime of OCT is usually restricted to superficial layers of scattering structures (unless special probes such as catheters are used), skin is a preferred candidate for OCT imaging. Using PS-OCT, dermal layers with different scattering and polarization properties can be observed, including stratum corneum, dermis, and epidermis (Figure 7) [50,64,172,173]. Oral and laryngeal tissue have also been imaged by PS-OCT. In the oropharyngeal tract, PS-OCT was demonstrated for investigating the mucosa of the vocal fold and for detecting lesions in the buccal mucosa based on increased birefringence [174,175].

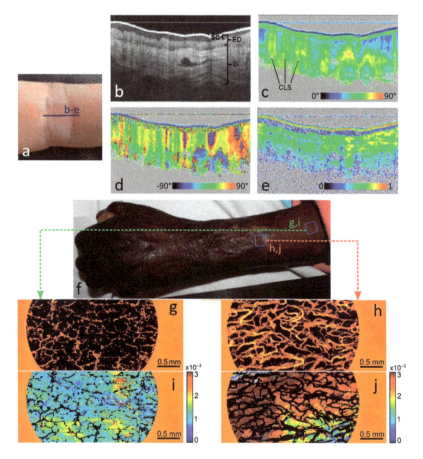

Figure 7. PS-OCT of skin. (**a–e**) Cross-sectional PS-OCT image of human skin (adapted with permission from [176], Optical Society of America, 2013). (**a**) Scan location at the proximal interphalangeal joint of middle finger; (**b**) Reflectivity image. SC stratum corneum, D dermis, ED epidermis; (**c**) Retardation image. CLS "column" like structure; (**d**) Axis orientation image; (**e**) DOPU image. (**f–j**) Birefringence and vascular imaging of a hypertrophic scar and adjacent normal skin (adapted with permission from [106], P. Gong, 2014). (**f**) Photograph. Locations of PS-OCT scans are indicated by two blue squares; (**g,h**) Vasculature maximum intensity projections of the normal skin and scar, respectively; (**i,j**) En face birefringence maps of the normal skin and scar, respectively.

Via dermal birefringence, PS-OCT provides access to tissue alterations caused by deformation, scarring, and burns [106,177–179]. Figure 7j shows an example of scarred skin exhibiting significantly higher birefringence than normal skin (Figure 7i). Additionally, wound healing processes including collagen restoration can be followed with PS-OCT [180,181]. Moreover, Stokes vector based depolarization imaging can reveal multiple scattering as well as pathological conditions in skin such as cancer [19,24,58], as will be discussed in the next section.

3.3. PS-OCT in Cancerous Tissue

Cancer alters tissue microstructure. This alteration can change the optical properties of affected tissues. PS-OCT has been applied for imaging cancerous tissues in several organs. Altered birefringence and depolarization characteristics enabled imaging and identification of skin lesions such as basal cell carcinoma (Figure 8a–d) [24,182,183].

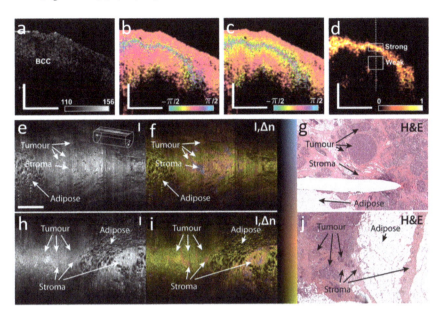

Figure 8. PS-OCT of cancerous tissue. (**a–d**) PS-OCT of basal cell carcinoma in human skin (adapted with permission from [24], Optical Society of America, 2016). (**a**) Reflectance B-scan image. BCC indicates the basal cell carcinoma tumor position; (**b**) Linear retardance along the *x–y* direction; (**c**) Linear retardance along the ±45° directions; (**d**) Normalized differential depolarization index image. Scale bar is 0.5(*x*) × 0.3(*z*) mm. (**e–j**) Needle-based PS-OCT of breast cancer (reproduced with permission from [112], M. Villiger et al., 2016). Imaging was performed in grade 1 invasive ductal carcinoma (**e–g**) and grade 2 invasive ductal carcinoma (**h,j**). (**e,h**) Structural intensity (I), (**f,i**) overlay of tissue birefringence and intensity (I, Δn), and (**g,j**) matching histological section stained with hematoxylin and eosin. Scale bar in (**e**) is 1 mm and applies to panels (**e–j**).

Further promising results of PS-OCT based cancer imaging were reported in larynx, ovaries, and bladder [32,184,185]. Several groups also successfully studied PS-OCT for imaging breast cancer [112,186,187]. Figure 8 shows exciting results of PS-OCT imaging, which enabled the differentiation of tumor from surrounding tissue. Using intraoperative scanning of excised tissue or in situ needle-based imaging (cf. Figure 8e–j), PS-OCT could represent a promising method for reliably demarking malignant breast tumors, thus reducing the re-excision rate due to positive margins.

3.4. PS-OCT in Muscles, Tendons, Cartilage, and Bone

Tendon was the first biological tissue imaged by PS-OCT [8]. Being collagen-rich structures, tendons and muscles exhibit strong birefringence, which enables an easy discrimination from surrounding supportive tissue by PS-OCT.

Since the integrity of collagen is an indicator for structural stability and pathologic state, PS-OCT was suggested for collagen assessment in tendons and ligaments [189]. Consequently, PS-OCT was used to visualize the evolution of the collagen fiber alignment via birefringence in tissue-engineered tendons in response to varying growth environments and to investigate degenerative changes related to rupture in Achilles tendons [190,191]. Lately, the influence of proteoglycans—which are essential components of the tendon extracellular matrix associated with tendinopathies—on the optical properties of tendons have been studied by PS-OCT [192]. In skeletal muscle of genetically-altered (mdx) mice, exercise-induced ultrastructural changes were detected by PS-OCT in in vivo animals [193]. Compared to wildtype controls, the highly birefringent properties of skeletal muscles markedly decreased in mdx mice, thus suggesting a relationship between the degree of birefringence detected using PS-OCT and the sarcomeric ultrastructure present within skeletal muscle. PS-OCT was also shown to be capable of detecting muscle necrosis in dystrophic mdx mice [194].

In cartilage, PS-OCT can detect areas of enhanced or reduced birefringence in hyaline cartilage (mostly composed of type-II collagen) and fibrocartilage (predominantly type-I collagen) related to degeneration and repair mechanisms [195]. In an in vivo study on human knee joints prior to partial or total joint replacement treatment, reduced birefringence was found in degenerated cartilage [196]. PS-OCT images of one proximal joint surface of bovine tibia are shown in Figure 9d [188]. PS-OCT using variable incidence angles was further used to investigate the 3D architecture of the collagen fiber network in cartilage [197,198]. Due to its high sensitivity to cartilage disorder, PS-OCT proved a promising tool for imaging cartilage in osteoarthritis in both humans and animal models [199,200].

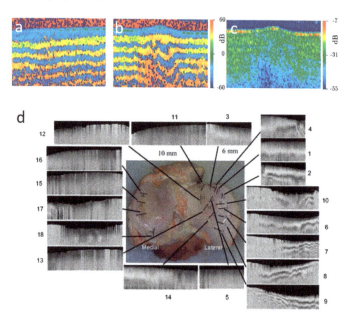

Figure 9. PS-OCT of tendon and cartilage. (**a–c**) First published PS-OCT images (adapted with permission from [8], Optical Society of America, 1997) showing birefringence of bovine tendon. (**a**) Fresh tendon; (**b**) Tendon after exposure to high-power laser irradiation. The banded structure appears disturbed compared to (**a**); (**c**) Color-coded intensity image; (**d**) PS-OCT optical phase retardation images at the various sites on a bovine tibia. The banded appearance in retardation images is indicative of high birefringence in some areas (adapted with permission from [188], G. M. Peavy, 2008).

3.5. PS-OCT in Vessels and Cardiac Tissue

Birefringence of vessel walls is a promising diagnostic parameter for arteriosclerotic vascular disease accessible by PS-OCT [201]. In atherosclerosis, artery walls locally thicken and may form atherosclerotic plaque lesions, which may be categorized into stable and unstable (called vulnerable) plaques. Ex vivo scanning compared to histology as well as catheter based PS-OCT imaging of atherosclerotic artery walls have revealed altered birefringence patterns in atherosclerotic plaques [202–204]. Examples of different plaques imaged by PS-OCT as well as corresponding histologic images are shown in Figure 10 [205].

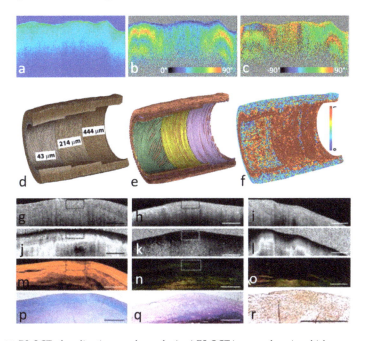

Figure 10. PS-OCT of cardiac tissue and vessels. (**a–c**) PS-OCT images of ex vivo chicken myocardium (adapted with permission from [14], Optical Society of America, 2001). (**a**) Intensity image; (**b**) Phase retardation image; (**c**) Image of fast axis distribution. (**d–f**) 3D visualization of a carotid arterial sample (adapted with permission from [70], Optical Society of America, 2015). (**d**) OCT intensity image; (**e**) Fiber orientation tractography; (**f**) Fiber alignment map in the vessel wall. Surface layers were partly removed in three sections to reveal OCT intensity and fiber orientation/alignment at three representative depths in the arterial wall. (**g–r**) PS-OCT images of fibrous plaques (adapted with permission from [205], Elsevier, 2007). (**g–i**) OCT intensity images; (**j,k**) Phase retardation images (scale range: 0–180°) showing high birefringence; (**l**) Retardation image of fibrous plaque showing black region corresponding to low birefringence below the luminal surface. (**m–o**) Picrosirius red stained histology section, showing orange-red fibers (thicker fibers), yellow-green (thinner fibers) and low collagen content in the plaque under polarized light microscopy, respectively; (**p,q**) Trichrome-stained histology images; (**r**) Corresponding histology section stained for α-smooth muscle actin shows numerous smooth muscle cells within the fibrous plaque. Scale bars are 500 μm.

Information on the birefringent axis orientation provides access to fiber alignment in fibrous tissue (Figure 10a–c) [14]. Tractographic PS-OCT imaging was performed in the walls of blood vessels (Figure 10d–f) and in the mouse heart, thereby revealing fibrous layers with varying fiber orientations [69,70]. In rabbit hearts, the geometry of the perfusion border zone was investigated using PS-OCT and tissue clearing [206], and tissue discrimination was enabled by PS-OCT in rat hearts where decreased birefringence was observed in infarcted hearts [207].

3.6. PS-OCT in Teeth

PS-OCT has been used for imaging dental structures for almost 20 years [208–210]. In teeth, PS-OCT provides contrast for dentin, enamel, as well as carious lesions. Dentin is a calcified tissue and is a central component of teeth. On the crown, it is covered by enamel, a highly mineralized substance. The apatite crystals in dental enamel are highly ordered and produce negative birefringence. In contrast, collagen makes dentin positively birefringent. Processes such as demineralization lead to birefringence changes which can be observed by PS-OCT.

PS-OCT was demonstrated for the assessment of early and advanced demineralization in dentin as well as in enamel (Figure 11d–e) [211–214]. Ablation of demineralized tooth structures was monitored by PS-OCT [215]. Demineralization was also investigated in tooth roots [216]. PS-OCT of enamel treated by CO_2 laser irradiation confirmed inhibited demineralization [217]. In particular, caries—characterized by mineral breakdown of teeth due to bacterial activity—has been an interesting target for PS-OCT imaging. Caries lesions in various conditions were investigated and their progression was followed longitudinally [209,210,218–220]. Consequently, also remineralization processes in enamel and dentin were imaged based on their birefringence [211,221]. Recently, an automated method for assessing remineralized lesions was developed based on PS-OCT [222].

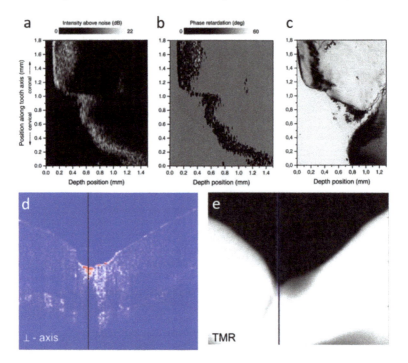

Figure 11. Dental PS-OCT. (a–c) PS-OCT tomograms of a carious lesion in enamel at the distal surface of a human molar (adapted with permission from [209], Karger Publishers, 1999). (a) Reflectivity image; (b) Phase retardation image; (c) Corresponding histologic section. (d–e) PS-OCT and transverse microradiography (TMR) of tooth demineralization (adapted with permission from [213], John Wiley and Sons, 2010). (d) PS-OCT image showing the cross-polarized signal; (e) Corresponding TMR. The area of demineralization has a strong cross-polarized signal in (d) and appears darker in (e).

3.7. PS-OCT in Nerves and Brain

Nerve fibers exhibit birefringence, an optical property that has been exploited for quantitative measurements in the retinal nerve fiber layer (see Section 3.1). Also nerve fibers in cerebral white matter or in peripheral nerves may be imaged by PS-OCT based on their birefringence [223].

Aside from neural structures in the central nervous system, peripheral nerves have also been imaged by PS-OCT. Improved delineation of the sciatic nerve boundaries to muscle and adipose tissues as well as quantitative birefringence measurements were enabled by the additional polarization contrast [224]. In an experimental nerve crush model, decreasing birefringence was observed in parallel to a loss of myelination (Figure 12a–d) [225]. The prostatic nerves—indiscernible from surrounding tissue by standard, intensity based OCT—were identified in prostates of rats and humans, thereby indicating the feasibility of PS-OCT as a method for intrasurgical imaging [226].

In the brain, PS-OCT was not only used to enhance the contrast of birefringent structures but also to trace white matter structures as shown in Figure 12e–h. PS-OCT based tractography provides images encoding the orientation of fiber tracts in different colors and can be used to verify diffusion tensor based MRI tractography images with micrometer scale resolution [51,227–229]. Lately, PS-OCT was also demonstrated for imaging a hallmark of Alzheimer's disease, namely neuritic amyloid-beta plaques as well as amyloidosis in cerebral vasculature [230]. PS-OCT images of birefringent neuritic plaques are shown in Figure 12i–k.

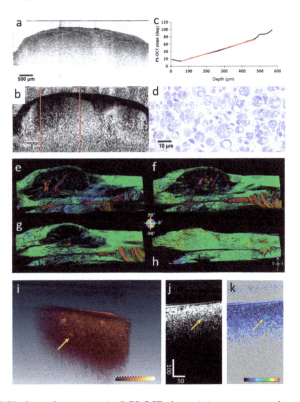

Figure 12. PS-OCT of neural structures. (**a–d**) PS-OCT of rat sciatic nerve two weeks post nerve crush (adapted with permission from [225], B. H. Park, 2015). (**a**) Intensity image; (**b**) PS-OCT retardation image; (**c**) Plot of average phase retardation per depth computed within the red lines shown in (**b**); (**d**) Histologic image. Toluidine blue staining for assessing the myelination state; (**e–h**) En face PS-OCT optic axis orientation maps quantitatively depict in-plane fiber orientations in the medulla (from [229] with permission by Elsevier). The color wheel shows the orientation values ranging between −90° and 90°. The brightness of colors in the images is determined by the en face retardance values. (**i–k**) PS-OCT imaging of neuritic plaques in post mortem cerebral cortex of an Alzheimer's disease patient. (**i**) Rendering of 3D retardation data showing increased retardation in plaques. Color map range: 7–41° (reproduced with permission from [230], B. Baumann et al., 2017). (**j**) Reflectivity B-scan image; (**k**) Retardation B-scan image. The location of one plaque is marked by an orange arrow.

3.8. Other Applications of PS-OCT

PS-OCT also found applications in biomedical fields other than those discussed in the previous sections as well as in non-medical fields. One exciting use of PS-OCT is in imaging applications relying on small particles as exogenous contrast agents. As such, plasmon-resonant nanoparticles like gold nanostars are popular contrast agents in biophotonic imaging. In order to increase detection sensitivity for single particles, their polarization-sensitive scattering signal in the near infrared can be modulated by an external oscillating magnetic field [231]. PS-OCT can be used to detect these dynamic scattering signals and was demonstrated for depth-resolved viscosity measurements based on the diffusion of gold nanorods [232]. Based on temporal changes in polarization contrast parameters, a method for differentiating light scatterers such as cells and gold nanorods was developed [233]. Recently, PS-OCT based detection of gold nanorods was proposed for detecting nanotopological changes in 3D tissue models of mammary extracellular matrix and pulmonary mucus (Figure 13d) [234]. Having a non-invasive imaging technique like PS-OCT for studies of such tissue models may help to interpret biophysical changes associated with disease progression.

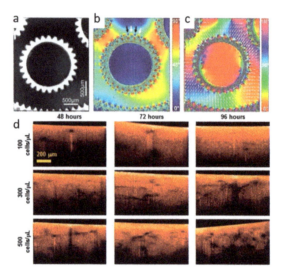

Figure 13. Nonmedical applications of PS-OCT. (**a–c**) Transversal ultrahigh resolution PS-OCT images of the resist-wafer interface of a photoresist mold for a micromechanical wheel (adapted with permission from [235], Optical Society of America, 2006). (**a**) Intensity image; (**b**) PS-OCT retardation image. The arrows indicate highly strained areas at the teeth of the wheel; (**c**) Orientation of the slow optic axis (color-coded and displayed as vector-field scaled by the magnitude of the retardation); (**d**) Diffusion of gold nanorods (GNRs) in in vitro extracellular matrix models (adapted with permission from [234], A. Oldenburg,2014). Representative PS-OCT B-scan images showing cross-polarized signal of fibroblasts in collagen I: Matrigel versus cell seed density and incubation time show how GNRs provide positive contrast within the matrix between cells, whereas cells appear dark.

Its sensitivity for microstructural changes affecting sample polarization properties along with its 3D imaging capabilities made PS-OCT also an interesting modality for materials science and nondestructive testing applications [236]. Using translucent glass-epoxy composite phantoms, the spatial distribution of mechanical stress was mapped by PS-OCT [237]. Strain mapping by also exploiting the optic axis orientation in PS-OCT images was demonstrated as a method for charting the directionality of strained sample areas (Figure 13a–c) [235]. Material dynamics were optically investigated by translating measured phase retardation into stress images [238]. Thereby, dynamics could be studied from the elastic regime over the deformation phase up to fracture.

4. Conclusions

PS-OCT is a versatile functional extension of OCT. As described in the section on the technical background, only few modifications to the standard OCT layout are necessary for polarization sensitivity. Of course, more sophisticated setups and advanced analysis methods were also developed. This review aimed to provide a concise introduction to the basic principles underlying PS-OCT and a crisp overview of advances in PS-OCT technology development. By highlighting research on state of the art PS-OCT applications based on the published literature, the obvious potential of this powerful technique for improved qualitative and quantitative imaging was portrayed. Given the achievements of the past 20 years discussed here, we are anticipating exciting new technological developments, advances of applied biomedical imaging, and potential applications in new fields for the next 20 years.

Acknowledgments: I would like to express my gratitude to Christoph K. Hitzenberger, Michael Pircher, and Erich Götzinger, Medical University of Vienna, Austria, who introduced me to the exciting world of PS-OCT more than 10 years ago. Also, I would like to thank my colleagues at the Center for Medical Physics and Biomedical Engineering (in particular Marco Augustin and Christoph Hitzenberger for proof-reading), at the Department of Ophthalmology, at the Division of Biomedical Research, at the Core Facility Imaging and at the Institute of Neurology at Medical University of Vienna as well as at the VetCore Facility for Research and Technology at the University of Veterinary Medicine Vienna for their continuous collaborative support, fruitful discussions, and creative feedback. Funding by the Austrian Science Fund (FWF grant P25823-B24) and the European Research Council (ERC Starting Grant 640396 OPTIMALZ) is gratefully acknowledged.

Conflicts of Interest: The author declares no conflict of interest.

References

1. Huang, D.; Swanson, E.A.; Lin, C.P.; Schuman, J.S.; Stinson, W.G.; Chang, W.; Hee, M.R.; Flotte, T.; Gregory, K.; Puliafito, C.A.; et al. Optical coherence tomography. *Science* **1991**, *254*, 1178–1181. [CrossRef] [PubMed]
2. Drexler, W.; Fujimoto, J.G. *Optical Coherence Tomography Technology and Applications*; Springer: Berlin/Heidelberg, Germany; New York, NY, USA, 2008.
3. Fercher, A.F. Optical coherence tomography—Development, principles, applications. *Zeitschrift Fur Medizinische Physik* **2010**, *20*, 251–276. [CrossRef] [PubMed]
4. Fercher, A.F.; Mengedoht, K.; Werner, W. Eye-length measurement by interferometry with partially coherent light. *Opt. Lett.* **1988**, *13*, 1867–1869. [CrossRef]
5. Fercher, A.F.; Hitzenberger, C.K.; Kamp, G.; Elzaiat, S.Y. Measurement of intraocular distances by backscattering spectral interferometry. *Opt. Commun.* **1995**, *117*, 43–48. [CrossRef]
6. Drexler, W.; Liu, M.; Kumar, A.; Kamali, T.; Unterhuber, A.; Leitgeb, R.A. Optical coherence tomography today: Speed, contrast, and multimodality. *J. Biomed. Opt.* **2014**, *19*, 071412. [CrossRef] [PubMed]
7. Tuchin, V.V. Polarized light interaction with tissues. *J. Biomed. Opt.* **2016**, *21*, 071114. [CrossRef] [PubMed]
8. De Boer, J.F.; Milner, T.E.; van Gemert, M.J.C.; Nelson, J.S. Two-dimensional birefringence imaging in biological tissue by polarization-sensitive optical coherence tomography. *Opt. Lett.* **1997**, *22*, 934–936. [CrossRef] [PubMed]
9. De Boer, J.F.; Milner, T.E. Review of polarization sensitive optical coherence tomography and Stokes vector determination. *J. Biomed. Opt.* **2002**, *7*, 359–371. [CrossRef] [PubMed]
10. De Boer, J.F.; Hitzenberger, C.K.; Yasuno, Y. Polarization sensitive optical coherence tomography—A review. *Biomed. Opt. Express* **2017**, *8*, 1838–1873. [CrossRef]
11. Jones, R.C. A new calculus for the treatment of optical systemsi. Description and discussion of the calculus. *J. Opt. Soc. Am.* **1941**, *31*, 488–493. [CrossRef]
12. Bickel, W.S.; Bailey, W.M. Stokes vectors, Mueller matrices, and polarized scattered light. *Am. J. Phys.* **1985**, *53*, 468–478. [CrossRef]
13. Hee, M.R.; Huang, D.; Swanson, E.A.; Fujimoto, J.G. Polarization-sensitive low-coherence reflectometer for birefringence characterization and ranging. *J. Opt. Soc. Am. B* **1992**, *9*, 903–908. [CrossRef]

14. Hitzenberger, C.K.; Gotzinger, E.; Sticker, M.; Pircher, M.; Fercher, A.F. Measurement and imaging of birefringence and optic axis orientation by phase resolved polarization sensitive optical coherence tomography. *Opt. Express* **2001**, *9*, 780–790. [CrossRef] [PubMed]

15. Park, B.H.; Pierce, M.C.; Cense, B.; de Boer, J.F. Jones matrix analysis for a polarization-sensitive optical coherence tomography system using fiber-optic components. *Opt. Lett.* **2004**, *29*, 2512–2514. [CrossRef] [PubMed]

16. Makita, S.; Yamanari, M.; Yasuno, Y. Generalized Jones matrix optical coherence tomography: Performance and local birefringence imaging. *Opt. Express* **2010**, *18*, 854–876. [CrossRef] [PubMed]

17. Todorovic, M.; Jiao, S.; Wang, L.V.; Stoica, G. Determination of local polarization properties of biological samples in the presence of diattenuation by use of Mueller optical coherence tomography. *Opt. Lett.* **2004**, *29*, 2402–2404. [CrossRef] [PubMed]

18. Braaf, B.; Vermeer, K.A.; de Groot, M.; Vienola, K.V.; de Boer, J.F. Fiber-based polarization-sensitive OCT of the human retina with correction of system polarization distortions. *Biomed. Opt. Express* **2014**, *5*, 2736–2758. [CrossRef] [PubMed]

19. Adie, S.G.; Hillman, T.R.; Sampson, D.D. Detection of multiple scattering in optical coherence tomography using the spatial distribution of Stokes vectors. *Opt. Express* **2007**, *15*, 18033–18049. [CrossRef] [PubMed]

20. Gotzinger, E.; Pircher, M.; Geitzenauer, W.; Ahlers, C.; Baumann, B.; Michels, S.; Schmidt-Erfurth, U.; Hitzenberger, C.K. Retinal pigment epithelium segmentation by polarization sensitive optical coherence tomography. *Opt. Express* **2008**, *16*, 16410–16422. [CrossRef] [PubMed]

21. Makita, S.; Hong, Y.-J.; Miura, M.; Yasuno, Y. Degree of polarization uniformity with high noise immunity using polarization-sensitive optical coherence tomography. *Opt. Lett.* **2014**, *39*, 6783–6786. [CrossRef] [PubMed]

22. Baumann, B.; Zotter, S.; Pircher, M.; Götzinger, E.; Rauscher, S.; Glösmann, M.; Lammer, J.; Schmidt-Erfurth, U.; Gröger, M.; Hitzenberger, C.K. Spectral degree of polarization uniformity for polarization-sensitive OCT. *J. Mod. Opt.* **2015**, *62*, 1758–1763. [CrossRef] [PubMed]

23. Lippok, N.; Villiger, M.; Bouma, B.E. Degree of polarization (uniformity) and depolarization index: Unambiguous depolarization contrast for optical coherence tomography. *Opt. Lett.* **2015**, *40*, 3954–3957. [CrossRef] [PubMed]

24. Ortega-Quijano, N.; Marvdashti, T.; Bowden, A.K.E. Enhanced depolarization contrast in polarization-sensitive optical coherence tomography. *Opt. Lett.* **2016**, *41*, 2350–2353. [CrossRef] [PubMed]

25. Baumann, B.; Baumann, S.O.; Konegger, T.; Pircher, M.; Götzinger, E.; Schlanitz, F.; Schütze, C.; Sattmann, H.; Litschauer, M.; Schmidt-Erfurth, U.; et al. Polarization sensitive optical coherence tomography of melanin provides intrinsic contrast based on depolarization. *Biomed. Opt. Express* **2012**, *3*, 1670–1683. [CrossRef] [PubMed]

26. Baumann, B.; Schirmer, J.; Rauscher, S.; Fialová, S.; Glösmann, M.; Augustin, M.; Pircher, M.; Gröger, M.; Hitzenberger, C.K. Melanin pigmentation in rat eyes: In vivo imaging by polarization-sensitive optical coherence tomography and comparison to histology. *Investig. Ophthalmol. Vis. Sci.* **2015**, *56*, 7462–7472. [CrossRef] [PubMed]

27. Zhang, E.Z.; Oh, W.-Y.; Villiger, M.L.; Chen, L.; Bouma, B.E.; Vakoc, B.J. Numerical compensation of system polarization mode dispersion in polarization-sensitive optical coherence tomography. *Opt. Express* **2013**, *21*, 1163–1180. [CrossRef] [PubMed]

28. Villiger, M.; Zhang, E.Z.; Nadkarni, S.; Oh, W.-Y.; Bouma, B.E.; Vakoc, B.J. Artifacts in polarization-sensitive optical coherence tomography caused by polarization mode dispersion. *Opt. Lett.* **2013**, *38*, 923–925. [CrossRef] [PubMed]

29. Villiger, M.; Zhang, E.Z.; Nadkarni, S.K.; Oh, W.-Y.; Vakoc, B.J.; Bouma, B.E. Spectral binning for mitigation of polarization mode dispersion artifacts in catheter-based optical frequency domain imaging. *Opt. Express* **2013**, *21*, 16353–16369. [CrossRef] [PubMed]

30. Ju, M.J.; Hong, Y.-J.; Makita, S.; Lim, Y.; Kurokawa, K.; Duan, L.; Miura, M.; Tang, S.; Yasuno, Y. Advanced multi-contrast Jones matrix optical coherence tomography for Doppler and polarization sensitive imaging. *Opt. Express* **2013**, *21*, 19412–19436. [CrossRef] [PubMed]

31. Wojtkowski, M.; Leitgeb, R.; Kowalczyk, A.; Bajraszewski, T.; Fercher, A.F. In vivo human retinal imaging by Fourier domain optical coherence tomography. *J. Biomed. Opt.* **2002**, *7*, 457–463. [CrossRef] [PubMed]

32. Kiseleva, E.; Kirillin, M.; Feldchtein, F.; Vitkin, A.; Sergeeva, E.; Zagaynova, E.; Streltzova, O.; Shakhov, B.; Gubarkova, E.; Gladkova, N. Differential diagnosis of human bladder mucosa pathologies in vivo with cross-polarization optical coherence tomography. *Biomed. Opt. Express* **2015**, *6*, 1464–1476. [CrossRef] [PubMed]

33. Gubarkova, E.V.; Kirillin, M.Y.; Dudenkova, V.V.; Timashev, P.S.; Kotova, S.L.; Kiseleva, E.B.; Timofeeva, L.B.; Belkova, G.V.; Solovieva, A.B.; Moiseev, A.A.; et al. Quantitative evaluation of atherosclerotic plaques using cross-polarization optical coherence tomography, nonlinear, and atomic force microscopy. *J. Biomed. Opt.* **2016**, *21*, 126010. [CrossRef] [PubMed]

34. Liu, B.; Harman, M.; Giattina, S.; Stamper, D.L.; Demakis, C.; Chilek, M.; Raby, S.; Brezinski, M.E. Characterizing of tissue microstructure with single-detector polarization-sensitive optical coherence tomography. *Appl. Opt.* **2006**, *45*, 4464–4479. [CrossRef] [PubMed]

35. Jiao, S.; Wang, L.V. Two-dimensional depth-resolved Mueller matrix of biological tissue measured with double-beam polarization-sensitive optical coherence tomography. *Opt. Lett.* **2002**, *27*, 101–103. [CrossRef] [PubMed]

36. Jiao, S.; Yao, G.; Wang, L.V. Depth-resolved two-dimensional Stokes vectors of backscattered light and Mueller matrices of biological tissue measured with optical coherence tomography. *Appl. Opt.* **2000**, *39*, 6318–6324. [CrossRef] [PubMed]

37. Jiao, S.; Yu, W.; Stoica, G.; Wang, L.V. Optical-fiber-based Mueller optical coherence tomography. *Opt. Lett.* **2003**, *28*, 1206–1208. [CrossRef] [PubMed]

38. Villiger, M.; Bouma, B.E. Practical decomposition for physically admissible differential Mueller matrices. *Opt. Lett.* **2014**, *39*, 1779–1782. [CrossRef] [PubMed]

39. Everett, M.J.; Schoenenberger, K.; Colston, B.W., Jr.; Da Silva, L.B. Birefringence characterization of biological tissue by use of optical coherence tomography. *Opt. Lett.* **1998**, *23*, 228–230. [CrossRef] [PubMed]

40. Pircher, M.; Goetzinger, E.; Leitgeb, R.; Hitzenberger, C.K. Transversal phase resolved polarization sensitive optical coherence tomography. *Phys. Med. Biol.* **2004**, *49*, 1257–1263. [CrossRef] [PubMed]

41. Götzinger, E.; Pircher, M.; Hitzenberger, C.K. High speed spectral domain polarization sensitive optical coherence tomography of the human retina. *Opt. Express* **2005**, *13*, 10217–10229. [CrossRef] [PubMed]

42. Schmoll, T.; Gotzinger, E.; Pircher, M.; Hitzenberger, C.K.; Leitgeb, R.A. Single-camera polarization-sensitive spectral-domain OCT by spatial frequency encoding. *Opt. Lett.* **2010**, *35*, 241–243. [CrossRef] [PubMed]

43. Fan, C.; Yao, G. Mapping local optical axis in birefringent samples using polarization-sensitive optical coherence tomography. *J. Biomed. Opt.* **2012**, *17*, 110501. [CrossRef] [PubMed]

44. Fan, C.; Yao, G. Mapping local retardance in birefringent samples using polarization sensitive optical coherence tomography. *Opt. Lett.* **2012**, *37*, 1415–1417. [CrossRef] [PubMed]

45. Al-Qaisi, M.K.; Akkin, T. Polarization-sensitive optical coherence tomography based on polarization-maintaining fibers and frequency multiplexing. *Opt. Express* **2008**, *16*, 13032–13041. [CrossRef] [PubMed]

46. Gotzinger, E.; Baumann, B.; Pircher, M.; Hitzenberger, C.K. Polarization maintaining fiber based ultra-high resolution spectral domain polarization sensitive optical coherence tomography. *Opt. Express* **2009**, *17*, 22704–22717. [CrossRef] [PubMed]

47. Al-Qaisi, M.K.; Akkin, T. Swept-source polarization-sensitive optical coherence tomography based on polarization-maintaining fiber. *Opt. Express* **2010**, *18*, 3392–3403. [CrossRef] [PubMed]

48. Wang, H.; Al-Qaisi, M.K.; Akkin, T. Polarization-maintaining fiber based polarization-sensitive optical coherence tomography in spectral domain. *Opt. Lett.* **2010**, *35*, 154–156. [CrossRef] [PubMed]

49. Zotter, S.; Pircher, M.; Torzicky, T.; Baumann, B.; Yoshida, H.; Hirose, F.; Roberts, P.; Ritter, M.; Schütze, C.; Götzinger, E.; et al. Large-field high-speed polarization sensitive spectral domain OCT and its applications in ophthalmology. *Biomed. Opt. Express* **2012**, *3*, 2720–2732. [CrossRef] [PubMed]

50. Bonesi, M.; Sattmann, H.; Torzicky, T.; Zotter, S.; Baumann, B.; Pircher, M.; Götzinger, E.; Eigenwillig, C.; Wieser, W.; Huber, R.; et al. High-speed polarization sensitive optical coherence tomography scan engine based on Fourier domain mode locked laser. *Biomed. Opt. Express* **2012**, *3*, 2987–3000. [CrossRef] [PubMed]

51. Wang, H.; Akkin, T.; Magnain, C.; Wang, R.P.; Dubb, J.; Kostis, W.J.; Yaseen, M.A.; Cramer, A.; Sakadzic, S.; Boas, D. Polarization sensitive optical coherence microscopy for brain imaging. *Opt. Lett.* **2016**, *41*, 2213–2216. [CrossRef] [PubMed]

52. Lin, H.; Kao, M.-C.; Lai, C.-M.; Huang, J.-C.; Kuo, W.-C. All fiber optics circular-state swept source polarization-sensitive optical coherence tomography. *J. Biomed. Opt.* **2013**, *19*, 021110. [CrossRef] [PubMed]

53. Lurie, K.L.; Moritz, T.J.; Ellerbee, A.K. Design considerations for polarization-sensitive optical coherence tomography with a single input polarization state. *Biomed. Opt. Express* **2012**, *3*, 2273–2287. [CrossRef] [PubMed]

54. Trasischker, W.; Zotter, S.; Torzicky, T.; Baumann, B.; Haindl, R.; Pircher, M.; Hitzenberger, C.K. Single input state polarization sensitive swept source optical coherence tomography based on an all single mode fiber interferometer. *Biomed. Opt. Express* **2014**, *5*, 2798–2809. [CrossRef] [PubMed]

55. Lippok, N.; Villiger, M.; Jun, C.; Bouma, B.E. Single input state, single-mode fiber-based polarization-sensitive optical frequency domain imaging by eigenpolarization referencing. *Opt. Lett.* **2015**, *40*, 2025–2028. [CrossRef] [PubMed]

56. Kemp, N.J.; Zaatari, H.N.; Park, J.; Rylander Iii, H.G.; Milner, T.E. Form-biattenuance in fibrous tissues measured with polarization-sensitive optical coherence tomography (PS-OCT). *Opt. Express* **2005**, *13*, 4611–4628. [CrossRef] [PubMed]

57. Pircher, M.; Götzinger, E.; Baumann, B.; Hitzenberger, C.K. Corneal birefringence compensation for polarization sensitive optical coherence tomography of the human retina. *J. Biomed. Opt.* **2007**, *12*, 041210. [CrossRef] [PubMed]

58. De Boer, J.F.; Milner, T.E.; Nelson, J.S. Determination of the depth-resolved Stokes parameters of light backscattered from turbid media by use of polarization-sensitive optical coherence tomography. *Opt. Lett.* **1999**, *24*, 300–302. [CrossRef] [PubMed]

59. Jiao, S.; Wang, L.V. Jones-matrix imaging of biological tissues with quadruple-channel optical coherence tomography. *J. Biomed. Opt.* **2002**, *7*, 350–358. [CrossRef] [PubMed]

60. Yasuno, Y.; Makita, S.; Endo, T.; Itoh, M.; Yatagai, T.; Takahashi, M.; Katada, C.; Mutoh, M. Polarization-sensitive complex Fourier domain optical coherence tomography for Jones matrix imaging of biological samples. *Appl. Phys. Lett.* **2004**, *85*, 3023–3025. [CrossRef]

61. Makita, S.; Yasuno, Y.; Endo, T.; Itoh, M.; Yatagi, T. Jones matrix imaging of biological samples using parallel-detecting polarization-sensitive Fourier domain optical coherence tomography. *Opt. Rev.* **2005**, *12*, 146–148. [CrossRef]

62. Park, J.; Kemp, N.J.; Zaatari, H.N.; Rylander, H.G.; Milner, T.E. Differential geometry of normalized Stokes vector trajectories in anisotropic media. *J. Opt. Soc. Am. A* **2006**, *23*, 679–690. [CrossRef]

63. Makita, S.; Yasuno, Y.; Endo, T.; Itoh, M.; Yatagai, T. Polarization contrast imaging of biological tissues by polarization-sensitive Fourier-domain optical coherence tomography. *Appl. Opt.* **2006**, *45*, 1142–1147. [CrossRef] [PubMed]

64. Saxer, C.E.; de Boer, J.F.; Park, B.H.; Zhao, Y.; Chen, Z.; Nelson, J.S. High-speed fiber-based polarization-sensitive optical coherence tomography of in vivo human skin. *Opt. Lett.* **2000**, *25*, 1355–1357. [CrossRef] [PubMed]

65. Yamanari, M.; Makita, S.; Madjarova, V.D.; Yatagai, T.; Yasuno, Y. Fiber-based polarization-sensitive Fourier domain optical coherence tomography using B-scan-oriented polarization modulation method. *Opt. Express* **2006**, *14*, 6502–6515. [CrossRef] [PubMed]

66. Park, B.H.; Saxer, C.; Srinivas, S.M.; Nelson, J.S.; de Boer, J.F. In vivo burn depth determination by high-speed fiber-based polarization sensitive optical coherence tomography. *J. Biomed. Opt.* **2001**, *6*, 474–479. [CrossRef] [PubMed]

67. Park, B.H.; Pierce, M.C.; Cense, B.; de Boer, J.F. Real-time multi-functional optical coherence tomography. *Opt. Express* **2003**, *11*, 782–793. [CrossRef] [PubMed]

68. Park, B.H.; Pierce, M.C.; Cense, B.; Yun, S.H.; Mujat, M.; Tearney, G.J.; Bouma, B.E.; de Boer, J.F. Real-time fiber-based multi-functional spectral-domain optical coherence tomography at 1.3 μm. *Opt. Express* **2005**, *13*, 3931–3944. [CrossRef] [PubMed]

69. Wang, Y.; Yao, G. Optical tractography of the mouse heart using polarization-sensitive optical coherence tomography. *Biomed. Opt. Express* **2013**, *4*, 2540–2545. [CrossRef] [PubMed]

70. Azinfar, L.; Ravanfar, M.; Wang, Y.; Zhang, K.; Duan, D.; Yao, G. High resolution imaging of the fibrous microstructure in bovine common carotid artery using optical polarization tractography. *J. Biophotonics* **2015**, *10*, 231–241. [CrossRef] [PubMed]

71. Oh, W.Y.; Yun, S.H.; Vakoc, B.J.; Shishkov, M.; Desjardins, A.E.; Park, B.H.; de Boer, J.F.; Tearney, G.J.; Bouma, E. High-speed polarization sensitive optical frequency domain imaging with frequency multiplexing. *Opt. Express* **2008**, *16*, 1096–1103. [CrossRef] [PubMed]

72. Oh, W.Y.; Vakoc, B.J.; Yun, S.H.; Tearney, G.J.; Bouma, B.E. Single-detector polarization-sensitive optical frequency domain imaging using high-speed intra a-line polarization modulation. *Opt. Lett.* **2008**, *33*, 1330–1332. [CrossRef] [PubMed]

73. Baumann, B.; Choi, W.; Potsaid, B.; Huang, D.; Duker, J.S.; Fujimoto, J.G. Swept source/Fourier domain polarization sensitive optical coherence tomography with a passive polarization delay unit. *Opt. Express* **2012**, *20*, 10229–10241. [CrossRef] [PubMed]

74. Lim, Y.; Hong, Y.-J.; Duan, L.; Yamanari, M.; Yasuno, Y. Passive component based multifunctional Jones matrix swept source optical coherence tomography for Doppler and polarization imaging. *Opt. Lett.* **2012**, *37*, 1958–1960. [CrossRef] [PubMed]

75. Lu, Z.; Kasaragod, D.K.; Matcher, S.J. Method to calibrate phase fluctuation in polarization-sensitive swept-source optical coherence tomography. *J. Biomed. Opt.* **2011**, *16*, 070502. [CrossRef] [PubMed]

76. Fan, C.; Yao, G. Imaging myocardial fiber orientation using polarization sensitive optical coherence tomography. *Biomed. Opt. Express* **2013**, *4*, 460–465. [CrossRef] [PubMed]

77. Lo, W.C.Y.; Villiger, M.; Golberg, A.; Broelsch, G.F.; Khan, S.; Lian, C.G.; Austen, W.G., Jr.; Yarmush, M.; Bouma, B.E. Longitudinal, 3D imaging of collagen remodeling in murine hypertrophic scars in vivo using polarization-sensitive optical frequency domain imaging. *J. Investig. Dermatol.* **2016**, *136*, 84–92. [CrossRef] [PubMed]

78. Li, E.; Makita, S.; Hong, Y.-J.; Kasaragod, D.; Yasuno, Y. Three-dimensional multi-contrast imaging of in vivo human skin by Jones matrix optical coherence tomography. *Biomed. Opt. Express* **2017**, *8*, 1290–1305. [CrossRef]

79. Cense, B.; Chen, T.C.; Park, B.H.; Pierce, M.C.; de Boer, J.F. In vivo depth-resolved birefringence measurements of the human retinal nerve fiber layer by polarization-sensitive optical coherence tomography. *Opt. Lett.* **2002**, *27*, 1610–1612. [CrossRef] [PubMed]

80. Pierce, M.C.; Park, B.H.; Cense, B.; de Boer, J.F. Simultaneous intensity, birefringence, and flow measurements with high-speed fiber-based optical coherence tomography. *Opt. Lett.* **2002**, *27*, 1534–1536. [CrossRef] [PubMed]

81. Leitgeb, R.; Hitzenberger, C.K.; Fercher, A.F. Performance of fourier domain vs. Time domain optical coherence tomography. *Opt. Express* **2003**, *11*, 889–894. [CrossRef] [PubMed]

82. De Boer, J.F.; Cense, B.; Park, B.H.; Pierce, M.C.; Tearney, G.J.; Bouma, B.E. Improved signal-to-noise ratio in spectral-domain compared with time-domain optical coherence tomography. *Opt. Lett.* **2003**, *28*, 2067–2069. [CrossRef] [PubMed]

83. Choma, M.A.; Sarunic, M.V.; Yang, C.H.; Izatt, J.A. Sensitivity advantage of swept source and Fourier domain optical coherence tomography. *Opt. Express* **2003**, *11*, 2183–2189. [CrossRef] [PubMed]

84. Baumann, B.; Gotzinger, E.; Pircher, M.; Hitzenberger, C.K. Single camera based spectral domain polarization sensitive optical coherence tomography. *Opt. Express* **2007**, *15*, 1054–1063. [CrossRef] [PubMed]

85. Cense, B.; Mujat, M.; Chen, T.C.; Park, B.H.; de Boer, J.F. Polarization-sensitive spectral-domain optical coherence tomography using a single line scan camera. *Opt. Express* **2007**, *15*, 2421–2431. [CrossRef] [PubMed]

86. Zhao, M.; Izatt, J.A. Single-camera sequential-scan-based polarization-sensitive SDOCT for retinal imaging. *Opt. Lett.* **2009**, *34*, 205–207. [CrossRef] [PubMed]

87. Fan, C.M.; Yao, G. Single camera spectral domain polarization-sensitive optical coherence tomography using offset B-scan modulation. *Opt. Express* **2010**, *18*, 7281–7287. [CrossRef] [PubMed]

88. Lee, S.W.; Jeong, H.W.; Kim, B.M. High-speed spectral domain polarization-sensitive optical coherence tomography using a single camera and an optical switch at 1.3 μm. *J. Biomed. Opt.* **2010**, *15*, 010501. [CrossRef] [PubMed]

89. Song, C.; Ahn, M.; Gweon, D. Polarization-sensitive spectral-domain optical coherence tomography using a multi-line single camera spectrometer. *Opt. Express* **2010**, *18*, 23805–23817. [CrossRef] [PubMed]

90. Lexer, F.; Hitzenberger, C.K.; Fercher, A.F.; Kulhavy, M. Wavelength-tuning interferometry of intraocular distances. *Appl. Opt.* **1997**, *36*, 6548–6553. [CrossRef] [PubMed]

91. Golubovic, B.; Bouma, B.E.; Tearney, G.J.; Fujimoto, J.G. Optical frequency-domain reflectometry using rapid wavelength tuning of a Cr^{4+}: Forsterite laser. *Opt. Lett.* **1997**, *22*, 1704–1706. [CrossRef] [PubMed]

92. Klein, T.; Huber, R. High-speed OCT light sources and systems. *Biomed. Opt. Express* **2017**, *8*, 828–859. [CrossRef] [PubMed]

93. Zhang, J.; Jung, W.; Nelson, J.S.; Chen, Z. Full range polarization-sensitive Fourier domain optical coherence tomography. *Opt. Express* **2004**, *12*, 6033–6039. [CrossRef] [PubMed]

94. Yamanari, M.; Makita, S.; Yasuno, Y. Polarization-sensitive swept-source optical coherence tomography with continuous source polarization modulation. *Opt. Express* **2008**, *16*, 5892–5906. [CrossRef] [PubMed]

95. Yasuno, Y.; Yamanari, M.; Kawana, K.; Oshika, T.; Miura, M. Investigation of post-glaucoma-surgery structures by three-dimensional and polarization sensitive anterior eye segment optical coherence tomography. *Opt. Express* **2009**, *17*, 3980–3996. [CrossRef] [PubMed]

96. Yamanari, M.; Makita, S.; Lim, Y.; Yasuno, Y. Full-range polarization-sensitive swept-source optical coherence tomography by simultaneous transversal and spectral modulation. *Opt. Express* **2010**, *18*, 13964–13980. [CrossRef] [PubMed]

97. Kim, K.H.; Park, B.H.; Tu, Y.; Hasan, T.; Lee, B.; Li, J.; de Boer, J.F. Polarization-sensitive optical frequency domain imaging based on unpolarized light. *Opt. Express* **2011**, *19*, 552–561. [CrossRef] [PubMed]

98. Torzicky, T.; Pircher, M.; Zotter, S.; Bonesi, M.; Götzinger, E.; Hitzenberger, C.K. Automated measurement of choroidal thickness in the human eye by polarization sensitive optical coherence tomography. *Opt. Express* **2012**, *20*, 7564–7574. [CrossRef] [PubMed]

99. Torzicky, T.; Marschall, S.; Pircher, M.; Baumann, B.; Bonesi, M.; Zotter, S.; Götzinger, E.; Trasischker, W.; Klein, T.; Wieser, W.; et al. Retinal polarization-sensitive optical coherence tomography at 1060 nm with 350 kHz a-scan rate using an Fourier domain mode locked laser. *J. Biomed. Opt.* **2013**, *18*, 026008. [CrossRef] [PubMed]

100. Hong, Y.-J.; Makita, S.; Sugiyama, S.; Yasuno, Y. Optically buffered Jones-matrix-based multifunctional optical coherence tomography with polarization mode dispersion correction. *Biomed. Opt. Express* **2015**, *6*, 225–243. [CrossRef] [PubMed]

101. Cho, H.S.; Oh, W.-Y. Polarization-sensitive OFDI using polarization-multiplexed wavelength-swept laser. *Opt. Lett.* **2014**, *39*, 4065–4067. [CrossRef] [PubMed]

102. Wang, Z.; Lee, H.-C.; Vermeulen, D.; Chen, L.; Nielsen, T.; Park, S.Y.; Ghaemi, A.; Swanson, E.; Doerr, C.; Fujimoto, J. Silicon photonic integrated circuit swept-source optical coherence tomography receiver with dual polarization, dual balanced, in-phase and quadrature detection. *Biomed. Opt. Express* **2015**, *6*, 2562–2574. [CrossRef] [PubMed]

103. Hong, Y.-J.; Miura, M.; Ju, M.J.; Makita, S.; Iwasaki, T.; Yasuno, Y. Simultaneous investigation of vascular and retinal pigment epithelial pathologies of exudative macular diseases by multifunctional optical coherence tomography. *Investig. Ophthalmol. Vis. Sci.* **2014**, *55*, 5016–5031. [CrossRef] [PubMed]

104. Kim, S.; Park, T.; Jang, S.-J.; Nam, A.S.; Vakoc, B.J.; Oh, W.-Y. Multi-functional angiographic OFDI using frequency-multiplexed dual-beam illumination. *Opt. Express* **2015**, *23*, 8939–8947. [CrossRef] [PubMed]

105. Augustin, M.; Fialová, S.; Himmel, T.; Glösmann, M.; Lengheimer, T.; Harper, D.J.; Plasenzotti, R.; Pircher, M.; Hitzenberger, C.K.; Baumann, B. Multi-functional OCT enables longitudinal study of retinal changes in a VLDLR knockout mouse model. *PLoS ONE* **2016**, *11*, e0164419. [CrossRef] [PubMed]

106. Gong, P.; Chin, L.; Es'haghian, S.; Liew, Y.M.; Wood, F.M.; Sampson, D.D.; McLaughlin, R.A. Imaging of skin birefringence for human scar assessment using polarization-sensitive optical coherence tomography aided by vascular masking. *J. Biomed. Opt.* **2014**, *19*, 126014. [CrossRef] [PubMed]

107. Pierce, M.C.; Shishkov, M.; Park, B.H.; Nassif, N.A.; Bouma, B.E.; Tearney, G.J.; de Boer, J.F. Effects of sample arm motion in endoscopic polarization-sensitive optical coherence tomography. *Opt. Express* **2005**, *13*, 5739–5749. [CrossRef] [PubMed]

108. Kim, K.H.; Park, B.H.; Maguluri, G.N.; Lee, T.W.; Rogomentich, F.J.; Bancu, M.G.; Bouma, B.E.; de Boer, J.F.; Bernstein, J.J. Two-axis magnetically-driven mems scanning catheter for endoscopic high-speed optical coherence tomography. *Opt. Express* **2007**, *15*, 18130–18140. [CrossRef] [PubMed]

109. Kim, K.H.; Burns, J.A.; Bernstein, J.J.; Maguluri, G.N.; Park, B.H.; de Boer, J.F. In vivo 3D human vocal fold imaging with polarization sensitive optical coherence tomography and a mems scanning catheter. *Opt. Express* **2010**, *18*, 14644–14653. [CrossRef] [PubMed]

110. Li, J.; Feroldi, F.; de Lange, J.; Daniels, J.M.A.; Grünberg, K.; de Boer, J.F. Polarization sensitive optical frequency domain imaging system for endobronchial imaging. *Opt. Express* **2015**, *23*, 3390–3402. [CrossRef] [PubMed]

111. Van der Sijde, J.N.; Karanasos, A.; Villiger, M.; Bouma, B.E.; Regar, E. First-in-man assessment of plaque rupture by polarization-sensitive optical frequency domain imaging in vivo. *Eur. Heart J.* **2016**, *37*, 1932. [CrossRef] [PubMed]

112. Villiger, M.; Lorenser, D.; McLaughlin, R.A.; Quirk, B.C.; Kirk, R.W.; Bouma, B.E.; Sampson, D.D. Deep tissue volume imaging of birefringence through fibre-optic needle probes for the delineation of breast tumour. *Sci. Rep.* **2016**, *6*, 28771. [CrossRef] [PubMed]

113. Li, B.H.; Leung, A.S.O.; Soong, A.; Munding, C.E.; Lee, H.; Thind, A.S.; Munce, N.R.; Wright, G.A.; Rowsell, C.H.; Yang, V.X.D.; et al. Hybrid intravascular ultrasound and optical coherence tomography catheter for imaging of coronary atherosclerosis. *Catheter. Cardiovasc. Interv.* **2013**, *81*, 494–507. [CrossRef] [PubMed]

114. Yoon, Y.; Jang, W.H.; Xiao, P.; Kim, B.; Wang, T.; Li, Q.; Lee, J.Y.; Chung, E.; Kim, K.H. In vivo wide-field reflectance/fluorescence imaging and polarization-sensitive optical coherence tomography of human oral cavity with a forward-viewing probe. *Biomed. Opt. Express* **2015**, *6*, 524–535. [CrossRef] [PubMed]

115. Liu, G.J.; Zhang, J.; Yu, L.F.; Xie, T.Q.; Chen, Z.P. Real-time polarization-sensitive optical coherence tomography data processing with parallel computing. *Appl. Opt.* **2009**, *48*, 6365–6370. [CrossRef] [PubMed]

116. Duan, L.; Yamanari, M.; Yasuno, Y. Automated phase retardation oriented segmentation of chorio-scleral interface by polarization sensitive optical coherence tomography. *Opt. Express* **2012**, *20*, 3353–3366. [CrossRef] [PubMed]

117. Kasaragod, D.; Makita, S.; Fukuda, S.; Beheregaray, S.; Oshika, T.; Yasuno, Y. Bayesian maximum likelihood estimator of phase retardation for quantitative polarization-sensitive optical coherence tomography. *Opt. Express* **2014**, *22*, 16472–16492. [CrossRef] [PubMed]

118. Cense, B.; Gao, W.; Brown, J.M.; Jones, S.M.; Jonnal, R.S.; Mujat, M.; Park, B.H.; de Boer, J.F.; Miller, D.T. Retinal imaging with polarization-sensitive optical coherence tomography and adaptive optics. *Opt. Express* **2009**, *17*, 21634–21651. [CrossRef] [PubMed]

119. Götzinger, E.; Pircher, M.; Baumann, B.; Schmoll, T.; Sattmann, H.; Leitgeb, R.A.; Hitzenberger, C.K. Speckle noise reduction in high speed polarization sensitive spectral domain optical coherence tomography. *Opt. Express* **2011**, *19*, 14568–14585. [CrossRef] [PubMed]

120. Sugita, M.; Zotter, S.; Pircher, M.; Makihira, T.; Saito, K.; Tomatsu, N.; Sato, M.; Roberts, P.; Schmidt-Erfurth, U.; Hitzenberger, C.K. Motion artifact and speckle noise reduction in polarization sensitive optical coherence tomography by retinal tracking. *Biomed. Opt. Express* **2014**, *5*, 106–122. [CrossRef] [PubMed]

121. Baumann, B.; Götzinger, E.; Pircher, M.; Sattmann, H.; Schütze, C.; Schlanitz, F.; Ahlers, C.; Schmidt-Erfurth, U.; Hitzenberger, C.K. Segmentation and quantification of retinal lesions in age-related macular degeneration using polarization-sensitive optical coherence tomography. *J. Biomed. Opt.* **2010**, *15*, 061704. [CrossRef] [PubMed]

122. Schlanitz, F.G.; Baumann, B.; Spalek, T.; Schütze, C.; Ahlers, C.; Pircher, M.; Götzinger, E.; Hitzenberger, C.K.; Schmidt-Erfurth, U. Performance of automated drusen detection by polarization-sensitive optical coherence tomography. *Investig. Ophthalmol. Vis. Sci.* **2011**, *52*, 4571–4579. [CrossRef] [PubMed]

123. Lammer, J.; Bolz, M.; Baumann, B.; Pircher, M.; Gerendas, B.; Schlanitz, F.; Hitzenberger, C.K.; Schmidt-Erfurth, U. Detection and analysis of hard exudates by polarization-sensitive optical coherence tomography in patients with diabetic maculopathy. *Investig. Ophthalmol. Vis. Sci.* **2014**, *55*, 1564–1571. [CrossRef] [PubMed]

124. Schlanitz, F.G.; Sacu, S.; Baumann, B.; Bolz, M.; Platzer, M.; Pircher, M.; Hitzenberger, C.K.; Schmidt-Erfurth, U. Identification of drusen characteristics in age-related macular degeneration by polarization-sensitive optical coherence tomography. *Am. J. Ophthalmol.* **2015**, *160*, 335–344. [CrossRef] [PubMed]

125. Sugita, M.; Pircher, M.; Zotter, S.; Baumann, B.; Saito, K.; Makihira, T.; Tomatsu, N.; Sato, M.; Hitzenberger, C.K. Analysis of optimum conditions of depolarization imaging by polarization-sensitive optical coherence tomography in the human retina. *J. Biomed. Opt.* **2015**, *20*, 016011. [CrossRef] [PubMed]

126. Miyazawa, A.; Yamanari, M.; Makita, S.; Miura, M.; Kawana, K.; Iwaya, K.; Goto, H.; Yasuno, Y. Tissue discrimination in anterior eye using three optical parameters obtained by polarization sensitive optical coherence tomography. *Opt. Express* **2009**, *17*, 17426–17440. [CrossRef] [PubMed]

127. Yasuno, Y.; Yamanari, M.; Kawana, K.; Miura, M.; Fukuda, S.; Makita, S.; Sakai, S.; Oshika, T. Visibility of trabecular meshwork by standard and polarization-sensitive optical coherence tomography. *J. Biomed. Opt.* **2010**, *15*, 061705. [CrossRef] [PubMed]

128. Fukuda, S.; Yamanari, M.; Lim, Y.; Hoshi, S.; Beheregaray, S.; Oshika, T.; Yasuno, Y. Keratoconus diagnosis using anterior segment polarization-sensitive optical coherence tomography. *Investig. Ophthalmol. Vis. Sci.* **2013**, *54*, 1384–1391. [CrossRef] [PubMed]

129. Duan, L.; Marvdashti, T.; Ellerbee, A.K. Polarization-sensitive interleaved optical coherence tomography. *Opt. Express* **2015**, *23*, 13693–13703. [CrossRef] [PubMed]

130. Drexler, W.; Fujimoto, J.G. State-of-the-art retinal optical coherence tomography. *Prog. Retin. Eye Res.* **2008**, *27*, 45–88. [CrossRef] [PubMed]

131. Pircher, M.; Hitzenberger, C.K.; Schmidt-Erfurth, U. Polarization sensitive optical coherence tomography in the human eye. *Prog. Retin. Eye Res.* **2011**, *30*, 431–451. [CrossRef] [PubMed]

132. Kingman, S. Glaucoma is second leading cause of blindness globally. *Bull. World Health Organ.* **2004**, *82*, 887–888. [PubMed]

133. Weinreb, R.N.; Dreher, A.W.; Coleman, A.; Quigley, H.; Shaw, B.; Reiter, K. Histopathologic validation of Fourier-ellipsometry measurements of retinal nerve fiber layer thickness. *Arch. Ophthalmol.* **1990**, *108*, 557–560. [CrossRef] [PubMed]

134. Huang, X.-R.; Knighton, R.W. Linear birefringence of the retinal nerve fiber layer measured in vitro with a multispectral imaging micropolarimeter. *J. Biomed. Opt.* **2002**, *7*, 199–204. [CrossRef] [PubMed]

135. Fortune, B.; Burgoyne, C.F.; Cull, G.; Reynaud, J.; Wang, L. Onset and progression of peripapillary retinal nerve fiber layer (RNFL) retardance changes occur earlier than RNFL thickness changes in experimental glaucoma. *Investig. Ophthalmol. Vis. Sci.* **2013**, *54*, 5653–5661. [CrossRef] [PubMed]

136. Bussel, I.I.; Wollstein, G.; Schuman, J.S. OCT for glaucoma diagnosis, screening and detection of glaucoma progression. *Br. J. Ophthalmol.* **2014**, *98*, ii15–ii19. [CrossRef] [PubMed]

137. Weinreb, R.N.; Zangwill, L.; Berry, C.C.; Bathija, R.; Sample, P.A. Detection of glaucoma with scanning laser polarimetry. *Arch. Ophthalmol.* **1998**, *116*, 1583–1589. [CrossRef] [PubMed]

138. Klemm, M.; Rumberger, E.; Walter, A.; Richard, G. Quantifizierung der retinalen nervenfaserschichtdicke ein vergleich von laser-scanning-ophthalmoskopie, polarimetrie und optischer kohärenztomographie bei gesunden und glaukomkranken augen. *Der Ophthalmologe* **2001**, *98*, 832–843. [CrossRef] [PubMed]

139. Knighton, R.W.; Huang, X.-R.; Greenfield, D.S. Analytical model of scanning laser polarimetry for retinal nerve fiber layer assessment. *Investig. Ophthalmol. Vis. Sci.* **2002**, *43*, 383–392.

140. Ducros, M.G.; Marsack, J.D.; Rylander, H.G.; Thomsen, S.L.; Milner, T.E. Primate retina imaging with polarization-sensitive optical coherence tomography. *J. Opt. Soc. Am. A* **2001**, *18*, 2945–2956. [CrossRef]

141. Cense, B.; Chen, T.C.; Park, B.H.; Pierce, M.C.; de Boer, J.F. Thickness and birefringence of healthy retinal nerve fiber layer tissue measured with polarization-sensitive optical coherence tomography. *Investig. Ophthalmol. Vis. Sci.* **2004**, *45*, 2606–2612. [CrossRef] [PubMed]

142. Yamanari, M.; Miura, M.; Makita, S.; Yatagai, T.; Yasuno, Y. Phase retardation measurement of retinal nerve fiber layer by polarization-sensitive spectral-domain optical coherence tomography and scanning laser polarimetry. *J. Biomed. Opt.* **2008**, *13*, 014013. [CrossRef] [PubMed]

143. Götzinger, E.; Pircher, M.; Baumann, B.; Hirn, C.; Vass, C.; Hitzenberger, C. Retinal nerve fiber layer birefringence evaluated with polarization sensitive spectral domain OCT and scanning laser polarimetry: A comparison. *J. Biophotonics* **2008**, *1*, 129–139. [CrossRef] [PubMed]

144. Götzinger, E.; Pircher, M.; Baumann, B.; Hirn, C.; Vass, C.; Hitzenberger, C.K. Analysis of the origin of atypical scanning laser polarimetry patterns by polarization-sensitive optical coherence tomography. *Investig. Ophthalmol. Vis. Sci.* **2008**, *49*, 5366–5372. [CrossRef] [PubMed]

145. Zotter, S.; Pircher, M.; Götzinger, E.; Torzicky, T.; Yoshida, H.; Hirose, F.; Holzer, S.; Kroisamer, J.; Vass, C.; Schmidt-Erfurth, U. Measuring retinal nerve fiber layer birefringence, retardation, and thickness using wide-field, high-speed polarization sensitive spectral domain OCT. *Investig. Ophthalmol. Vis. Sci.* **2013**, *54*, 72–84. [CrossRef] [PubMed]

146. Pocock, G.M.; Aranibar, R.G.; Kemp, N.J.; Specht, C.S.; Markey, M.K.; Rylander, H.G. The relationship between retinal ganglion cell axon constituents and retinal nerve fiber layer birefringence in the primate. *Investig. Ophthalmol. Vis. Sci.* **2009**, *50*, 5238–5246. [CrossRef] [PubMed]

147. Dwelle, J.; Liu, S.; Wang, B.; McElroy, A.; Ho, D.; Markey, M.K.; Milner, T.; Rylander, H.G. Thickness, phase retardation, birefringence, and reflectance of the retinal nerve fiber layer in normal and glaucomatous non-human primates. *Investig. Ophthalmol. Vis. Sci.* **2012**, *53*, 4380–4395. [CrossRef] [PubMed]

148. Fialová, S.; Augustin, M.; Fischak, C.; Schmetterer, L.; Handschuh, S.; Glösmann, M.; Pircher, M.; Hitzenberger, C.K.; Baumann, B. Posterior rat eye during acute intraocular pressure elevation studied using polarization sensitive optical coherence tomography. *Biomed. Opt. Express* **2017**, *8*, 298–314. [CrossRef] [PubMed]

149. Sugita, M.; Pircher, M.; Zotter, S.; Baumann, B.; Roberts, P.; Makihira, T.; Tomatsu, N.; Sato, M.; Vass, C.; Hitzenberger, C.K. Retinal nerve fiber bundle tracing and analysis in human eye by polarization sensitive OCT. *Biomed. Opt. Express* **2015**, *6*, 1030–1054. [CrossRef] [PubMed]

150. Pircher, M.; Götzinger, E.; Leitgeb, R.; Sattmann, H.; Findl, O.; Hitzenberger, C.K. Imaging of polarization properties of human retina in vivo with phase resolved transversal PS-OCT. *Opt. Express* **2004**, *12*, 5940–5951. [CrossRef] [PubMed]

151. Pircher, M.; Götzinger, E.; Findl, O.; Michels, S.; Geitzenauer, W.; Leydolt, C.; Schmidt-Erfurth, U.; Hitzenberger, C.K. Human macula investigated in vivo with polarization-sensitive optical coherence tomography. *Investig. Ophthalmol. Vis. Sci.* **2006**, *47*, 5487–5494. [CrossRef] [PubMed]

152. Baumann, B.; Götzinger, E.; Pircher, M.; Hitzenberger, C.K. Measurements of depolarization distribution in the healthy human macula by polarization sensitive OCT. *J. Biophotonics* **2009**, *2*, 426–434. [CrossRef] [PubMed]

153. Schütze, C.; Ritter, M.; Blum, R.; Zotter, S.; Baumann, B.; Pircher, M.; Hitzenberger, C.K.; Schmidt-Erfurth, U. Retinal pigment epithelium findings in patients with albinism using wide-field polarization-sensitive optical coherence tomography. *Retina* **2014**, *34*, 2208–2217. [CrossRef] [PubMed]

154. Ahlers, C.; Götzinger, E.; Pircher, M.; Golbaz, I.; Prager, F.; Schütze, C.; Baumann, B.; Hitzenberger, C.K.; Schmidt-Erfurth, U. Imaging of the retinal pigment epithelium in age-related macular degeneration using polarization-sensitive optical coherence tomography. *Investig. Ophthalmol. Vis. Sci.* **2010**, *51*, 2149–2157. [CrossRef] [PubMed]

155. Michels, S.; Pircher, M.; Geitzenauer, W.; Simader, C.; Götzinger, E.; Findl, O.; Schmidt-Erfurth, U.; Hitzenberger, C. Value of polarisation-sensitive optical coherence tomography in diseases affecting the retinal pigment epithelium. *Br. J. Ophthalmol.* **2008**, *92*, 204–209. [CrossRef] [PubMed]

156. Schlanitz, F.G.; Baumann, B.; Kundi, M.; Sacu, S.; Baratsits, M.; Scheschy, U.; Shahlaee, A.; Mittermüller, T.J.; Montuoro, A.; Roberts, P. Drusen volume development over time and its relevance to the course of age-related macular degeneration. *Br. J. Ophthalmol.* **2016**, *101*, 198–203. [CrossRef] [PubMed]

157. Schütze, C.; Bolz, M.; Sayegh, R.; Baumann, B.; Pircher, M.; Götzinger, E.; Hitzenberger, C.K.; Schmidt-Erfurth, U. Lesion size detection in geographic atrophy by polarization-sensitive optical coherence tomography and correlation to conventional imaging techniques. *Investig. Ophthalmol. Vis. Sci.* **2013**, *54*, 739–745. [CrossRef] [PubMed]

158. Sayegh, R.G.; Zotter, S.; Roberts, P.K.; Kandula, M.M.; Sacu, S.; Kreil, D.P.; Baumann, B.; Pircher, M.; Hitzenberger, C.K.; Schmidt-Erfurth, U. Polarization-sensitive optical coherence tomography and conventional retinal imaging strategies in assessing foveal integrity in geographic atrophy. *Investig. Ophthalmol. Vis. Sci.* **2015**, *56*, 5246–5255. [CrossRef] [PubMed]

159. Roberts, P.; Sugita, M.; Deák, G.; Baumann, B.; Zotter, S.; Pircher, M.; Sacu, S.; Hitzenberger, C.K.; Schmidt-Erfurth, U. Automated identification and quantification of subretinal fibrosis in neovascular age-related macular degeneration using polarization-sensitive OCT. *Investig. Ophthalmol. Vis. Sci.* **2016**, *57*, 1699–1705. [CrossRef] [PubMed]

160. Roberts, P.; Baumann, B.; Lammer, J.; Gerendas, B.; Kroisamer, J.; Bühl, W.; Pircher, M.; Hitzenberger, C.K.; Schmidt-Erfurth, U.; Sacu, S. Retinal pigment epithelial features in central serous chorioretinopathy identified by polarization-sensitive optical coherence tomography. *Investig. Ophthalmol. Vis. Sci.* **2016**, *57*, 1595–1603. [CrossRef] [PubMed]

161. Miura, M.; Yamanari, M.; Iwasaki, T.; Elsner, A.E.; Makita, S.; Yatagai, T.; Yasuno, Y. Imaging polarimetry in age-related macular degeneration. *Investig. Ophthalmol. Vis. Sci.* **2008**, *49*, 2661–2667. [CrossRef] [PubMed]

162. Schütze, C.; Wedl, M.; Baumann, B.; Pircher, M.; Hitzenberger, C.K.; Schmidt-Erfurth, U. Progression of retinal pigment epithelial atrophy in antiangiogenic therapy of neovascular age-related macular degeneration. *Am. J. Ophthalmol.* **2015**, *159*, 1100–1114. [CrossRef] [PubMed]

163. Ritter, M.; Zotter, S.; Schmidt, W.M.; Bittner, R.E.; Deak, G.G.; Pircher, M.; Sacu, S.; Hitzenberger, C.K.; Schmidt-Erfurth, U.M. Characterization of stargardt disease using polarization-sensitive optical coherence tomography and fundus autofluorescence imaging. *Investig. Ophthalmol. Vis. Sci.* **2013**, *54*, 6416–6425. [CrossRef] [PubMed]

164. Schütze, C.; Ahlers, C.; Pircher, M.; Baumann, B.; Götzinger, E.; Prager, F.; Matt, G.; Sacu, S.; Hitzenberger, C.K.; Schmidt-Erfurth, U. Morphologic characteristics of idiopathic juxtafoveal telangiectasia using spectral-domain and polarization-sensitive optical coherence tomography. *Retina* **2012**, *32*, 256–264. [CrossRef] [PubMed]

165. Götzinger, E.; Pircher, M.; Dejaco-Ruhswurm, I.; Kaminski, S.; Skorpik, C.; Hitzenberger, C.K. Imaging of birefringent properties of keratoconus corneas by polarization-sensitive optical coherence tomography. *Investig. Ophthalmol. Vis. Sci.* **2007**, *48*, 3551–3558. [CrossRef] [PubMed]

166. Ju, M.J.; Tang, S. Usage of polarization-sensitive optical coherence tomography for investigation of collagen cross-linking. *J. Biomed. Opt.* **2015**, *20*, 046001. [CrossRef] [PubMed]

167. Fukuda, S.; Beheregaray, S.; Kasaragod, D.; Hoshi, S.; Kishino, G.; Ishii, K.; Yasuno, Y.; Oshika, T. Noninvasive evaluation of phase retardation in blebs after glaucoma surgery using anterior segment polarization-sensitive optical coherence tomography. *Investig. Ophthalmol. Vis. Sci.* **2014**, *55*, 5200–5206. [CrossRef] [PubMed]

168. Miura, M.; Kawana, K.; Iwasaki, T.; Kiuchi, T.; Oshika, T.; Mori, H.; Yamanari, M.; Makita, S.; Yatagai, T.; Yasuno, Y. Three-dimensional anterior segment optical coherence tomography of filtering blebs after trabeculectomy. *J. Glaucoma* **2008**, *17*, 193–196. [CrossRef] [PubMed]

169. Lim, Y.; Yamanari, M.; Fukuda, S.; Kaji, Y.; Kiuchi, T.; Miura, M.; Oshika, T.; Yasuno, Y. Birefringence measurement of cornea and anterior segment by office-based polarization-sensitive optical coherence tomography. *Biomed. Opt. Express* **2011**, *2*, 2392–2402. [CrossRef] [PubMed]

170. Iwasaki, T. Polarization-sensitive optical coherence tomography of necrotizing scleritis. *Ophthalmic Surg. Lasers Imaging Retin.* **2009**, *40*, 607.

171. Yamanari, M.; Nagase, S.; Fukuda, S.; Ishii, K.; Tanaka, R.; Yasui, T.; Oshika, T.; Miura, M.; Yasuno, Y. Scleral birefringence as measured by polarization-sensitive optical coherence tomography and ocular biometric parameters of human eyes in vivo. *Biomed. Opt. Express* **2014**, *5*, 1391–1402. [CrossRef] [PubMed]

172. Pircher, M.; Goetzinger, E.; Leitgeb, R.; Hitzenberger, C.K. Three dimensional polarization sensitive OCT of human skin in vivo. *Opt. Express* **2004**, *12*, 3236–3244. [CrossRef] [PubMed]

173. Mogensen, M.; Morsy, H.A.; Thrane, L.; Jemec, G.B.E. Morphology and epidermal thickness of normal skin imaged by optical coherence tomography. *Dermatology* **2008**, *217*, 14–20. [CrossRef] [PubMed]

174. Lee, A.M.D.; Cahill, L.; Liu, K.; MacAulay, C.; Poh, C.; Lane, P. Wide-field in vivo oral OCT imaging. *Biomed. Opt. Express* **2015**, *6*, 2664–2674. [CrossRef] [PubMed]

175. Burns, J.A.; Zeitels, S.M.; Anderson, R.R.; Kobler, J.B.; Pierce, M.C.; de Boer, J.F. Imaging the mucosa of the human vocal fold with optical coherence tomography. *Ann. Otol. Rhinol. Laryngol.* **2005**, *114*, 671–676. [CrossRef] [PubMed]

176. Bonesi, M.; Sattmann, H.; Torzicky, T.; Zotter, S.; Baumann, B.; Pircher, M.; Götzinger, E.; Eigenwillig, C.; Wieser, W.; Huber, R. High-speed polarization sensitive optical coherence tomography scan engine based on Fourier domain mode locked laser: Erratum. *Biomed. Opt. Express* **2013**, *4*, 241–244. [CrossRef]

177. Sakai, S.; Yamanari, M.; Lim, Y.; Nakagawa, N.; Yasuno, Y. In vivo evaluation of human skin anisotropy by polarization-sensitive optical coherence tomography. *Biomed. Opt. Express* **2011**, *2*, 2623–2631. [CrossRef] [PubMed]

178. Jiao, S.; Yu, W.; Stoica, G.; Wang, L.V. Contrast mechanisms in polarization-sensitive Mueller-matrix optical coherence tomography and application in burn imaging. *Appl. Opt.* **2003**, *42*, 5191–5197. [CrossRef] [PubMed]

179. Pierce, M.C.; Sheridan, R.L.; Park, B.H.; Cense, B.; de Boer, J.F. Collagen denaturation can be quantified in burned human skin using polarization-sensitive optical coherence tomography. *Burns* **2004**, *30*, 511–517. [CrossRef] [PubMed]

180. Oh, J.-T.; Lee, S.-W.; Kim, Y.-S.; Suhr, K.-B.; Kim, B.-M. Quantification of the wound healing using polarization-sensitive optical coherence tomography. *J. Biomed. Opt.* **2006**, *11*, 041124. [CrossRef] [PubMed]

181. Sahu, K.; Verma, Y.; Sharma, M.; Rao, K.; Gupta, P. Non-invasive assessment of healing of bacteria infected and uninfected wounds using optical coherence tomography. *Skin Res. Technol.* **2010**, *16*, 428–437. [CrossRef] [PubMed]

182. Strasswimmer, J.; Pierce, M.C.; Park, B.H.; Neel, V.; de Boer, J.F. Polarization-sensitive optical coherence tomography of invasive basal cell carcinoma. *J. Biomed. Opt.* **2004**, *9*, 292–298. [CrossRef] [PubMed]

183. Duan, L.; Marvdashti, T.; Lee, A.; Tang, J.Y.; Ellerbee, A.K. Automated identification of basal cell carcinoma by polarization-sensitive optical coherence tomography. *Biomed. Opt. Express* **2014**, *5*, 3717–3729. [CrossRef] [PubMed]

184. Burns, J.A.; Kim, K.H.; deBoer, J.F.; Anderson, R.R.; Zeitels, S.M. Polarization-sensitive optical coherence tomography imaging of benign and malignant laryngeal lesions: An in vivo study. *Otolaryngol. Head Neck Surg.* **2011**, *145*, 91–99. [CrossRef] [PubMed]

185. Wang, T.; Yang, Y.; Zhu, Q. A three-parameter logistic model to characterize ovarian tissue using polarization-sensitive optical coherence tomography. *Biomed. Opt. Express* **2013**, *4*, 772–777. [CrossRef] [PubMed]

186. Patel, R.; Khan, A.; Quinlan, R.; Yaroslavsky, A.N. Polarization-sensitive multimodal imaging for detecting breast cancer. *Cancer Res.* **2014**, *74*, 4685–4693. [CrossRef] [PubMed]

187. South, F.A.; Chaney, E.J.; Marjanovic, M.; Adie, S.G.; Boppart, S.A. Differentiation of ex vivo human breast tissue using polarization-sensitive optical coherence tomography. *Biomed. Opt. Express* **2014**, *5*, 3417–3426. [CrossRef] [PubMed]

188. Xie, T.; Xia, Y.; Guo, S.; Hoover, P.; Chen, Z.; Peavy, G.M. Topographical variations in the polarization sensitivity of articular cartilage as determined by polarization-sensitive optical coherence tomography and polarized light microscopy. *J. Biomed. Opt.* **2008**, *13*, 054034. [CrossRef] [PubMed]

189. Martin, S.; Patel, N.; Adams, S.; Roberts, M.; Plummer, S.; Stamper, D.; Fujimoto, J.; Brezinski, M. New technology for assessing microstructural components of tendons and ligaments. *Int. Orthop.* **2003**, *27*, 184–189. [PubMed]

190. Ahearne, M.; Bagnaninchi, P.O.; Yang, Y.; El Haj, A.J. Online monitoring of collagen fibre alignment in tissue-engineered tendon by PSOCT. *J. Tissue Eng. Regen. Med.* **2008**, *2*, 521–524. [CrossRef] [PubMed]

191. Bagnaninchi, P.; Yang, Y.; Bonesi, M.; Maffulli, G.; Phelan, C.; Meglinski, I.; El Haj, A.; Maffulli, N. In-depth imaging and quantification of degenerative changes associated with achilles ruptured tendons by polarization-sensitive optical coherence tomography. *Phys. Med. Biol.* **2010**, *55*, 3777. [CrossRef] [PubMed]

192. Yang, Y.; Rupani, A.; Bagnaninchi, P.; Wimpenny, I.; Weightman, A. Study of optical properties and proteoglycan content of tendons by polarization sensitive optical coherence tomography. *J. Biomed. Opt.* **2012**, *17*, 081417. [CrossRef] [PubMed]

193. Pasquesi, J.J.; Schlachter, S.C.; Boppart, M.D.; Chaney, E.; Kaufman, S.J.; Boppart, S.A. In vivo detection of exercise-induced ultrastructural changes in genetically-altered murine skeletal muscle using polarization-sensitive optical coherence tomography. *Opt. Express* **2006**, *14*, 1547–1556. [PubMed]

194. Yang, X.; Chin, L.; Klyen, B.R.; Shavlakadze, T.; McLaughlin, R.A.; Grounds, M.D.; Sampson, D.D. Quantitative assessment of muscle damage in the mdx mouse model of duchenne muscular dystrophy using polarization-sensitive optical coherence tomography. *J. Appl. Physiol.* **2013**, *115*, 1393–1401. [CrossRef] [PubMed]

195. Matcher, S.J. What can biophotonics tell us about the 3D microstructure of articular cartilage? *Quant. Imaging Med. Surg.* **2015**, *5*, 143–158. [PubMed]

196. Chu, C.R.; Izzo, N.J.; Irrgang, J.J.; Ferretti, M.; Studer, R.K. Clinical diagnosis of potentially treatable early articular cartilage degeneration using optical coherence tomography. *J. Biomed. Opt.* **2007**, *12*, 051703. [CrossRef] [PubMed]

197. Ugryumova, N.; Jacobs, J.; Bonesi, M.; Matcher, S.J. Novel optical imaging technique to determine the 3-d orientation of collagen fibers in cartilage: Variable-incidence angle polarization-sensitive optical coherence tomography. *Osteoarthr. Cartil.* **2009**, *17*, 33–42. [CrossRef] [PubMed]

198. Kasaragod, D.K.; Lu, Z.; Jacobs, J.; Matcher, S.J. Experimental validation of an extended Jones matrix calculus model to study the 3D structural orientation of the collagen fibers in articular cartilage using polarization-sensitive optical coherence tomography. *Biomed. Opt. Express* **2012**, *3*, 378–387. [CrossRef] [PubMed]

199. Patel, N.A.; Zoeller, J.; Stamper, D.L.; Fujimoto, J.G.; Brezinski, M.E. Monitoring osteoarthritis in the rat model using optical coherence tomography. *IEEE Trans. Med. Imaging* **2005**, *24*, 155–159. [CrossRef] [PubMed]

200. Li, X.; Martin, S.; Pitris, C.; Ghanta, R.; Stamper, D.L.; Harman, M.; Fujimoto, J.G.; Brezinski, M.E. High-resolution optical coherence tomographic imaging of osteoarthritic cartilage during open knee surgery. *Arthritis Res. Ther.* **2005**, *7*, R318. [CrossRef] [PubMed]

201. Nadkarni, S.K. Optical measurement of arterial mechanical properties: From atherosclerotic plaque initiation to rupture. *J. Biomed. Opt.* **2013**, *18*, 121507. [CrossRef] [PubMed]

202. Kuo, W.-C.; Hsiung, M.-W.; Shyu, J.-J.; Chou, N.-K.; Yang, P.-N. Assessment of arterial characteristics in human atherosclerosis by extracting optical properties from polarization-sensitive optical coherence tomography. *Opt. Express* **2008**, *16*, 8117–8125. [CrossRef] [PubMed]

203. Giattina, S.D.; Courtney, B.K.; Herz, P.R.; Harman, M.; Shortkroff, S.; Stamper, D.L.; Liu, B.; Fujimoto, J.G.; Brezinski, M.E. Assessment of coronary plaque collagen with polarization sensitive optical coherence tomography (PS-OCT). *Int. J. Cardiol.* **2006**, *107*, 400–409. [CrossRef] [PubMed]

204. Nadkarni, S.K.; Bouma, B.E.; de Boer, J.; Tearney, G.J. Evaluation of collagen in atherosclerotic plaques: The use of two coherent laser-based imaging methods. *Lasers Med. Sci.* **2009**, *24*, 439–445. [CrossRef] [PubMed]

205. Nadkarni, S.K.; Pierce, M.C.; Park, B.H.; de Boer, J.F.; Whittaker, P.; Bouma, B.E.; Bressner, J.E.; Halpern, E.; Houser, S.L.; Tearney, G.J. Measurement of collagen and smooth muscle cell content in atherosclerotic plaques using polarization-sensitive optical coherence tomography. *J. Am. Coll. Cardiol.* **2007**, *49*, 1474–1481. [CrossRef] [PubMed]

206. Smith, R.M.; Black, A.J.; Velamakanni, S.S.; Akkin, T.; Tolkacheva, E.G. Visualizing the complex 3D geometry of the perfusion border zone in isolated rabbit heart. *Appl. Opt.* **2012**, *51*, 2713–2721. [CrossRef] [PubMed]

207. Sun, C.-W.; Wang, Y.-M.; Lu, L.-S.; Lu, C.-W.; Hsu, I.-J.; Tsai, M.-T.; Yang, C.-C.; Kiang, Y.-W.; Wu, C.-C. Myocardial tissue characterization based on a polarization-sensitive optical coherence tomography system with an ultrashort pulsed laser. *J. Biomed. Opt.* **2006**, *11*, 054016. [CrossRef] [PubMed]

208. Baumgartner, A.; Hitzenberger, C.K.; Dichtl, S.; Sattmann, H.; Moritz, A.; Sperr, W.; Fercher, A.F. Optical Coherence Tomography of Dental Structures. *SPIE Proc.* **1998**, *3248*, 130–136.

209. Baumgartner, A.; Dichtl, S.; Hitzenberger, C.; Sattmann, H.; Robl, B.; Moritz, A.; Fercher, A.; Sperr, W. Polarization–sensitive optical coherence tomography of dental structures. *Caries Res.* **1999**, *34*, 59–69. [CrossRef]

210. Everett, M.J.; Colston, B.W., Jr.; Sathyam, U.S.; Da Silva, L.B.; Fried, D.; Featherstone, J.D. Noninvasive diagnosis of early caries with polarization-sensitive optical coherence tomography (PS-OCT). *SPIE Proc.* **1999**, *3593*, 177–182.

211. Manesh, S.K.; Darling, C.L.; Fried, D. Nondestructive assessment of dentin demineralization using polarization-sensitive optical coherence tomography after exposure to fluoride and laser irradiation. *J. Biomed. Mater. Res. Part B Appl. Biomater.* **2009**, *90*, 802–812. [CrossRef] [PubMed]

212. Le, M.H.; Darling, C.L.; Fried, D. Automated analysis of lesion depth and integrated reflectivity in PS-OCT scans of tooth demineralization. *Lasers Surg. Med.* **2010**, *42*, 62–68. [CrossRef] [PubMed]

213. Louie, T.; Lee, C.; Hsu, D.; Hirasuna, K.; Manesh, S.; Staninec, M.; Darling, C.L.; Fried, D. Clinical assessment of early tooth demineralization using polarization sensitive optical coherence tomography. *Lasers Surg. Med.* **2010**, *42*, 898–905. [CrossRef] [PubMed]

214. Tom, H.; Simon, J.C.; Chan, K.H.; Darling, C.L.; Fried, D. Near-infrared imaging of demineralization under sealants. *J. Biomed. Opt.* **2014**, *19*, 077003. [CrossRef] [PubMed]

215. Tom, H.; Chan, K.H.; Darling, C.L.; Fried, D. Near-IR image-guided laser ablation of demineralization on tooth occlusal surfaces. *Lasers Surg. Med.* **2016**, *48*, 52–61. [CrossRef] [PubMed]

216. Lee, C.; Darling, C.L.; Fried, D. Polarization-sensitive optical coherence tomographic imaging of artificial demineralization on exposed surfaces of tooth roots. *Dent. Mater.* **2009**, *25*, 721–728. [CrossRef] [PubMed]

217. Can, A.M.; Darling, C.L.; Ho, C.; Fried, D. Non-destructive assessment of inhibition of demineralization in dental enamel irradiated by a $\lambda = 9.3$-μm CO_2 laser at ablative irradiation intensities with PS-OCT. *Lasers Surg. Med.* **2008**, *40*, 342–349. [CrossRef] [PubMed]

218. Fried, D.; Xie, J.; Shafi, S.; Featherstone, J.D.; Breunig, T.M.; Le, C. Imaging caries lesions and lesion progression with polarization sensitive optical coherence tomography. *J. Biomed. Opt.* **2002**, *7*, 618–627. [CrossRef] [PubMed]

219. Jones, R.S.; Staninec, M.; Fried, D. Imaging artificial caries under composite sealants and restorations. *J. Biomed. Opt.* **2004**, *9*, 1297–1304. [CrossRef] [PubMed]

220. Simon, J.C.; Lucas, S.A.; Lee, R.C.; Darling, C.L.; Staninec, M.; Vaderhobli, R.; Pelzner, R.; Fried, D. Near-infrared imaging of secondary caries lesions around composite restorations at wavelengths from 1300–1700-nm. *Dent. Mater.* **2016**, *32*, 587–595. [CrossRef] [PubMed]

221. Jones, R.; Fried, D. Remineralization of enamel caries can decrease optical reflectivity. *J. Dent. Res.* **2006**, *85*, 804–808. [CrossRef] [PubMed]

222. Lee, R.C.; Kang, H.; Darling, C.L.; Fried, D. Automated assessment of the remineralization of artificial enamel lesions with polarization-sensitive optical coherence tomography. *Biomed. Opt. Express* **2014**, *5*, 2950–2962. [CrossRef] [PubMed]

223. De Boer, J.F.; Srinivas, S.M.; Park, B.H.; Pham, T.H.; Chen, Z.; Milner, T.E.; Nelson, J.S. Polarization effects in optical coherence tomography of various biological tissues. *IEEE J. Sel. Top. Quantum Electron.* **1999**, *5*, 1200–1204. [CrossRef] [PubMed]

224. Islam, M.S.; Oliveira, M.C.; Wang, Y.; Henry, F.P.; Randolph, M.A.; Park, B.H.; de Boer, J.F. Extracting structural features of rat sciatic nerve using polarization-sensitive spectral domain optical coherence tomography. *J. Biomed. Opt.* **2012**, *17*, 056012. [CrossRef] [PubMed]

225. Henry, F.P.; Wang, Y.; Rodriguez, C.L.R.; Randolph, M.A.; Rust, E.A.Z.; Winograd, J.M.; de Boer, J.F.; Park, B.H. In vivo optical microscopy of peripheral nerve myelination with polarization sensitive-optical coherence tomography. *J. Biomed. Opt.* **2015**, *20*, 046002. [CrossRef] [PubMed]

226. Yoon, Y.; Jeon, S.H.; Park, Y.H.; Jang, W.H.; Lee, J.Y.; Kim, K.H. Visualization of prostatic nerves by polarization-sensitive optical coherence tomography. *Biomed. Opt. Express* **2016**, *7*, 3170–3183. [CrossRef] [PubMed]

227. Wang, H.; Black, A.J.; Zhu, J.; Stigen, T.W.; Al-Qaisi, M.K.; Netoff, T.I.; Abosch, A.; Akkin, T. Reconstructing micrometer-scale fiber pathways in the brain: Multi-contrast optical coherence tomography based tractography. *Neuroimage* **2011**, *58*, 984–992. [CrossRef] [PubMed]

228. Wang, H.; Zhu, J.; Akkin, T. Serial optical coherence scanner for large-scale brain imaging at microscopic resolution. *Neuroimage* **2014**, *84*, 1007–1017. [CrossRef] [PubMed]

229. Wang, H.; Zhu, J.F.; Reuter, M.; Vinke, L.N.; Yendiki, A.; Boas, D.A.; Fischl, B.; Akkin, T. Cross-validation of serial optical coherence scanning and diffusion tensor imaging: A study on neural fiber maps in human medulla oblongata. *Neuroimage* **2014**, *100*, 395–404. [CrossRef] [PubMed]

230. Baumann, B.; Woehrer, A.; Ricken, G.; Augustin, M.; Mitter, C.; Pircher, M.; Kovacs, G.G.; Hitzenberger, C.K. Visualization of neuritic plaques in Alzheimer's disease by polarization-sensitive optical coherence microscopy. *Sci. Rep.* **2017**, *7*, 43477. [CrossRef] [PubMed]

231. Wei, Q.; Song, H.-M.; Leonov, A.P.; Hale, J.A.; Oh, D.; Ong, Q.K.; Ritchie, K.; Wei, A. Gyromagnetic imaging: Dynamic optical contrast using gold nanostars with magnetic cores. *J. Am. Chem. Soc.* **2009**, *131*, 9728–9734. [CrossRef] [PubMed]

232. Chhetri, R.K.; Kozek, K.A.; Johnston-Peck, A.C.; Tracy, J.B.; Oldenburg, A.L. Imaging three-dimensional rotational diffusion of plasmon resonant gold nanorods using polarization-sensitive optical coherence tomography. *Phys. Rev. E* **2011**, *83*, 040903. [CrossRef] [PubMed]

233. Oldenburg, A.L.; Chhetri, R.K.; Cooper, J.M.; Wu, W.-C.; Troester, M.A.; Tracy, J.B. Motility-, autocorrelation-, and polarization-sensitive optical coherence tomography discriminates cells and gold nanorods within 3D tissue cultures. *Opt. Lett.* **2013**, *38*, 2923–2926. [CrossRef] [PubMed]

234. Chhetri, R.K.; Blackmon, R.L.; Wu, W.-C.; Hill, D.B.; Button, B.; Casbas-Hernandez, P.; Troester, M.A.; Tracy, J.B.; Oldenburg, A.L. Probing biological nanotopology via diffusion of weakly constrained plasmonic nanorods with optical coherence tomography. *Proc. Natl. Acad. Sci. USA* **2014**, *111*, E4289–E4297. [CrossRef] [PubMed]

235. Wiesauer, K.; Pircher, M.; Goetzinger, E.; Hitzenberger, C.; Engelke, R.; Ahrens, G.; Gruetzner, G.; Stifter, D. Transversal ultrahigh-resolution polarization-sensitive optical coherence tomography for strain mapping in materials. *Opt. Express* **2006**, *14*, 5945–5953. [CrossRef] [PubMed]

236. Stifter, D. Beyond biomedicine: A review of alternative applications and developments for optical coherence tomography. *Appl. Phys. B* **2007**, *88*, 337–357. [CrossRef]

237. Oh, J.-T.; Kim, S.-W. Polarization-sensitive optical coherence tomography for photoelasticity testing of glass/epoxy composites. *Opt. Express* **2003**, *11*, 1669–1676. [CrossRef] [PubMed]

238. Stifter, D.; Leiss-Holzinger, E.; Major, Z.; Baumann, B.; Pircher, M.; Götzinger, E.; Hitzenberger, C.K.; Heise, B. Dynamic optical studies in materials testing with spectral-domain polarization-sensitive optical coherence tomography. *Opt. Express* **2010**, *18*, 25712–25725. [CrossRef] [PubMed]

Review

Can OCT Angiography Be Made a Quantitative Blood Measurement Tool?

Jun Zhu [1], Conrad W. Merkle [1], Marcel T. Bernucci [1], Shau Poh Chong [1] and Vivek J. Srinivasan [1,2,*]

1   Department of Biomedical Engineering, University of California Davis, Davis, CA 95616, USA;
     znjzhu@ucdavis.edu (J.Z.); cwmerkle@ucdavis.edu (C.W.M.); mtbernucci@ucdavis.edu (M.T.B.);
     spchong@ucdavis.edu (S.P.C.)
2   Department of Ophthalmology and Vision Science, School of Medicine, University of California Davis,
     Sacramento, CA 95817, USA
*   Correspondence: vjsriniv@ucdavis.edu; Tel.: +1-530-752-9277

Academic Editor: Michael Pircher
Received: 11 May 2017; Accepted: 19 June 2017; Published: 4 July 2017

Featured Application: Optical Coherence Tomography Angiography (OCTA) is a technique for label-free vascular imaging in fields such as ophthalmology, gastroenterology, cancer biology, and neuroscience. Here, we discuss advances that relate OCTA more rigorously to underlying blood physiology and hemodynamics, which promise to make OCTA an even more powerful quantitative tool.

Abstract: Optical Coherence Tomography Angiography (OCTA) refers to a powerful class of OCT scanning protocols and algorithms that selectively enhance the imaging of blood vessel lumens, based mainly on the motion and scattering of red blood cells (RBCs). Though OCTA is widely used in clinical and basic science applications for visualization of perfused blood vessels, OCTA is still primarily a *qualitative* tool. However, more quantitative hemodynamic information would better delineate disease mechanisms, and potentially improve the sensitivity for detecting early stages of disease. Here, we take a broader view of OCTA in the context of microvascular hemodynamics and light scattering. Paying particular attention to the unique challenges presented by capillaries versus larger supplying and draining vessels, we critically assess opportunities and challenges in making OCTA a *quantitative* tool.

Keywords: optical coherence tomography; angiography; scattering; red blood cells; rheology; imaging; hemodynamics; blood flow

---

1. Introduction

The microcirculation comprises a network of blood vessels that delivers oxygen and nutrients to surrounding tissues, removes waste products and heat, and otherwise supports tissue viability [1–3]. Red blood cells (RBCs) are the main carriers of oxygen in blood. "Optical Coherence Tomography Angiography (OCTA)" is a term for the specialized Optical Coherence Tomography (OCT) scanning protocols and post-processing algorithms that mainly enhance the motion contrast of red blood cells (RBCs) in OCT images to selectively highlight these vessels. By enabling the visualization of cell-perfused vasculature without an exogenous contrast agent, OCT angiography has generated enormous interest in ophthalmology [4–11], gastroenterology [12,13], cancer biology [14,15], and neuroscience [16,17] over the past decade. It has been particularly useful in studying diseases where the microvascular morphology or presence of perfusion changes over time. However, with few exceptions [18–20], the majority of published studies have used OCT angiography *qualitatively*, primarily as a means of visualization. Here, we review the relevant basic hemodynamic principles, fundamentals of OCTA, categories of OCTA scanning protocols, and classes of OCTA algorithms. We argue that a rigorous and model-based relationship between hemodynamic parameters, light scattering theory, and measurement observables [21] in OCT angiography will pave the way towards more *quantitative* imaging of hemodynamics by OCTA and related methods, with the potential to enhance all applications.

2. OCTA Fundamentals

A unifying feature of all OCTA algorithms is that they visualize objects that are both moving and backscattering. Hence, we begin our review with a discussion of hemodynamics and light scattering properties of blood. Importantly, we distinguish between capillaries (<10 μm in diameter), where RBCs flow in a line and hematocrits are low, and macrovasculature, where RBCs flow side-by-side and hematocrits approach systemic levels, with the understanding that non-capillary microvessels (10–100 μm in diameter) represent an intermediate case between the two extremes discussed here [2,3].

2.1. Hemodynamic Parameters

What are the main hemodynamic parameters that impact observed OCTA signals? In capillaries (Figure 1A), the RBC flow is single-file, with plasma gaps in between [22,23]. RBC speed (distance/time), flux (#/time), and linear density (#/distance) are thus primary hemodynamic parameters in capillaries. Due to the plasma gaps between cells, flux can often be determined by imaging individual capillaries and counting RBCs traversing a single location [23]. Assuming single-file capillary flow, microvascular tube hematocrit (H_{tube}), or RBC volume fraction, is related to linear density (ϱ) by $\varrho = H_{tube}A/V_{RBC}$, where V_{RBC} is the red blood cell volume and A is the vessel cross-sectional area. Capillary tube hematocrit is generally a factor of ~2–3× lower than systemic levels [24], but hematocrit can vary considerably between capillaries. In macrovessels (Figure 1B), which include supplying arteries and draining veins, the blood velocity varies across the vessel cross-section. In contrast to microvessels, macrovascular hematocrit approaches systemic levels of ~40–45% [25]. Flow is typically laminar with some degree of blunting [24], with the largest shear rate, or velocity gradient, at the edge of the vessel. In macrovessels, RBC velocity or speed (distance/time), flow rate (volume/time), and hematocrit (volume/volume) are the primary hemodynamic parameters. All hemodynamic parameters vary over time with respiration and the heartbeat of the subject [24].

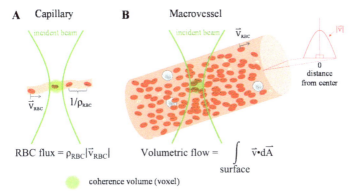

A Capillary   **B** Macrovessel

RBC flux = $\rho_{RBC}|\vec{v}_{RBC}|$ Volumetric flow = $\int_{surface} \vec{v}\cdot d\vec{A}$

coherence volume (voxel)

Figure 1. (A) Flow in capillaries (microvessels with diameters of <10 μm) is single-file, usually with highly variable hematocrits that fall below systemic levels; **(B)** On the other hand, macrovascular flow often follows a blunted laminar profile at near-systemic hematocrits. Consequently, different approaches are required to quantify microvascular versus macrovascular hemodynamics via OCTA (Optical Coherence Tomography Angiography) imaging.

2.2. Light Scattering from Red Blood Cells

What are the physical properties of RBCs that enable their detection by OCTA? RBC scattering and absorption properties derive from the presence of hemoglobin and its complex refractive index [26]. Major absorption bands of hemoglobin, related to the imaginary part of the complex refractive index, predominate at visible and shorter wavelengths, while hemoglobin absorption becomes negligible at near-infrared wavelengths, where scattering dominates. The light scattering properties of individual RBCs are determined by the refractive index contrast with respect to the surrounding plasma, as well as their shape and size relative to the medium wavelength. The real part of the complex refractive index, or refractive index, of hemoglobin is larger by ~3–6% relative to the surrounding plasma [26–28]. RBCs are biconcave disks (Figure 2), with a diameter of 6–8 μm and thickness of ~2 μm, although their shape changes under external stress. Due to the large volume fraction of RBCs and their refractive index mismatch relative to plasma, RBCs are the main scattering constituent in blood [27,29,30].

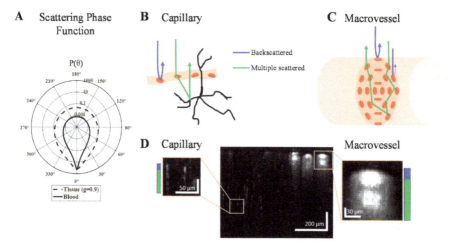

Figure 2. Single and multiple scattering in the OCTA of capillaries versus macrovasculature. (**A**) Blood has a high scattering anisotropy, leading to a high probability of detecting multiple scattered light paths; (**B**) For capillaries, dynamic RBC (red blood cell) forward scattering precedes or follows static tissue backscattering, which leads to "multiple scattering" tails; (**C**) In large vessels, the backscattering cross-section is determined by the shear-induced orientation of RBCs with their flat face parallel to the shear force. If the vessel lumen exceeds a scattering length, multiple intravascular dynamic scattering events (green) before detection are likely; (**D**) Cross-sectional OCT (Optical Coherence Tomography) angiogram of the mouse brain at 1300 nm (complex interframe subtraction method) with a qualitative colorbar showing the balance of backscattered light (blue) and multiple scattered light (green) in a capillary (left) and macrovessel (right).

The scattering properties of both individual RBCs and ensembles of RBCs are important in OCTA. Due to their irregular shape, the probability of light scattering in a given direction for a particular RBC depends on both its orientation and the direction of incident light. An ensemble of RBCs with different orientations can be characterized by a scattering coefficient (μ_s), the scattering probability per unit distance; a scattering phase function ($P(\theta)$), the probability of scattering in a given elevation direction θ per unit solid angle; and a scattering anisotropy ($g = E[\cos(\theta)]$), the expectation or average (E[ ]) of $\cos(\theta)$ over solid angle. These parameters characterize scattering of whole blood, which comprises an ensemble volume of RBCs with random orientations. In whole blood, empirically measured g and μ_s include dependent and multiple scattering effects [31]. With a hematocrit of around 45%, whole blood is found to be highly forward scattering between 750 and 950 nm, with a scattering coefficient

(μ_s) between 65 and 80 mm^{-1}, and anisotropy (g) between 0.97 and 0.99 [29,32–34]. Exemplary phase functions [30,32,35] for tissue (Henyey-Greenstein with g = 0.9) and blood (Gegenbauer-Kernel with g = 0.972 and α = 0.49 [36]) are shown in Figure 2A on a logarithmic scale. Tissue has a higher probability of back scattering than blood, while blood is considerably more forward scattering.

In OCTA (Figure 2B,C), detected light ideally results from paths with single RBC backscattering (θ = 180°) events (blue). However, the high RBC anisotropy (Figure 2A) makes detection of multiple scattered light (green) likely. Probable light paths can be understood through the principles of radiative transport. In capillaries, where RBC flow is single-file, light forward scattered from RBCs is also backscattered from extravascular tissue (Figure 2B), creating axial multiple scattering tails (Figure 2D left box). In macrovessels, there are two important effects. First, RBCs tend to align their flat face parallel to the shear force, i.e., facing outwards along the vessel circumference (Figure 2C). The largest backscattering cross-section occurs when the shortest RBC dimension is aligned with the incident light. Therefore, the signal is enhanced at the top and bottom of the vessel lumen and reduced at the side (Figure 2D right box) [37]. At higher shear rates, RBCs elongate and the backscattering pattern disappears [38]. Second, for vessel lumens larger than a scattering length ($1/\mu_s$), multiple intravascular dynamic scattering events (green) can occur before detection.

As OCTA images are created by post-processing OCT data, OCTA has an image penetration depth comparable to or less than OCT. This is typically ~0.5–1.5 mm in most tissues, depending on the source wavelength and the sample optical properties [39,40]. It is important to note that while OCTA visualizes blood vessels, the penetration depth of OCTA may be determined by the attenuation of both intravascular and extravascular tissue.

3. OCTA Signal

In this section, we provide a unifying framework for the OCT signal to facilitate the discussion of OCTA algorithms in Section 4. Commonly-used symbols or variables and their definitions are summarized in Table 1, while other symbols are defined in the text.

Table 1. Symbols or variables used and their meaning.

Symbol	Meaning
S	Complex OCT signal/field
|S|	Amplitude of the OCT signal
$I = \|S\|^2$	Intensity of the OCT signal
$\varnothing$	Phase of the OCT signal
S_m	OCT field from one scatterer
SV	Speckle variance
cmOCT	Correlation mapping OCT signal
PV	Phase variance
$\Delta\varnothing$	Phase difference
ΔS	Complex field difference
CDV	Complex differential variance
R	Autocorrelation function
P	Power spectral density

All standard OCTA algorithms [41,42] start from the complex OCT signal. The complex, depth-resolved OCT signal can be expressed as:

$$S(x, z, t) = |S(x, z, t)| \exp\{i\varnothing(x, z, t)\}. \tag{1}$$

Note that $S(x, z, t)$ is related to the depth-resolved optical field, integrated over a resolution element (coherence volume). Therefore, the depth-resolved intensity, $I(x, z, t)$, is equivalent to the magnitude square of the field, i.e., $I(x, z, t) = |S(x, z, t)|^2$. OCTA algorithms may operate on either $S(x, z, t)$, $\varnothing(x, z, t)$, or $I(x, z, t)$ as the "signal", and accordingly, can be categorized into complex field-based techniques, phase-based techniques, and intensity-based techniques. In its simplest form, OCTA employs differences between OCT signals at the same spatial position over a series of time points to highlight scatterer motion. As discussed in Section 2.2, RBCs are the main blood scattering component. Due to the dynamic motion of RBCs, the overall field, phase, and intensity fluctuate. For the field, these variations are determined, in a statistical sense, by the first-order field autocorrelation function, $r(\tau)$, in which $r(\tau) = R(\tau)/R(0)$ and $R(\tau) = E[S(x, z, t + \tau)S^*(x, z, t)]$, where $E[\ ]$ represents expectation and τ is the time lag. Under some circumstances, all other signal variations, including those of the intensity and phase, derive their statistical properties from the field autocorrelation [43].

OCT complex signal dynamics are illustrated in Figure 3. The complex signal is treated as a complex summation of backscattered fields from individual scatterers within the coherence volume. The coherence volume is defined by the beam waist in the transverse direction and the coherence length in the axial direction. Changes in the fields from individual scatterers over time leads to changes in the total signal over time (Figure 3A,B). In many practical situations, scatterers may be further classified as "dynamic" and "static" depending on whether they move or not, with both scatterer types contributing to the signal in the same coherence volume (Figure 3C,D).

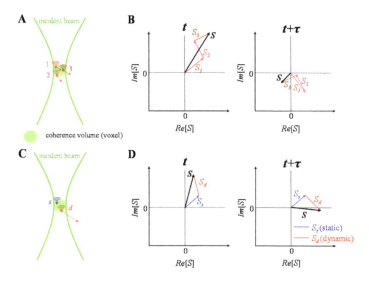

Figure 3. The motion of scatterers in a coherence volume gives rise to complex field fluctuations that form the basis for OCTA signals. The contributions to the complex field are shown at two different points in time (t and t + τ). (**A,B**) Field fluctuations due to dynamic scatterers in a coherence volume. (**C,D**) Field fluctuations due to a combination of static (blue) and dynamic (red) scatterers in a coherence volume.

The nature of scatterer dynamics plays a major role in determining the OCT signal changes (Figure 4). Generally, scatterer motion is accompanied by both a Doppler shift and decorrelation [44]. When the scatterer has an axial velocity component, moving towards or away from the incident beam, the complex field rotates, tracing a helix over time (Figure 4A–C). This effect can be described as a linear phase shift over time due to the Doppler effect, or a "Doppler phase shift". When the scatterer is undergoing a dynamic conformational change, rotation, or translational motion through the coherence volume, both the OCT signal intensity and phase change randomly (Figure 4D–F). This random change of the complex field is known as decorrelation. Doppler shifts are associated with a change in the phase of the complex field autocorrelation, while decorrelation is associated with a decrease in the magnitude of the complex field autocorrelation, $|R(\tau)|$, with increasing τ.

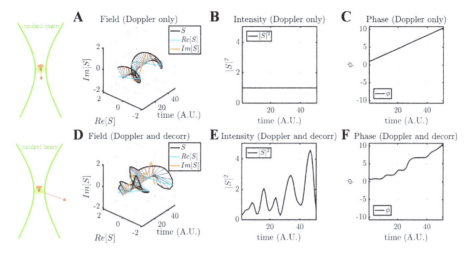

Figure 4. The major categories of OCT signal fluctuations are Doppler shifts and decorrelation. Comparison of complex field, intensity, and phase time courses, for the case of a pure Doppler shift (**A**–**C**) and a Doppler shift with decorrelation (**D**–**F**). For a pure Doppler shift, the field traces out a helical pattern (**A**), whereas decorrelation introduces random deviations from this pattern (**D**).

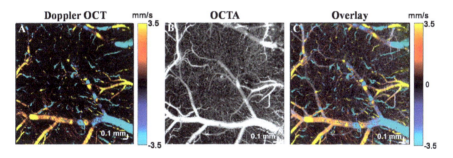

Figure 5. Doppler OCT and OCTA in the mouse brain. (**A**) Doppler OCT can visualize flow based on Doppler shifts, caused by motion in the axial direction, towards or away from the probe beam. On the other hand, OCTA visualizes flow based on decorrelation, usually caused by translational motion through the coherence volume, as well as Doppler shifts. The overlay of both methods (**C**) shows that Doppler OCT is mainly limited to ascending venules or descending arterioles, where Doppler shifts dominate. On the other hand, OCTA, which is sensitive to decorrelation, more comprehensively shows vasculature, including regions with predominantly transverse flow. A standard Kasai algorithm was used on transversally oversampled images for (**A**) and a complex interframe subtraction method was used on rapidly acquired repeated cross-sectional images for (**B**).

Both Doppler shifts and decorrelation are present to varying degrees in all vasculature. Note that a Doppler shift due to translational axial motion through the coherence volume implies decorrelation. On the other hand, decorrelation occurs even for transverse motion or rotation, and does not necessarily imply a Doppler shift. To illustrate this, Figure 5 shows a comparison between Doppler OCT and OCTA of mouse brain microvasculature. Doppler OCT detects phase changes caused by translational axial motion [45]. The requirement for axial phase shifts renders Doppler OCT only sensitive to motion parallel to the incident beam. Doppler shifts predominate in larger microvessels which are ascending or descending (Figure 5A); hence when used for angiography, the Doppler effect provides only a partial microvascular map. By comparison, decorrelation involves random deviations of the complex field and predominates in vessels with transverse flow. Thus, OCTA, which senses decorrelation via intensity and/or phase, more comprehensively shows the vasculature (Figure 5B,C).

Finally, it should be noted that the presence of static scattering can significantly alter time courses. The OCT field, intensity, and phase time courses due to dynamic scattering in the presence of a static scatterer are shown in Figure 6. As suggested by Figure 3D, the presence of static scattering confines the field fluctuations to a portion of the complex plane (Figure 6A). As will be discussed in Section 7.1, the possible presence of static scatterer(s) must be considered in order to recover quantitative information about the Doppler phase shift or the decorrelation rate in OCTA.

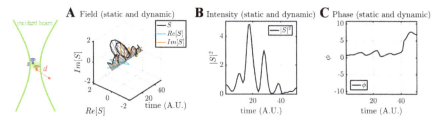

Figure 6. Comparison of complex field (**A**), intensity (**B**), and phase (**C**) time courses when both static and dynamic scattering are present in a coherence volume (with Doppler shift and decorrelation of the dynamic component). Such coherence volumes are present at the edges of vessels.

4. OCTA Algorithms

The previous section showed that OCTA signals depend on the type of dynamics (Doppler shift or decorrelation), the observed parameter (intensity, phase, or field), and the possible presence of static scattering in the coherence volume. With this discussion in mind, we now present the main classes of angiography algorithms.

4.1. Intensity- or Amplitude-Based OCTA Algorithms

Intensity-based OCTA algorithms use $I(x, z, t) = |S(x, z, t)|^2$, while amplitude-based OCTA algorithms use $|S(x, z, t)|$ in Equation (1).

The first class of intensity-based OCTA algorithms is the speckle variance method. Speckle [46] can be described as the random interference of scattering fields (indexed by m) that cannot be resolved within a coherence volume:

$$S(z) = \sum_m S_m(z),\tag{2}$$

S_m represents the fields within a coherence volume, each weighted according to the point spread function at the scatterer location (Figure 3). The intensity (as well as the phase and field) changes over time as the configuration of scatterers changes, causing decorrelation (Figure 4D–F). Decorrelation can occur as RBCs pass through a coherence volume, but may also occur due to rotational motion or diffusion. In 2005, Barton and Stromski showed the feasibility of flow speed measurement without phase information by evaluating speckle pattern changes [47]. In 2008, Mariampillai et al. [48] used interframe speckle variance to visualize microcirculation. In [48], speckle variance was defined as:

$$SV(x, z) = \frac{1}{N} \sum_{t=0}^{(N-1)T} [I(x, z, t) - \overline{I(x, z)}]^2,\tag{3}$$

where $t = 0, T, 2T, \ldots, (N-1)T$ represents the OCT acquisition time; T is the time interval; N is the total number of acquisitions at the same position; x and z denote lateral and depth indices, respectively; and $\overline{I(x, z)} = \overline{|S(x, z)|^2}$ is the time-averaged intensity at position (x, z). This is a temporally averaged, variance-based algorithm without normalization. By ignoring the phase in Equation (1), the method is not sensitive to pure Doppler shifts. Consequently, speckle variance is not susceptible to phase noise. However, the speckle variance method may be compromised due to interframe bulk tissue motion. While in-plane (xz) motion can be compensated in principle, out-of-plane motion is more challenging to correct. To minimize motion effects, later in 2010, Mariampillai et al. [49] optimized the frame number and frame rate for a given level of bulk tissue motion, through maximizing the speckle variance signal-to-noise ratio (SNR) between a "dynamic" and "static" pixel. Speckle variance SNR is calculated as:

$$SV_{SNR}(N, \overline{I(x,z)}) = \frac{SV_{dynamic}(N, \overline{I(x,z)}) - SV_{static}(N, \overline{I(x,z)})}{\sqrt{\sigma^2_{dynamic}(N, \overline{I(x,z)}) + \sigma^2_{static}(N, \overline{I(x,z)})}}, \tag{4}$$

where $\overline{I(x,z)}$ is the time-averaged intensity for both "dynamic" and "static" pixels, $SV_{dynamic}$ and SV_{static} are speckle variances calculated from Equation (3), and $\sigma^2_{dynamic}$ and σ^2_{static} are variances of $SV_{dynamic}$ and SV_{static}, respectively. By optimizing the frame number under conditions of low tissue bulk motion, capillaries can be reliably detected [49].

As the dynamic tissue signal has a lower temporal correlation, at a given time lag, than static tissue, correlation has been investigated as a parameter for angiography. In 2011, Enfield et al. [50] demonstrated in vivo human volar forearm imaging of the capillary density and vessel diameter with correlation mapping optical coherence tomography (cmOCT). The correlation between OCT frames acquired at time t and t + T at the same position is:

$$cmOCT(x,z) = \sum_{p=0}^{V} \sum_{q=0}^{W} \frac{[I(x+p, z+q, t) - \overline{I(t)}][I(x+p, z+q, t+T) - \overline{I(t+T)}]}{\sqrt{[I(x+p, z+q, t) - \overline{I(t)}]^2 + [I(x+p, z+q, t+T) - \overline{I(t+T)}]^2}}, \tag{5}$$

where V and W define the extent of the spatial region for correlation calculation, and $\overline{I(t)}$ denotes the spatially averaged intensity over this region. This is a spatially averaged, correlation-based algorithm with normalization. After this calculation, a 2D correlation map can be formed by applying a threshold to binarize the image into static and dynamic regions. In 2012, Jia et al. [51] proposed split-spectrum amplitude-decorrelation angiography (SSADA) to image the human macula and optic nerve head. Ensuring a nearly isotropic coherence volume size by splitting the spectrum to degrade the axial resolution to equal the transverse resolution, they then applied a method similar to cmOCT.

4.2. Phase-Based OCTA Algorithms

Phase-based OCTA algorithms rely on $\varnothing(x, z, t)$ in Equation (1) to distinguish dynamic and static tissue. Doppler OCT, a category of phase-based OCTA, uses a deterministic Doppler phase shift for in vivo blood flow measurements [52,53]. While Doppler OCT can quantify flow, visualization applications are limited due to its angle dependence (Figure 5A). For instance, retinal blood vessels are nearly perpendicular to the optic axis, particularly outside of the optic nerve head, yielding insufficient phase shifts for Doppler measurements [42]. Power Doppler [54,55] and phase variance imaging [56] represent alternative approaches that are sensitive to decorrelation, or random non-deterministic Doppler shifts. In 2007, Fingler et al. [57] proposed phase variance for motion contrast. In [58], the phase variance at position (x, z) is defined as:

$$PV(x,z) = \frac{1}{N-1} \sum_{t=0}^{(N-2)T} [\Delta\varnothing(x,z,t) - \overline{\Delta\varnothing(x,z)}]^2. \tag{6}$$

The phase difference at a given location is given by:

$$\Delta\varnothing(x,z,t) = \varnothing(x,z,t+T) - \varnothing(x,z,t), \tag{7}$$

where T is the time lag. Equation (6) is a temporally averaged, variance-based algorithm without normalization. Phase-based OCTA algorithms are advantageous over amplitude- and intensity-based algorithms if phase changes but intensity and amplitude do not. However, phase-based OCTA loses information about the OCT signal amplitude and intensity. Moreover, phase-based OCTA may not detect changes in the presence of a large static scattering component. Similar to amplitude- and intensity-based algorithms, phase-based OCTA is sensitive to decorrelation (Figure 4E,F). However, as phase is particularly sensitive to axial motion, additional bulk motion phase correction is typically required. In [57], before phase variance analysis, Fingler et al. removed the bulk motion phase change:

$$\Delta\varnothing^{corr}(x, z, t) = \Delta\varnothing(x, z, t) - \Delta\varnothing^{bulk}(x, t),\tag{8}$$

where $\Delta\varnothing^{corr}(x, z, t)$ denotes the corrected phase change, and $\Delta\varnothing^{bulk}(x, t)$ represents the phase change due to bulk motion, estimated as:

$$\Delta\varnothing^{bulk}(x, t) = \frac{\sum_{z=a}^{b}\left[|S(x, z, t)|\Delta\varnothing(x, z, t)\right]}{\sum_{z=a}^{b}\left[|S(x, z, t)|\right]}\tag{9}$$

The phase change due to bulk motion is thus calculated by a weighted mean from $z = a$ to $z = b$ in one A-scan. Note that bulk phase change estimation based on cross-correlation is also possible [16,59].

4.3. Complex Signal-Based OCTA Algorithms

Complex signal-based OCTA algorithms use $S(x, z, t)$, the complex field, which includes both the intensity/amplitude and phase in Equation (1). As both intensity and phase fluctuations (Figure 4B,C,E,F) arise from field fluctuations (Figure 4A,D), we assert that the complex field is more fundamental than either the intensity or phase. In particular, the static component can be readily handled in the complex domain (Figure 6). Also, unlike intensity-based OCTA, complex signal-based OCTA is sensitive to slow flow with only phase changes [60]. In 2007, Wang et al. [61] demonstrated complex signal-based OCT angiography, also called optical microangiography (OMAG), for the first time, while interframe complex OCTA was introduced later [16,62]. The most basic complex OCTA algorithm is based on subtraction,

$$\Delta S(x, z, t) = |S(x, z, t+T) - S(x, z, t)|,\tag{10}$$

where $S(x, z, t+T)$ and $S(x, z, t)$ are complex OCT signals acquired at the same position separated by a time lag T. This is a difference-based algorithm without normalization. Spatial or temporal averaging may be applied as needed. This expression may also be generalized as a variance calculation (or high-pass filter [16]) that eliminates static scattering:

$$\Delta S(x, z) = \frac{1}{N}\sum_{t=0}^{(N-1)T}\left|S(x, z, t) - \overline{S(x, z)}\right|^{2}.\tag{11}$$

This is a temporally averaged, variance-based algorithm without normalization. As static and dynamic scatterer fields add in the complex domain (Figure 6), the above expression correctly eliminates static scattering to quantify the dynamic scattering signal.

Using a complex signal-based algorithm, several applications of OCTA are demonstrated here. Figure 7A,B shows OCTA graphing of the mouse brain vasculature in vivo. Longitudinal monitoring of recovery in the mouse brain, one week after an experimental ischemic stroke, is shown in Figure 7C. Note the presence of vascular remodeling (yellow arrows). Figure 8 shows OCTA of a rodent eye in vivo. Figure 9 presents OCTA of pig ear skin, including a cross-sectional intensity image (Figure 9A), cross-sectional angiogram image (Figure 9B), color-coded angiogram of superficial and deep vasculature (Figure 9C), and angiograms centered at different depths (Figure 9D–I). All figures employ a complex interframe subtraction algorithm for angiography.

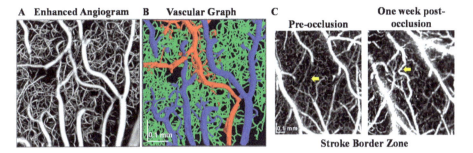

Figure 7. OCTA microscopy of the mouse brain enables an assessment of vascular connectivity (**A**,**B**) and longitudinal monitoring of microvascular remodeling (**C**) one week after distal middle cerebral artery occlusion (yellow arrow).

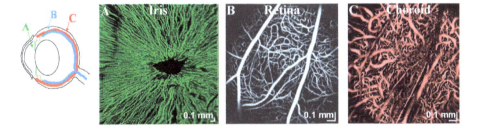

Figure 8. Ocular OCTA of iris (**A**), retina (**B**), and choroid (**C**). Hessian vesselness enhancement was applied to retinal and choroidal vasculature before display. Note that the pupil was dilated prior to OCTA acquisition for (**B**,**C**).

In 2014, Nam et al. [63] proposed a complex differential variance (CDV) algorithm. This differential variance algorithm, applied to the OCT signal at a position (x, z), is:

$$\mathrm{CDV}(x,z) = \sqrt{1 - \frac{\sum_{t=0}^{(N-2)T} |\sum_{k=-L}^{L} w_k S(x, z - k, t) S^*(x, z - k, t + T)|}{\sum_{t=0}^{(N-2)T} \sum_{k=-L}^{L} w_k \frac{1}{2}[I(x, z - k, t) + I(x, z - k, t + T)]}},\tag{12}$$

where w_k is a depth-dependent window function of length $2L + 1$. Though it is referred to as a "variance" method, this algorithm is actually a spatially and temporally averaged, correlation-based method with normalization (see the discussion of variance versus correlation in Section 4.4). The correlation is estimated by averaging on a complex basis axially (in z) and a magnitude basis over time. Also note that the correlation definition is the complex conjugate of that used elsewhere in this paper, though due to the absolute value operation, this minor discrepancy has no effect on the final CDV.

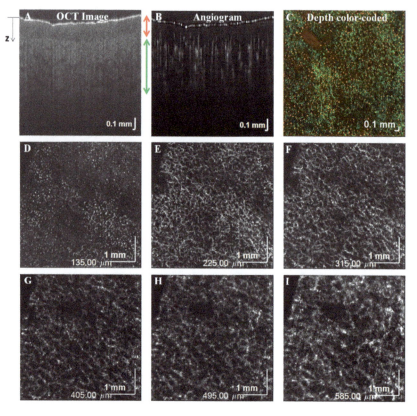

Figure 9. OCTA of the skin on a pig ear. OCT cross-sectional intensity image (**A**) and angiogram (**B**) determined by complex subtraction. (**C**) Overlay of superficial vessels in the epidermis (red) with deeper vasculature in the dermis (green). (**D–I**) Maximum intensity projections centered at different axial (z) positions relative to the surface.

4.4. Classification of Present OCTA Algorithms

Historically, all of the OCTA algorithms described above were novel at the time they were introduced. However, with the benefit of hindsight, we propose basic categories to classify OCTA algorithms in Table 2.

Table 2. Classification of OCTA algorithms.

Category	Classification
OCT signal	Field vs. Intensity/Amplitude vs. Phase
Calculation	Variance/Difference vs. Correlation
Averaging method	Temporal vs. Spatial vs. Spectral
Normalization	Normalized vs. Non-normalized

The primary distinction between algorithms, discussed in Section 4.1, Section 4.2, Section 4.3, is the OCT signal(s) employed. The second distinction, which is emphasized in the literature, is between variance/difference-based methods and correlation-based methods. However, here we argue that in some cases, this distinction is meaningless. Difference-based methods are actually estimating the following:

$$D(T) = E[|X_{t+T} - X_t|^2], \tag{13}$$

where $D(T)$ denotes the difference at a time lag of T and X_t can be the OCT field, intensity, or amplitude at time t. Variance-based methods are estimating the following:

$$V = E[|X_t - E(X_t)|^2]. \tag{14}$$

On the other hand, the un-normalized autocorrelation is defined as:

$$R(T) = E[X_t^* X_{t+T}]. \tag{15}$$

Further expanding Equation (13), the difference can be written in terms of the autocorrelation:

$$D(T) = E[|X_{t+T}|^2] + E[|X_t|^2] - 2\text{Re}\{E[X_t^* X_{t+T}]\} = 2R(0) - 2\text{Re}\{R(T)\}. \tag{16}$$

Therefore, difference and correlation methods are very closely connected if $R(T)$ is real. If $R(T)$ is complex, as would be the case if X_t represented the field and Doppler shifting were present, the difference $D(T)$ depends only on the real part of $R(T)$. From Equations (14) and (15), it can be readily shown that $R(0) = V$ if $E[X_t] = 0$. Thus, every difference method corresponds to an equivalent correlation method via Equation (16).

The third distinction between algorithms is the way that the expectation, $E[\ ]$, is realized in practice. One method of realizing the expectation is by averaging over time. Another way is by averaging over space, at different tissue locations. Yet another way is spectral or optical wavelength averaging, employed in split-spectrum methods [51]. Under the assumption of ergodicity [64], all averaging methods are asymptotically equivalent, and in practice, all can be used to some degree. However, note that averaging over one dimension will automatically degrade the resolution in that dimension.

Fourth, OCTA methods can be distinguished by the use of normalization. The normalized correlation is divided by the signal power, $R(0)$:

$$r(T) = R(T)/E[|X_t|^2] = R(T)/R(0) \tag{17}$$

For the complex signal, the power $R(0)$ is related to the total scattering within a coherence volume. In a vessel, this depends on the backscattering cross-section of RBCs (which depends on orientation according to Section 2.2), and the RBC density (hematocrit). As discussed further in Section 7.1, $|R(\tau)|$ is a monotonically decreasing function under certain conditions, with the decorrelation rate, or rate of autocorrelation decay, being proportional to speed. As difference methods depend on $R(0)$ and $R(T)$, there are two regimes to consider in understanding Equation (16). The first is when T is much longer than the intrinsic decorrelation time. In this case, $R(T) \ll R(0)$, and $D(T)$ is proportional to the signal power $R(0)$, typically related to backscattering (RBC density and orientation). If T is on the order of the intrinsic decorrelation time, the difference $D(T)$ depends on both the signal power $R(0)$ and the decorrelation rate. In this case, the interpretation of the difference $D(T)$ becomes more ambiguous, and it can be affected by the signal power or decorrelation rate, which can be impacted by the RBC density, orientation, and speed. With the normalization in Equation (17), $r(T)$ is more directly related to the rate of decorrelation, and hence, the RBC speed. However, to rigorously account for the possible presence of static scattering, measurements at several time lags [18] are required.

5. OCTA Scanning Protocols

The efficiency and sensitivity of OCTA measurements are determined by the OCTA scanning protocol. At a fundamental level, scanning protocols can be categorized based on whether the analysis is performed on consecutive A-scans, frames, or volumes. Figure 10 shows the so-called MB-scan,

BM-scan and intervolume scanning methods. In Figure 10, the cube represents the imaged object, and t_1, t_2, t_3 are the first, second, and third OCT scanning time scales, respectively, with $t_3 > t_2 > t_1$. Each protocol can be characterized by the time duration for which a single location is observed.

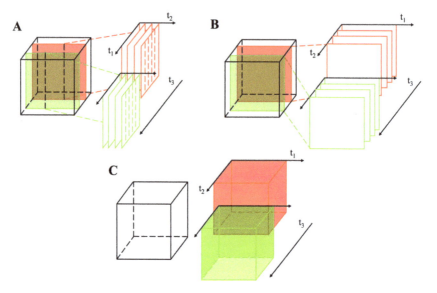

Figure 10. Volumetric OCTA scanning protocols can operate with respect to A-scan (**A**), frame (**B**), or volume (**C**). A cubic volume is scanned with time scales t_1, t_2, and t_3. Data acquired sequentially along time scale t_3 are shown in red and green. (**A**) MB-scan: multiple A-scans are obtained at one lateral position before switching to the next lateral position; (**B**) BM-scan or interframe scan: multiple B-scans are obtained at one cross-sectional location before switching to the next location; (**C**) intervolume scan: successive scans of the whole volume. Each scan achieves a progressively larger observation time for a single spatial position ($t_3 > t_2 > t_1$).

To our knowledge, Fingler et al. [57] were the first to rigorously compare different OCTA scanning patterns. They compared the MB-scan (Figure 10A) and the BM-scan (Figure 10B), using a phase contrast algorithm. An M-scan is a repeated zero-dimensional scan at a single position, while a B-scan is a one-dimensional scan along a single axis. An MB-scan comprises multiple A-scans taken at one lateral position before switching to the next position (Figure 10A), while a BM-scan comprises repetitive B-scans taken along the same cross-section (Figure 10B). According to [57], the advantages and disadvantages of the two scanning methods are described here.

The MB-scan is an extension of Doppler OCT protocols. By increasing N, the number of A-scans per M-scan, the dynamic range for the measurement increases. However, the MB-scan is not time-efficient, because the total observation time for a single location is ~t_1. Unless the dwell time is very long, $|r(t_1)| \sim 1$; thus it is challenging to observe decorrelation. However, due to the rapid repeated sampling of the same position, the MB-scan can sample fast Doppler velocities [57] without aliasing.

On the contrary, a BM-scan compares consecutive frames, thereby more efficiently utilizing the total acquisition time. With a BM-scan, the total observation time for a single location is ~t_2. In [57], the BM-scan was able to acquire data 200 times faster than an MB-scan of the same size. Even when using fast systems, the BM-scan may suffer from aliasing of fast Doppler velocities; however, the decorrelation rate can be obtained if the interframe time is short enough, i.e., $|r(t_1)| > 0$, and provided that t_2 exceeds the intrinsic decorrelation time.

In a logical extension of the above two scanning methods, in 2016, Wei et al. [65] proposed a volumetric optical microangiography method (Figure 10C) which used intervolume OCT scans to extract dynamic changes. The total observation time for a single location is ~t_3. However, in this volumetric protocol, all information about the decorrelation rate is lost as speckles decorrelate between volumes (i.e., $|r(t_2)| \sim 0$) for all but the slowest flows. Nevertheless, the volumetric OCTA is likely to become more prevalent as imaging speeds continue to improve [66].

6. Empirical Validation of OCTA

A major question in quantitative OCTA is the degree to which the measured signals are affected by the RBC speed versus density or orientation. Several authors have attempted to answer this question empirically. In 2016, Choi et al. [67] investigated the relationship between OMAG (complex difference OCTA) signals and capillary flow. They proposed an analytic model that expressed OMAG signals as a function of time interval between successive B-scan frames, particle speed, and concentration (the last two determine flux). Based on this model, they performed simulations, as well as phantom experiments, using microfluidic channels filled with diluted Intralipid solution to model blood vessels. It was shown that OMAG signal increases with flow speed within a certain range that depends on the time interval between successive B-scan frames, as expected based on Equation (16). Furthermore, OMAG signal increased with particle concentration, but was not strictly linear. One limitation of this study is that the Intralipid solution and blood possess very different scattering properties [68,69]. Su et al. [70] used blood samples in microfluidic channels to demonstrate the relationship between SSADA decorrelation signal and the flow speed and channel width. They concluded that before saturation, the decorrelation rate was proportional to the blood flow speed when the channel width was fixed.

Even if flow velocities, channel widths, and particle/cell concentrations are realistic, controlled ex vivo experiments are limited in how well they can model the range of phenomena that are present in vivo. These include effects such as static scattering and multiple scattering involving extravascular tissue (Figure 2B), RBC orientation and transit deformation, vascular compliance, and cell-endothelium interactions. So, in vitro experiments may verify algorithms under model conditions, but the model might only partially capture the range of rheological and hemodynamic phenomena present in vivo.

One proposed in vivo benchmark for OCTA is fluorescence angiography (FA), which is a gold standard method for perfusion imaging [7,42]. Comparative OCTA-FA studies [7,55] have suggested that the presence of moving blood cells is a prerequisite for detection by OCTA. The threshold red blood cell density and speed required for OCTA detection are usually determined by the algorithm sensitivity. While FA shows plasma perfusion, limited depth resolution and lack of three-dimensional data and quantitative flow information make FA a less-than-ideal technique for OCTA validation.

The gold standard for single vessel hemodynamic imaging in deep tissue is multiphoton microscopy (MPM) [71,72]. In the simplest implementation, a fluorescent label is injected into the bloodstream and volumetric two-photon microscopy (TPM) is performed to acquire an angiogram. Vakoc et al. [15] showed that OCTA and two-photon microscopy angiogram morphologies correlate well for vessels larger than capillaries, and that OCTA is not confounded by dye leakage, which can impair TPM. Aside from morphology, TPM line scans enable red blood cell imaging in individual capillaries [22], measuring in vivo speed, flux, and linear density quantitatively. In 2012, Srinivasan et al. [18] performed OCTA and TPM line scans sequentially in the same vessels in vivo, showing that OCTA decorrelation rate increases with RBC speed measured by TPM. Later in 2014, Wang et al. [73] validated OMAG (complex difference OCTA) with TPM, finding no significant difference between the respective vessel densities derived from OMAG and TPM, up to the penetration depth of TPM.

When comparing OCTA and TPM, it is important to recognize that their contrast mechanisms are complementary. As OCTA measures RBC scattering and TPM measures plasma tracer fluorescence, measurements of vessel diameter must disagree in small vessels due to the plasma only, cell-free layer [2]. Moreover, typically, OCTA has a worse volumetric resolution than TPM, and asynchrony in measurements [18] can additionally confound comparisons between modalities unless physiology is carefully maintained. Thus, rigorous verification of OCTA with simultaneous TPM is a promising topic for further investigation.

The most appealing and direct validation approach is to use another OCT modality or algorithm to cross-validate OCTA. In 2012, Ren et al. [74] noticed that the passage of a red blood cell through the OCT coherence volume led to phase and intensity transients. Based on this insight, they developed a particle counting method for measuring the flux, speed, and linear density in a capillary. Using particle counting, they developed and validated a phase intensity mapping (PIM) algorithm for measuring quantitative cerebral blood flow (CBF) [75]. It remains unclear whether individual red blood cell passage can be measured at all locations in an image, or whether these results are merely anecdotal. Moreover, the intensity pattern created by decorrelation can create random transients that could be easily mistaken for RBC passage (e.g., Figures 4E and 6B). Still, particle counting remains an attractive approach for validating OCTA in stable preparations.

7. OCTA Measurements of Hemodynamics

Based on dynamic changes in intensity, phase, or complex signal, OCTA algorithms can distinguish dynamic tissue from static tissue. Thus, while OCTA can answer the question "where is there flow?", it cannot yet reliably answer the question "how much flow is there?". In recent years, several attempts have been made to further quantify OCTA signals. Many of these efforts are based on estimating the autocorrelation function. While the autocorrelation function can be estimated, to date, there is no rigorous theory or model for recovering RBC flow or speed from OCTA signals. Here, we summarize some promising work towards these goals.

7.1. Flow Quantification

In 2010, Wang et al. [19] made an early effort at providing an autocorrelation model to measure transverse particle flow speed. Though they focused on intensity transients, here we generalize their initial work. The basic principle of their model is that when particles pass through the imaging beam, they create OCT signal transients that may provide information about the speed of the underlying particles. However, with a large coherence volume, the individual transients may overlap in time. The complex signal at position (x, z) is expressed as a superposition of particle contributions:

$$S(x, z, t) = \sum_{k=1}^{G(x,z)} M_k(x, z) REC(x, z, t - t_k), \tag{18}$$

$$REC(x, z, t) = \begin{cases} 1, & 0 \leq t \leq \tau_0(x, z) \\ 0, & \text{otherwise} \end{cases}, \tag{19}$$

where k is index of the kth particle, $G(x, z)$ is the total number of particles passing through the imaging beam within the signal acquisition period, $M_k(x, z)$ is the complex amplitude of the kth particle transient, t_k denotes the time when a particle begins to pass through the beam, and $\tau_0(x, z)$ is the position-dependent transit time of the particle.

After expressing the complex OCT signal in terms of particle contributions, the normalized autocorrelation function of $S(x, z, t)$ is given by:

$$\frac{R(x, z, \tau)}{R(x, z, 0)} = \begin{cases} 1 - \frac{\tau}{\tau_0(x,z)}, & \tau \leq \tau_0(x, z) \\ 0, & \tau \geq \tau_0(x, z) \end{cases}, \tag{20}$$

where $R(x, z, \tau)$ is the autocorrelation function of $S(x, z, t)$ with time lag τ. Note that this is equivalent to the normalized autocorrelation of $REC(t)$. The slope of the normalized autocorrelation function in Equation (20) is proportional to the transverse speed ($\sim 1/\tau_0$). Note that Equation (20) can be further generalized to accommodate other transient shapes.

In 2012, Srinivasan et al. [18] proposed an alternative model to relate the autocorrelation to speed. For small particles undergoing isotropic motion through a coherence volume, they proposed that the autocorrelation decay is determined by axial and transverse point spread functions, while for large particles, the spatial characteristics of the particles themselves dominate the autocorrelation as described above. In [76], for small particles, the autocorrelation function at time lag τ in cylindrical coordinates (ϱ, φ, z) is:

$$R_d(\tau) = \frac{2|K|^2}{\pi^2 w_\varrho^4} \sqrt{\frac{\pi}{\left(\frac{v_\varrho^2}{w_\varrho^2} + \frac{v_z^2}{w_z^2}\right)}} P_A \exp\left[-\frac{(v_\varrho\tau)^2}{w_\varrho^2} - \frac{(v_z\tau)^2}{w_z^2}\right] \exp[i(\frac{4\pi n}{\lambda_0})v_z\tau], \quad (21)$$

where w_ϱ is the transverse beam profile, w_z is the axial resolution, K is an arbitrary complex constant [77], P_A is the power in the random process which describes the field, v is the particle's speed, n denotes the refractive index, and λ_0 is the central wavelength. The power spectral density, P_d, derived from the temporal autocorrelation function, is expressed as:

$$P_d(f) = \frac{2|K|^2}{\pi w_\varrho^4 \left(\frac{v_\varrho^2}{w_\varrho^2} + \frac{v_z^2}{w_z^2}\right)} P_A \exp\left[-\frac{\pi^2 \left(f - \frac{2nv_z}{\lambda_0}\right)^2}{\left(\frac{v_\varrho^2}{w_\varrho^2} + \frac{v_z^2}{w_z^2}\right)}\right]. \quad (22)$$

In the presence of static scattering (Figure 6), the autocorrelation takes the form:

$$R(\tau) = R_d(\tau) + R_s(\tau), \quad (23)$$

where $R_s(\tau)$ is the autocorrelation of the static component, with a much longer decorrelation time than the autocorrelation of the dynamic component, $R_d(\tau)$. In practice, $R_s(\tau)$ is usually constant over time scales of interest. Aside from the Doppler shift, the un-normalized autocorrelation $R_d(\tau)$ provides two essential observables: the decorrelation rate, which is sensitive to speed, and power (P_A), which is sensitive to the RBC density. Recent work has proposed to quantify OCTA using difference algorithms measured at several time delays [20,78], providing the ability to measure blood flow speed. Since difference and correlation algorithms are related by Equation (16), these algorithms essentially estimate the un-normalized autocorrelation. As highlighted in Equation (23), static scattering, if present, must also be taken into account in parametric estimations based on the autocorrelation.

Finally, a major limitation of existing models is that they do not account for multiple scattering. In particular, multiple dynamic scattering events (Figure 2C, green) increase the decorrelation rate relative to the single scattering models described above, as each dynamic scattering event causes momentum transfer [79]. In such cases, the decorrelation rate depends on the number of scattering events, which in turn is impacted by the RBC density. Thus, with multiple intravascular scattering events, decorrelation rate is not a "pure" metric of speed. Therefore, decorrelation rate is not a good metric of speed within macrovessels where multiple scattering dominates, but may perform better in capillaries where hematocrits are lower and singly backscattered light prevails (Figure 2B).

7.2. Hematocrit Quantification

Since OCTA signal depends on the RBC density, can OCTA be used to quantify hematocrit? The differences in rheology, geometry, and light scattering in capillaries versus macrovessels suggest different approaches for each. In macrovessels, backscattering or attenuation (signal slope) are possible observables which may help to determine hematocrit. However, due to the high scattering coefficient and anisotropy of RBCs, multiple scattering events are very likely, except at superficial path lengths (Figure 2C,D). In particular, at physiological hematocrits, dependent scattering and shadowing effects lead to a highly nonlinear relationship between the RBC concentration and scattering coefficient [80,81]. This nonlinear relationship hampers efforts at quantifying hematocrit based on light scattering and the signal slope alone. Additionally, the oxygen saturation dependence of hemoglobin refractive index and RBC scattering further complicate efforts to measure hematocrit based on attenuation [26]. The orientation-dependence of light scattering from RBCs (Figure 2C) makes quantifying hematocrit from backscattering alone challenging. Thus, quantification is challenging in macrovessels.

The single file flow and relatively lower hematocrit in capillaries makes multiple scattering within these vessels less problematic. However, RBCs may re-orient themselves and possibly deform as they squeeze through the smallest diameter capillaries, thereby changing their backscattering cross-sections. Moreover, measuring backscattering directly would need absolute calibration, which can be difficult in vivo. However, backscattering may still measure relative changes in the red blood cell content in capillaries [82] and, possibly, at the surfaces of macrovessels over time. Thus, while quantification of hematocrit changes is possible in capillaries, absolute measurements of hematocrit with conventional OCTA are currently challenging.

8. Can OCTA Be Made a Quantitative Tool?

OCTA systems can observe dynamic signal power (variance) and decorrelation rate [19,83], based on the dynamics of light scattering. As algorithms, imaging system performance, and motion tracking/compensation continue to improve, OCTA observables, particularly decorrelation rate, can be precisely and accurately measured. These observables may generate useful diagnostic information, even if their underlying hemodynamic correlates remain unclear. However, if OCTA observables can be directly linked to hemodynamic parameters such as blood flow, volume, hematocrit, and speed, OCTA diagnostics could aid understanding of pathogenesis. This effort requires an appropriate model to describe OCTA signals. The model may be empirical (Section 6), but ideally, should have a theoretical foundation (Section 7.1). Current theoretical models are very simple, and only account for single scattering [18,19,83,84]. Improvements in OCTA theory to include multiple scattering [85] and orientation effects [37] are needed. Empirical models have been developed for flow phantoms [67], but they may be limited to in vitro conditions, and their applicability in vivo remains uncertain. Better in vivo validation experiments, perhaps in well-controlled and stable animal preparations, are needed. Last, due to differences in light scattering and hemodynamics (Figures 1 and 2), models for capillaries and macrovessels must be developed independently.

In spite of these proposed efforts, the inherent complexity of the rheology and light transport in microvasculature may prevent reliable quantification of OCTA. Therefore, we propose that alternative optical properties (aside from light scattering) may enable more quantitative OCTA. For instance, visible light OCTA [86] enables direct absorption-based measurements of hemoglobin concentration, which is expected to correlate well with hematocrit (RBC volume fraction) under most conditions [87]. Yet another way to circumvent the pitfalls of RBC scattering is to introduce an exogenous contrast agent with more desirable scattering properties into the bloodstream [88]. If a more isotropically scattering contrast agent such as Intralipid® [89,90] is used, angiograms derived from the contrast agent signal alone do not suffer from multiple scattering tails [89,90]. Microbubbles [91,92] are promising for enhancing intravascular scattering signals, and may present more well-defined decorrelation characteristics than blood. Moreover, if the contrast agent behaves like plasma and the signal can be calibrated and related to concentration [89,90], plasma flow, transit time, and volume can all be measured.

9. Conclusions

Despite recent strides in OCTA imaging speed, field-of-view, and measurement of OCTA observables, OCTA remains a qualitative tool at present. The obstacles to quantification include the irregular shape of RBCs, the consequent orientation-dependence of RBC backscattering, and the high anisotropy of the RBC scattering phase function, which leads to multiple scattering in large vessels. Quantification of OCTA signals can be achieved only through a rigorous understanding of the relationship between hemodynamics, rheology, and light scattering of RBCs. Improvements in theoretical models, validated in microvasculature in vivo against gold standard techniques and possibly in simulation, may help to improve this understanding. Finally, alternative measurements, based on absorption or exogenous contrast agents, may help to alleviate some of the confounds associated with RBC scattering and enhance the quantitative information provided by OCTA. More quantitative interpretation of OCTA would aid the application of this promising technique to study pathophysiology, and also potentially enhance the clinical impact of OCTA, making this endeavor well worth the effort.

Acknowledgments: We acknowledge support from the National Institutes of Health (EB023591, AG010129, and NS094681) and the Glaucoma Research Foundation Catalyst for a Cure.

Author Contributions: Jun Zhu and Vivek J. Srinivasan conceived the paper. Conrad W. Merkle, Marcel T. Bernucci, and Shau Poh Chong generated data for the figures. Conrad W. Merkle and Vivek J. Srinivasan processed the data. Jun Zhu, Conrad W. Merkle, Shau Poh Chong, and Vivek J. Srinivasan made the figures for the paper. All authors wrote and reviewed the paper.

Conflicts of Interest: Vivek J. Srinivasan receives royalties from Optovue, Inc.

References

1. Fedosov, D.A.; Noguchi, H.; Gompper, G. Multiscale modeling of blood flow: From single cells to blood rheology. *Biomech. Model. Mechanobiol.* **2014**, *13*, 239–258. [CrossRef] [PubMed]
2. Secomb, T.W. Blood flow in the microcirculation. *Annu. Rev. Fluid Mech.* **2017**, *49*, 443–461. [CrossRef]
3. Fung, Y.-C. *Biomechanics: Motion, Flow, Stress, and Growth*; Springer Science & Business Media: New York, NY, USA, 2013.
4. Choi, W.; Mohler, K.J.; Potsaid, B.; Lu, C.D.; Liu, J.J.; Jayaraman, V.; Cable, A.E.; Duker, J.S.; Huber, R.; Fujimoto, J.G. Choriocapillaris and choroidal microvasculature imaging with ultrahigh speed oct angiography. *PLoS ONE* **2013**, *8*, e81499. [CrossRef] [PubMed]
5. Jia, Y.; Bailey, S.T.; Wilson, D.J.; Tan, O.; Klein, M.L.; Flaxel, C.J.; Potsaid, B.; Liu, J.J.; Lu, C.D.; Kraus, M.F.; et al. Quantitative optical coherence tomography angiography of choroidal neovascularization in age-related macular degeneration. *Ophthalmology* **2014**, *121*, 1435–1444. [CrossRef] [PubMed]
6. Jia, Y.; Bailey, S.T.; Hwang, T.S.; McClintic, S.M.; Gao, S.S.; Pennesi, M.E.; Flaxel, C.J.; Lauer, A.K.; Wilson, D.J.; Hornegger, J.; et al. Quantitative optical coherence tomography angiography of vascular abnormalities in the living human eye. *Proc. Natl. Acad. Sci. USA* **2015**, *112*, 2395–2402. [CrossRef] [PubMed]
7. Ishibazawa, A.; Nagaoka, T.; Takahashi, A.; Omae, T.; Tani, T.; Sogawa, K.; Yokota, H.; Yoshida, A. Optical coherence tomography angiography in diabetic retinopathy: A prospective pilot study. *Am. J. Ophthalmol.* **2015**, *160*, 35–44. [CrossRef] [PubMed]
8. Jia, Y.; Wei, E.; Wang, X.; Zhang, X.; Morrison, J.C.; Parikh, M.; Lombardi, L.H.; Gattey, D.M.; Armour, R.L.; Edmunds, B.; et al. Optical coherence tomography angiography of optic disc perfusion in glaucoma. *Ophthalmology* **2014**, *121*, 1322–1332. [CrossRef] [PubMed]
9. Spaide, R.F. Optical coherence tomography angiography signs of vascular abnormalization with antiangiogenic therapy for choroidal neovascularization. *Am. J. Ophthalmol.* **2015**, *160*, 6–16. [CrossRef] [PubMed]
10. Kim, D.Y.; Fingler, J.; Zawadzki, R.J.; Park, S.S.; Morse, L.S.; Schwartz, D.M.; Fraser, S.E.; Werner, J.S. Optical imaging of the chorioretinal vasculature in the living human eye. *Proc. Natl. Acad. Sci. USA* **2013**, *110*, 14354–14359. [CrossRef] [PubMed]
11. Talisa, E.; Bonini Filho, M.A.; Chin, A.T.; Adhi, M.; Ferrara, D.; Baumal, C.R.; Witkin, A.J.; Reichel, E.; Duker, J.S.; Waheed, N.K. Spectral-domain optical coherence tomography angiography of choroidal neovascularization. *Ophthalmology* **2015**, *122*, 1228–1238.
12. Tsai, T.H.; Ahsen, O.O.; Lee, H.C.; Liang, K.; Figueiredo, M.; Tao, Y.K.; Giacomelli, M.G.; Potsaid, B.M.; Jayaraman, V.; Huang, Q.; et al. Endoscopic optical coherence angiography enables 3-dimensional visualization of subsurface microvasculature. *Gastroenterology* **2014**, *147*, 1219–1221. [CrossRef] [PubMed]

13. Lee, H.C.; Ahsen, O.O.; Liang, K.; Wang, Z.; Cleveland, C.; Booth, L.; Potsaid, B.; Jayaraman, V.; Cable, A.E.; Mashimo, H.; et al. Circumferential optical coherence tomography angiography imaging of the swine esophagus using a micromotor balloon catheter. *Biomed. Opt. Express* **2016**, *7*, 2927–2942. [CrossRef] [PubMed]

14. Jung, Y.; Zhi, Z.; Wang, R.K. Three-dimensional optical imaging of microvascular networks within intact lymph node in vivo. *J. Biomed. Opt.* **2010**, *15*, 050501–050503. [CrossRef] [PubMed]

15. Vakoc, B.J.; Lanning, R.M.; Tyrrell, J.A.; Padera, T.P.; Bartlett, L.A.; Stylianopoulos, T.; Munn, L.L.; Tearney, G.J.; Fukumura, D.; Jain, R.K.; et al. Three-dimensional microscopy of the tumor microenvironment in vivo using optical frequency domain imaging. *Nat. Med.* **2009**, *15*, 1219–1223. [CrossRef] [PubMed]

16. Srinivasan, V.J.; Jiang, J.Y.; Yaseen, M.A.; Radhakrishnan, H.; Wu, W.; Barry, S.; Cable, A.E.; Boas, D.A. Rapid volumetric angiography of cortical microvasculature with optical coherence tomography. *Opt. Lett.* **2010**, *35*, 43–45. [CrossRef] [PubMed]

17. Srinivasan, V.J.; Atochin, D.N.; Radhakrishnan, H.; Jiang, J.Y.; Ruvinskaya, S.; Wu, W.; Barry, S.; Cable, A.E.; Ayata, C.; Huang, P.L.; et al. Optical coherence tomography for the quantitative study of cerebrovascular physiology. *J. Cereb. Blood Flow Metab.* **2011**, *31*, 1339–1345. [CrossRef] [PubMed]

18. Srinivasan, V.J.; Radhakrishnan, H.; Lo, E.H.; Mandeville, E.T.; Jiang, J.Y.; Barry, S.; Cable, A.E. Oct methods for capillary velocimetry. *Biomed. Opt. Express* **2012**, *3*, 612–629. [CrossRef] [PubMed]

19. Wang, Y.; Wang, R. Autocorrelation optical coherence tomography for mapping transverse particle-flow velocity. *Opt. Lett.* **2010**, *35*, 3538–3540. [CrossRef] [PubMed]

20. Choi, W.; Moult, E.M.; Waheed, N.K.; Adhi, M.; Lee, B.; Lu, C.D.; de Carlo, T.E.; Jayaraman, V.; Rosenfeld, P.J.; Duker, J.S.; et al. Ultrahigh-speed, swept-source optical coherence tomography angiography in nonexudative age-related macular degeneration with geographic atrophy. *Ophthalmology* **2015**, *122*, 2532–2544. [CrossRef] [PubMed]

21. Duncan, D.D.; Kirkpatrick, S.J. Can laser speckle flowmetry be made a quantitative tool? *J. Opt. Soc. Am. A Opt. Image Sci. Vis.* **2008**, *25*, 2088–2094. [CrossRef] [PubMed]

22. Kleinfeld, D.; Mitra, P.P.; Helmchen, F.; Denk, W. Fluctuations and stimulus-induced changes in blood flow observed in individual capillaries in layers 2 through 4 of rat neocortex. *Proc. Natl. Acad. Sci. USA* **1998**, *95*, 15741–15746. [CrossRef] [PubMed]

23. Kamoun, W.S.; Chae, S.S.; Lacorre, D.A.; Tyrrell, J.A.; Mitre, M.; Gillissen, M.A.; Fukumura, D.; Jain, R.K.; Munn, L.L. Simultaneous measurement of rbc velocity, flux, hematocrit and shear rate in vascular networks. *Nat. Method.* **2010**, *7*, 655–660. [CrossRef] [PubMed]

24. Santisakultarm, T.P.; Cornelius, N.R.; Nishimura, N.; Schafer, A.I.; Silver, R.T.; Doerschuk, P.C.; Olbricht, W.L.; Schaffer, C.B. In vivo two-photon excited fluorescence microscopy reveals cardiac- and respiration-dependent pulsatile blood flow in cortical blood vessels in mice. *Am. J. Physiol. Heart Circ. Phys.* **2012**, *302*, H1367–H1377. [CrossRef] [PubMed]

25. Desjardins, C.; Duling, B.R. Microvessel hematocrit: Measurement and implications for capillary oxygen transport. *Am. J. Physiol. Heart Circ. Physiol.* **1987**, *252*, H494–H503.

26. Faber, D.J.; Aalders, M.C.; Mik, E.G.; Hooper, B.A.; van Gemert, M.J.; van Leeuwen, T.G. Oxygen saturation-dependent absorption and scattering of blood. *Phys. Rev. Lett.* **2004**, *93*, 028102. [CrossRef] [PubMed]

27. Meinke, M.; Müller, G.; Helfmann, J.; Friebel, M. Optical properties of platelets and blood plasma and their influence on the optical behavior of whole blood in the visible to near infrared wavelength range. *J. Biomed. Opt.* **2007**, *12*, 014024–014029. [CrossRef] [PubMed]

28. Sydoruk, O.; Zhernovaya, O.; Tuchin, V.; Douplik, A. Refractive index of solutions of human hemoglobin from the near-infrared to the ultraviolet range: Kramers-kronig analysis. *J. Biomed. Opt.* **2012**, *17*, 115002. [CrossRef] [PubMed]

29. Bosschaart, N.; Edelman, G.J.; Aalders, M.C.; van Leeuwen, T.G.; Faber, D.J. A literature review and novel theoretical approach on the optical properties of whole blood. *Lasers Med. Sci.* **2014**, *29*, 453–479. [CrossRef] [PubMed]

30. Yaroslavsky, A.N.; Yaroslavsky, I.V.; Goldbach, T.; Schwarzmaier, H.-J. Optical properties of blood in the near-infrared spectral range. Proceedings of Photonics West 1996, San Jose, CA, USA, 17 May 1996; pp. 314–324.

31. Simon, J.-C. Dependent scattering and radiative transfer in dense inhomogeneous media. *Phys. A Stat. Mech. Appl.* **1997**, *241*, 77–81. [CrossRef]

32. Roggan, A.; Friebel, M.; Dörschel, K.; Hahn, A.; Muller, G. Optical properties of circulating human blood in the wavelength range 400–2500 nm. *J. Biomed. Opt.* **1999**, *4*, 36–46. [CrossRef] [PubMed]

33. Friebel, M.; Roggan, A.; Müller, G.; Meinke, M. Determination of optical properties of human blood in the spectral range 250 to 1100 nm using monte carlo simulations with hematocrit-dependent effective scattering phase functions. *J. Biomed. Opt.* **2006**, *11*, 034021. [CrossRef] [PubMed]

34. Meinke, M.; Müller, G.; Helfmann, J.; Friebel, M. Empirical model functions to calculate hematocrit-dependent optical properties of human blood. *Appl. Opt.* **2007**, *46*, 1742–1753. [CrossRef] [PubMed]

35. Henyey, L.G.; Greenstein, J.L. Diffuse radiation in the galaxy. *Astrophys. J.* **1941**, *93*, 70–83. [CrossRef]

36. Hammer, M.; Yaroslavsky, A.N.; Schweitzer, D. A scattering phase function for blood with physiological haematocrit. *Phys. Med. Biol.* **2001**, *46*, N65. [CrossRef] [PubMed]

37. Cimalla, P.; Walther, J.; Mittasch, M.; Koch, E. Shear flow-induced optical inhomogeneity of blood assessed in vivo and in vitro by spectral domain optical coherence tomography in the 1.3 μm wavelength range. *J. Biomed. Opt.* **2011**, *16*, 116020. [CrossRef] [PubMed]

38. Friebel, M.; Helfmann, J.; Müller, G.; Meinke, M. Influence of shear rate on the optical properties of human blood in the spectral range 250 to 1100 nm. *J. Biomed. Opt.* **2007**, *12*, 054005–054008. [CrossRef] [PubMed]

39. Prati, F.; Regar, E.; Mintz, G.S.; Arbustini, E.; Di Mario, C.; Jang, I.-K.; Akasaka, T.; Costa, M.; Guagliumi, G.; Grube, E. Expert review document on methodology, terminology, and clinical applications of optical coherence tomography: Physical principles, methodology of image acquisition, and clinical application for assessment of coronary arteries and atherosclerosis. *Eur. Heart J.* **2010**, *31*, 401–415. [CrossRef] [PubMed]

40. Bigio, I.J.; Fantini, S. *Quantitative Biomedical Optics: Theory, Methods, and Applications*; Cambridge University Press: Cambridge, UK, 2016.

41. Chen, C.-L.; Wang, R.K. Optical coherence tomography based angiography [invited]. *Biomed. Opt. Express* **2017**, *8*, 1056. [CrossRef] [PubMed]

42. De Carlo, T.E.; Romano, A.; Waheed, N.K.; Duker, J.S. A review of optical coherence tomography angiography (octa). *Int. J. Retina Vitreous* **2015**, *1*. [CrossRef] [PubMed]

43. Siegert, A. *On the Fluctuations in Signals Returned by Many Independently Moving Scatterers*; Massachusetts Institute of Technology: Cambridge, MA, USA, 1943.

44. Srinivasan, V.J.; Chan, A.C.; Lam, E.Y. *Doppler OCT and OCT Angiography for In Vivo Imaging of Vascular Physiology*; INTECH Open Access Publisher: Rijeka, Croatia, 2012.

45. Leitgeb, R.A.; Werkmeister, R.M.; Blatter, C.; Schmetterer, L. Doppler optical coherence tomography. *Prog. Retinal Eye Res.* **2014**, *41*, 26–43. [CrossRef] [PubMed]

46. Schmitt, J.M.; Xiang, S.; Yung, K.M. Speckle in optical coherence tomography. *J. Biomed. Opt.* **1999**, *4*, 95–105. [CrossRef] [PubMed]

47. Barton, J.K.; Stromski, S. Flow measurement without phase information in optical coherence tomography images. *Opt. Express* **2005**, *13*, 5234–5239. [CrossRef] [PubMed]

48. Mariampillai, A.; Standish, B.A.; Moriyama, E.H.; Khurana, M.; Munce, N.R.; Leung, M.K.; Jiang, J.; Cable, A.; Wilson, B.C.; Vitkin, I.A. Speckle variance detection of microvasculature using swept-source optical coherence tomography. *Opt. Lett.* **2008**, *33*, 1530–1532. [CrossRef] [PubMed]

49. Mariampillai, A.; Leung, M.K.; Jarvi, M.; Standish, B.A.; Lee, K.; Wilson, B.C.; Vitkin, A.; Yang, V.X. Optimized speckle variance oct imaging of microvasculature. *Opt. Lett.* **2010**, *35*, 1257–1259. [CrossRef] [PubMed]

50. Enfield, J.; Jonathan, E.; Leahy, M. In vivo imaging of the microcirculation of the volar forearm using correlation mapping optical coherence tomography (cmoct). *Biomed. Opt. Express* **2011**, *2*, 1184–1193. [CrossRef] [PubMed]

51. Jia, Y.; Tan, O.; Tokayer, J.; Potsaid, B.; Wang, Y.; Liu, J.J.; Kraus, M.F.; Subhash, H.; Fujimoto, J.G.; Hornegger, J. Split-spectrum amplitude-decorrelation angiography with optical coherence tomography. *Opt. Express* **2012**, *20*, 4710–4725. [CrossRef] [PubMed]

52. Leitgeb, R.A.; Schmetterer, L.; Drexler, W.; Fercher, A.; Zawadzki, R.; Bajraszewski, T. Real-time assessment of retinal blood flow with ultrafast acquisition by color doppler fourier domain optical coherence tomography. *Opt. Express* **2003**, *11*, 3116–3121. [CrossRef] [PubMed]

53. Leitgeb, R.A.; Schmetterer, L.; Hitzenberger, C.K.; Fercher, A.F.; Berisha, F.; Wojtkowski, M.; Bajraszewski, T. Real-time measurement of in vitro flow by fourier-domain color doppler optical coherence tomography. *Opt. Lett.* **2004**, *29*, 171–173. [CrossRef] [PubMed]

54. Zhao, Y.; Chen, Z.; Saxer, C.; Shen, Q.; Xiang, S.; de Boer, J.F.; Nelson, J.S. Doppler standard deviation imaging for clinical monitoring of in vivo human skin blood flow. *Opt. Lett.* **2000**, *25*, 1358–1360. [CrossRef] [PubMed]

55. Makita, S.; Hong, Y.; Yamanari, M.; Yatagai, T.; Yasuno, Y. Optical coherence angiography. *Opt. Express* **2006**, *14*, 7821–7840. [CrossRef] [PubMed]

56. Park, B.H.; Pierce, M.C.; Cense, B.; Yun, S.-H.; Mujat, M.; Tearney, G.J.; Bouma, B.E.; de Boer, J.F. Real-time fiber-based multi-functional spectral-domain optical coherence tomography at 1.3 μm. *Opt. Express* **2005**, *13*, 3931–3944. [CrossRef] [PubMed]

57. Fingler, J.; Schwartz, D.; Yang, C.; Fraser, S.E. Mobility and transverse flow visualization using phase variance contrast with spectral domain optical coherence tomography. *Opt. Express* **2007**, *15*, 12636–12653. [CrossRef] [PubMed]

58. Kim, D.Y.; Fingler, J.; Werner, J.S.; Schwartz, D.M.; Fraser, S.E.; Zawadzki, R.J. In vivo volumetric imaging of human retinal circulation with phase-variance optical coherence tomography. *Biomed. Opt. Express* **2011**, *2*, 1504–1513. [CrossRef] [PubMed]

59. Lee, J.; Srinivasan, V.; Radhakrishnan, H.; Boas, D.A. Motion correction for phase-resolved dynamic optical coherence tomography imaging of rodent cerebral cortex. *Opt. Express* **2011**, *19*, 21258–21270. [CrossRef] [PubMed]

60. Zhang, A.; Zhang, Q.; Chen, C.-L.; Wang, R.K. Methods and algorithms for optical coherence tomography-based angiography: A review and comparison. *J. Biomed. Opt.* **2015**, *20*, 100901. [CrossRef] [PubMed]

61. Wang, R.K.; Jacques, S.L.; Ma, Z.; Hurst, S.; Hanson, S.R.; Gruber, A. Three dimensional optical angiography. *Opt. Express* **2007**, *15*, 4083–4097. [CrossRef] [PubMed]

62. An, L.; Qin, J.; Wang, R.K. Ultrahigh sensitive optical microangiography for in vivo imaging of microcirculations within human skin tissue beds. *Opt. Express* **2010**, *18*, 8220–8228. [CrossRef] [PubMed]

63. Nam, A.S.; Chico-Calero, I.; Vakoc, B.J. Complex differential variance algorithm for optical coherence tomography angiography. *Biomed. Opt. Express* **2014**, *5*, 3822–3832. [CrossRef] [PubMed]

64. Goodman, J.W. *Statistical Optics*; Wiley: New York, NY, USA, 2000.

65. Wei, W.; Xu, J.; Baran, U.; Song, S.; Qin, W.; Qi, X.; Wang, R.K. Intervolume analysis to achieve four-dimensional optical microangiography for observation of dynamic blood flow. *J. Biomed. Opt.* **2016**, *21*, 36005. [CrossRef] [PubMed]

66. Klein, T.; Wieser, W.; Reznicek, L.; Neubauer, A.; Kampik, A.; Huber, R. Multi-mhz retinal oct. *Biomed. Opt. Express* **2013**, *4*, 1890–1908. [CrossRef] [PubMed]

67. Choi, W.J.; Qin, W.; Chen, C.L.; Wang, J.; Zhang, Q.; Yang, X.; Gao, B.Z.; Wang, R.K. Characterizing relationship between optical microangiography signals and capillary flow using microfluidic channels. *Biomed. Opt. Express* **2016**, *7*, 2709–2728. [CrossRef] [PubMed]

68. Kinnunen, M.; Myllylä, R. Effect of glucose on photoacoustic signals at the wavelengths of 1064 and 532 nm in pig blood and intralipid. *J. Phys. D Appl. Phys.* **2005**, *38*, 2654. [CrossRef]

69. Van Staveren, H.J.; Moes, C.J.; van Marie, J.; Prahl, S.A.; Van Gemert, M.J. Light scattering in lntralipid-10% in the wavelength range of 400–1100 nm. *Appl. Opt.* **1991**, *30*, 4507–4514. [CrossRef] [PubMed]

70. Su, J.P.; Chandwani, R.; Gao, S.S.; Pechauer, A.D.; Zhang, M.; Wang, J.; Jia, Y.; Huang, D.; Liu, G. Calibration of optical coherence tomography angiography with a microfluidic chip. *J. Biomed. Opt.* **2016**, *21*, 86015. [CrossRef] [PubMed]

71. Denk, W.; Strickler, J.H.; Webb, W.W. Two-photon laser scanning fluorescence microscopy. *Science* **1990**, *248*, 73–76. [CrossRef] [PubMed]

72. Zipfel, W.R.; Williams, R.M.; Webb, W.W. Nonlinear magic: Multiphoton microscopy in the biosciences. *Nat. Biotechnol.* **2003**, *21*, 1369–1377. [CrossRef] [PubMed]

73. Wang, H.; Baran, U.; Li, Y.; Qin, W.; Wang, W.; Zeng, H.; Wang, R.K. Does optical microangiography provide accurate imaging of capillary vessels?: Validation using multiphoton microscopy. *J. Biomed. Opt.* **2014**, *19*, 106011. [CrossRef] [PubMed]

74. Ren, H.; Du, C.; Park, K.; Volkow, N.D.; Pan, Y. Quantitative imaging of red blood cell velocity invivo using optical coherence doppler tomography. *Appl. Phys. Lett.* **2012**, *100*, 233702. [CrossRef] [PubMed]

75. Ren, H.; Du, C.; Yuan, Z.; Park, K.; Volkow, N.D.; Pan, Y. Cocaine-induced cortical microischemia in the rodent brain: Clinical implications. *Mol. Psychiatry* **2012**, *17*, 1017–1025. [CrossRef] [PubMed]

76. Chan, A.C.; Merkle, C.W.; Lam, E.Y.; Srinivasan, V.J. *Maximum Likelihood Estimation of Blood Velocity Using Doppler Optical Coherence Tomography*; SPIE BiOS: Bellingham, WA, USA, 2014; p. 89349.

77. Srinivasan, V.J.; Sakadžić, S.; Gorczynska, I.; Ruvinskaya, S.; Wu, W.; Fujimoto, J.G.; Boas, D.A. Quantitative cerebral blood flow with optical coherence tomography. *Opt. Express* **2010**, *18*, 2477–2494. [CrossRef] [PubMed]

78. Ploner, S.B.; Moult, E.M.; Choi, W.; Waheed, N.K.; Lee, B.; Novais, E.A.; Cole, E.D.; Potsaid, B.; Husvogt, L.; Schottenhamml, J. Toward quantitative optical coherence tomography angiography: Visualizing blood flow speeds in ocular pathology using variable interscan time analysis. *Retina* **2016**, *36*, S118–S126. [CrossRef] [PubMed]

79. Bonner, R.; Nossal, R. Model for laser doppler measurements of blood flow in tissue. *Appl. Opt.* **1981**, *20*, 2097–2107. [CrossRef] [PubMed]

80. Faber, D.J.; van der Meer, F.J.; Aalders, M.C.; van Leeuwen, T.G. Hematocrit-dependence of the scattering coefficient of blood determined by optical coherence tomography. In Proceedings of the Saratov Fall Meeting 2005: Optical Technologies in Biophysics and Medicine VII, Saratov, Russia, 27–30 October 2005; SPIE: Bellingham, WA, USA, 2006; p. 61639.

81. Faber, D.J.; van Leeuwen, T.G. Are quantitative attenuation measurements of blood by optical coherence tomography feasible? *Opt Lett.* **2009**, *34*, 1435–1437. [CrossRef] [PubMed]

82. Srinivasan, V.J.; Radhakrishnan, H. Optical coherence tomography angiography reveals laminar microvascular hemodynamics in the rat somatosensory cortex during activation. *NeuroImage* **2014**, *102*, 393–406. [CrossRef] [PubMed]

83. Tokayer, J.; Jia, Y.; Dhalla, A.-H.; Huang, D. Blood flow velocity quantification using split-spectrum amplitude-decorrelation angiography with optical coherence tomography. *Biomed. Opt. Express* **2013**, *4*, 1909–1924. [CrossRef] [PubMed]

84. Choi, W.J.; Li, Y.; Qin, W.; Wang, R.K. Cerebral capillary velocimetry based on temporal oct speckle contrast. *Biomed. Opt. Express* **2016**, *7*, 4859–4873. [CrossRef] [PubMed]

85. Hartinger, A.E.; Nam, A.S.; Chico-Calero, I.; Vakoc, B.J. Monte carlo modeling of angiographic optical coherence tomography. *Biomed. Opt. Express* **2014**, *5*, 4338–4349. [CrossRef] [PubMed]

86. Yi, J.; Chen, S.; Backman, V.; Zhang, H.F. In vivo functional microangiography by visible-light optical coherence tomography. *Biomed. Opt. Express* **2014**, *5*, 3603–3612. [CrossRef] [PubMed]

87. Chong, S.P.; Merkle, C.W.; Leahy, C.; Radhakrishnan, H.; Srinivasan, V.J. Quantitative microvascular hemoglobin mapping using visible light spectroscopic optical coherence tomography. *Biomed. Opt. Express* **2015**, *6*, 1429–1450. [CrossRef] [PubMed]

88. Pan, Y.; You, J.; Volkow, N.D.; Park, K.; Du, C. Ultrasensitive detection of 3d cerebral microvascular network dynamics in vivo. *Neuroimage* **2014**, *103*, 492–501. [CrossRef] [PubMed]

89. Merkle, C.W.; Leahy, C.; Srinivasan, V.J. Dynamic contrast optical coherence tomography images transit time and quantifies microvascular plasma volume and flow in the retina and choriocapillaris. *Biomed. Opt. Express* **2016**, *7*, 4289–4312. [CrossRef] [PubMed]

90. Merkle, C.W.; Srinivasan, V.J. Laminar microvascular transit time distribution in the mouse somatosensory cortex revealed by dynamic contrast optical coherence tomography. *Neuroimage* **2016**, *125*, 350–362. [CrossRef] [PubMed]

91. Assadi, H.; Demidov, V.; Karshafian, R.; Douplik, A.; Vitkin, I.A. Microvascular contrast enhancement in optical coherence tomography using microbubbles. *J. Biomed. Opt.* **2016**, *21*, 076014. [CrossRef] [PubMed]

92. Barton, J.K.; Hoying, J.B.; Sullivan, C.J. Use of microbubbles as an optical coherence tomography contrast agent. *Acad. Radiol.* **2002**, *9*, S52–S55. [CrossRef]

Review

Dental Applications of Optical Coherence Tomography (OCT) in Cariology

Hartmut Schneider *, Kyung-Jin Park, Matthias Häfer, Claudia Rüger, Gerhard Schmalz, Felix Krause, Jana Schmidt, Dirk Ziebolz and Rainer Haak

Department of Cariology, Endodontology and Periodontology, University of Leipzig, Liebigstraße 12, 04103 Leipzig, Germany; kyungjin.park@medizin.uni-leipzig.de (K.-J.P.); matthias.haefer@medizin.uni-leipzig.de (M.H.); claudia.rueger@medizin.uni-leipzig.de (C.R.); gerhard.schmalz@medizin.uni-leipzig.de (G.S.); felix.krause@medizin.uni-leipzig.de (F.K.); jana.schmidt@medizin.uni-leipzig.de (J.S.); dirk.ziebolz@medizin.uni-leipzig.de (D.Z.); rainer.haak@medizin.uni-leipzig.de (R.H.)
* Correspondence: hartmut.schneider@medizin.uni-leipzig.de; Tel.: +49-341-97-21263

Academic Editor: Michael Pircher
Received: 10 February 2017; Accepted: 26 April 2017; Published: 3 May 2017

Abstract: Across all medical disciplines, therapeutic interventions are based on previously acquired diagnostic information. In cariology, which includes the detection and assessment of the disease "caries" and its lesions, as well as non-invasive to invasive treatment and caries prevention, visual inspection and radiology are routinely used as diagnostic tools. However, the specificity and sensitivity of these standard methods are still unsatisfactory and the detection of defects is often afflicted with a time delay. Numerous novel methods have been developed to improve the unsatisfactory diagnostic possibilities in this specialized medical field. These newer techniques have not yet found widespread acceptance in clinical practice, which might be explained by the generated numerical or color-coded output data that are not self-explanatory. With optical coherence tomography (OCT), an innovative image-based technique has become available that has considerable potential in supporting the routine assessment of teeth in the future. The received cross-sectional images are easy to interpret and can be processed. In recent years, numerous applications of OCT have been evaluated in cariology beginning with the diagnosis of different defects up to restoration assessment and their monitoring, or the visualization of individual treatment steps. Based on selected examples, this overview outlines the possibilities and limitations of this technique in cariology and restorative dentistry, which pertain to the most clinical relevant fields of dentistry.

Keywords: optical coherence tomography; cariology; caries prevention; caries diagnosis; caries therapy; restoration assessment; process monitoring; in vitro, ex vivo and in vivo evaluation

1. Introduction

Caries is still the most common chronic disease worldwide. Although prevention is generally possible, carious lesions are still one of the main reasons for dental treatment. In order to have an approach to prevent or causally treat caries, a clear diagnosis of the relevant findings in the oral cavity is necessary. Currently, visual and radiologic inspections are two standard techniques for the clinical detection, assessment and monitoring of carious lesions and dental restorations [1,2]. Although the "International Caries Detection and Assessment System" (ICDAS II) has helped to standardize visual diagnosis [3], the validity and reproducibility of lesion assessment is in need of improvement [4,5]. In particular, very early incipient demineralizations within outer enamel layers remain invisible, or cannot be correctly assessed, and radiography generally fails (Figures 1 and 2).

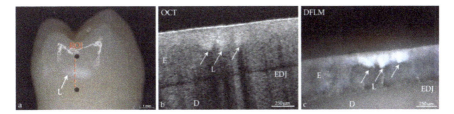

Figure 1. Conformance of visual assessment with optical coherence tomography. (**a**) Extracted human premolar with a carious lesion (L) of International Caries Detection and Assessment System (ICDAS II) Code 1. The lesion is visually detectable only after drying. Marks define the region of interest (ROI); (**b**) In the cross-sectional image of the ROI, generated using spectral domain OCT, bright shadowed spots reveal the lesion area is limited to the first half of the enamel (E); (**c**) After sectioning, demineralization can be imaged using dark-field light microscopy (DFLM) corresponding to the lesion details in the OCT image. D: dentin, EDJ: enamel-dentin junction.

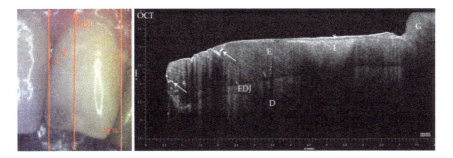

Figure 2. Screenshot, in vivo image of a human incisor. Partial conformance of visual assessment with OCT. A carious lesion (L) of ICDAS II Code 2 can be detected visually at the air-dried enamel surface (red arrow). In the SD-OCT B-scan, this lesion appears as a bright shadowed area without cavitation. Compared to visual assessment, the OCT signal reveals an extended lesion body with dentin (D) involvement and a mineral-rich and porous surface layer (*). In addition, defects, or cracks in the enamel (E) are seen to cause shadowing (white arrows). G: gingiva, EDJ: enamel-dentin junction.

Caries is still one of the main reasons for invasive restorative treatment [6]. Today, it is a primary objective in oral health care to preserve as much of the healthy tooth structure as possible. Therefore, early non- or minimally-invasive treatment options are becoming more important, thus placing higher demands on diagnostic tools as it is no longer sufficient to detect already extended carious lesions or cavitations. Rather, a goal is to recognize lesions, or other defects at their earliest possible stage, and to assess reproducibly whether they progress or not. Early lesion manifestations could be stabilized, or remineralization could even be initiated [7–9]. Traditional caries diagnostic methods like visual-tactile detection and radiography cannot achieve these objectives. X-ray diagnosis, in particular, has limitations and drawbacks: the exposure to ionizing radiation, the interpretation failures due to superpositioning effects, and a low sensitivity [10] that hinders the early detection and monitoring of minor pathological changes. Recently, innovative methods based on fluorescence [11], impedance spectroscopy, or digital infrared-transillumination have been implemented in caries diagnosis [12]. However, even these techniques were not able to solve the referred challenges and showed deficits in both sensitivity and specificity [1,13]. Therefore, there is still a medical need for procedures which could display early caries lesions and their progression. Initial findings concerning OCT suggested that different expressions of lesions, starting with very early demineralizations, could be detected. A threshold independent assessment of the validity by ROC analysis is aimed for the future to clarify whether a separation between sound and diseased is possible over the entire spectrum of carious lesions [14].

Furthermore, there is a substantial need for the assessment of dental restorations, which is technically demanding. Challenges include quality evaluation in terms of material homogeneity (bubble formation within material or at increments), or the detection of material overhangs at restoration margins. Currently there are no suitable methods available to enable dentists to self-check the quality of intermediate steps during restoration placement. Moreover, it is important to have the possibility to assess the bond at restoration margins and along the tooth-restoration interface, as well as to detect carious lesions adjacent to restorations (secondary caries) and compare different restoration materials in clinical trials or in in vitro studies. Today, the in vivo evaluation of marginal integrity can only be performed visually [15,16] and by a sensitive technique with a time-consuming quantitative margin analysis using scanning electron microscopy (SEM) [17]. However, the detection of early lesions adjacent to restorations is still a challenge [1].

Against the background of these demands, optical coherence tomography seems to be a promising technique. The method, which is based on low coherence interferometry, was first presented by Huang et al. [18]. In medicine, OCT is currently primarily used in ophthalmology for evaluating the pathological changes in retinal layers and the optic nerve [19–22]. Further applications are found in dermatology [23,24] and cardiology [25]. The first images of human dental tissues were presented by Colston et al. [26,27] and Feldchtein et al. [28]. Since then, the number of studies using OCT have increased markedly [29,30]. Seven features which make it particularly suitable for clinical application are non-invasive action avoiding radiation-induced tissue damage; easy to understand cross-sectional and 3D-images of soft and hard tissues; as well as defects at a very early stage with high spatial resolution. Structural changes can be obtained qualitatively as well as quantitatively. In particular, high speed data acquisition allows real-time imaging in vitro, ex vivo and in vivo [9,30].

Although OCT has been widely used in dental research, it nevertheless stands at the beginning of its application career. Numerous functional OCT systems [31,32], methodological extensions (e.g., polarization sensitive OCT [10]), as well as evaluation methods have been described [31–33], and also show the potential of this technology in characterizing specific material characteristics [33], or to improve the optical performance of the method, such as detection sensitivity or depth penetration [31,34]. This review presents applications of optical coherence tomography in the field of cariology, which includes subsections on caries prevention, detection, and assessment of carious lesions, as well as caries therapy. Particular focus is placed on restoration assessment and monitoring of the tooth-restoration bond.

2. Optical Coherence Tomography—The Methods

In biomedical and clinical research, the Fourier-domain optical coherence tomography (FD-OCT) is the technique currently applied and has two approaches: swept source OCT (SS-OCT) and the spectrometer-based system (spectral domain OCT, SD-OCT). Both approaches, which are based on low-coherence interferometry, have been well described in numerous publications [8,26,31,32,35]. The light from the source is split into a sample and a reference arm (Michelson interferometer configuration). The back reflected light from the sample (sample arm) and the reference mirror interfere with each other, and the resulting signal (spectrum) is then recorded by the detector.

In contrast, SS-OCT uses a frequency sweeping laser source combined with a photo diode; and SD-OCT applies a wideband laser source in combination with a spectrometer to split the interference signal into single wavelengths. After Fourier transform of the signal, a depth profile of backscattering along a perpendicular line to the object surface is generated (A-scan). The point-by-point scanning of the OCT beam across the sample produces 2D cross-sectional images (B-scans), and the line-by-line scanning generates a series of 2D images from which 3D image stacks can be created (Figure 3). OCT enables images to be generated from the different absorption and scattering of light of various material components in hard and soft tissues. Image contrast arises in areas with structures of different refractive index and light absorption, such as tooth-restoration interfaces, gaps, bubbles, material cracks, or porous areas in carious lesions [8,29,31].

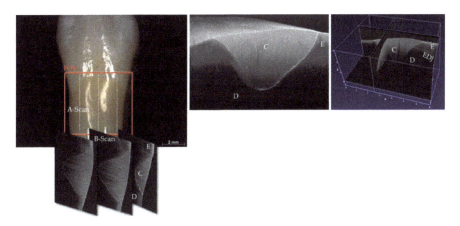

Figure 3. SD-OCT imaging (2D, 3D) of an extracted human premolar restored with composite (C). FOV: field of view, E: enamel, D: dentin, EDJ: enamel-dentin junction.

The scenarios presented in this article are based on both SS- and SD-OCT. The techniques produce images non-invasively and at high-speed, allowing real-time imaging, and provide high axial and lateral resolution in the micrometer range. Dental structures and restorations can be imaged up to a depth of 2–2.5 mm thanks to the translucency and specific refractive indices of dental hard tissues (n_{dentin}: 1.48–1.80; n_{enamel}: 1.45–1.61, own measurements) and composite restoration materials ($n_{composite}$: 1.42–1.63, own readings) in the range 1325 nm of near infrared light. In this wavelength range the transparency of enamel is highest [31] with high axial resolution at the same time. The refractive indices of lesions vary widely depending on the demineralization degree such as the values of natural carious enamel lesions. Usually these exceeded the values of the adjacent sound enamel up to +16% (own data).

The following setups were used in our lab:

SD-OCT: TELESTO SP II or SP 5, Thorlabs GmbH, Dachau, Germany (Figure 4a).

Recording parameters in vitro, ex vivo and in vivo were: center wavelength 1310 ± 107 nm; sensitivity ≤106 dB; axial/lateral resolution <7.5 (air)/15 µm; field of view maximum 9 mm × 9 mm × 3.5 mm (air, pixel size 700 × 700 × 512 or 1024); imaging speed 76/91 kHz (in vivo) or 48–91 kHz (in vitro, ex vivo); A-Scan average 1 (in vivo) or 1–5 (in vitro, ex vivo); spot size 20 µm; and power on sample 3 mW.

SS-OCT: SS-OCT, OCS 1300SS, Thorlabs Inc., Newton, NJ, USA (Figure 4b).

The parameters of the OCT equipment in vitro, ex vivo and in vivo were: center wavelength 1325 ± 100 nm; sensitivity 100 dB; axial/lateral resolution 12 (air)/25 µm; field of view maximum 10 mm × 10 mm × 3 mm (pixel size 512 × 512 × 512); imaging speed 16 kHz; A-Scan average 1; and power on sample 4.5 mW.

In vitro/ex vivo studies: OCT signals were verified within a predefined region of interest (ROI) by comparison with corresponding X-ray microtomography images (µCT, Skyscan 1172-100-50, Bruker MicroCT, Kontich, Belgium); and after sectioning specimens with images from the reference methods transverse microradiography (TMR, PW 3830/40, Philips, Amsterdam, The Netherlands); scanning electron microscopy (SEM, 5 kV, Phenom, Phenom World, Eindhoven, The Netherlands); spinning disc confocal laser microscopy (SDCM, Smartproof 5, Carl Zeiss Microscopy, Oberkochen Germany); and different modes of light microscopy.

In vitro/ex vivo studies were performed on extracted human teeth. The teeth were sound or had caries lesions, were unrestored or restored with composite. The focus on the specimen was set on a ROI and the surface was dried using a cotton pellet to leave it moist with no visible water droplets (controlled hydrated condition). 2D cross-sectional images of the ROI and 3D image stacks were recorded and images processed (ImageJ from version 1.45S; Wayne Rasband, NIH, Bethesda, MD, USA).

In vivo studies: Because of the geometric dimensions of the probes (diameter 34 mm, working distance about 42 mm), the imaging was limited to vestibular smooth surfaces of incisors, canines and premolars, after restorations and tooth surfaces were cleaned. The handheld and mechanically stabilized scanning probe was positioned at a rough right-angle to the restoration surface (Figure 4c). The restoration area was brought into focus and within 6 s, 2D cross-sectional images and 3D image stacks were generated and processed (ImageJ from version 1.45S).

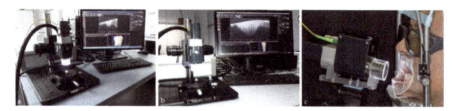

Figure 4. (**a**) SD-OCT Telesto SP II; and (**b**) SSOCT OCS1300SS; (**c**) In vivo imaging of vestibular smooth surfaces of incisors, canines and premolars with the handheld and mechanically stabilized scanning probe (P).

The studies were conducted in accordance with the Declaration of Helsinki, and the protocols were approved by the Ethics Committee of the University of Leipzig. The extracted teeth were used with patients' approvals (informed consents, protocol no. 299-10-04102010), and all subjects gave their informed consent for inclusion before they participated in the study (protocols no. 087/2003, 131-11-18042011, 196-14-14042014, 294-15-13072015).

With OCT, in vivo imaging is a challenge if dental hard tissues (enamel, dentin), which induce more scattering and containing less water, are to be displayed together with soft tissues (gingiva, pulp), which contain more water. The optical properties of these tissues, such as light scattering and absorption, are very diverse and also vary with imaging wavelength. One limitation of OCT is the shallow penetration depth, caused by signal attenuation due to scattering and absorption. Depth penetration can be enhanced by optimizing the center wavelength. In particular, in the presence of structures such as carious lesions which induce increased scattering, multiple scattering can further reduce imaging quality at deeper penetration depths. As well as the technical parameters of the OCT equipment, the center wavelength of the light source and the optical bandwidth are key determinants for axial resolution. For soft and hard tissues, there are characteristic dependences of wavelength of OCT signal attenuation [36]. Some authors suggested that an imaging window at 1700 nm offers an advantage over shorter wavelengths by increasing penetration depth as well as enhancing the image contrast at deeper penetration depths by reducing multiple scattering [37]. For biological tissues, it is expected an increase in OCT imaging depth at 1600 nm compared to 1300 nm on samples with high scattering power and low water content (enamel) [38]. Otherwise at longer wavelengths, absorption increases with greater water content, which has a counterbalancing effect [37]. Also, as in the case of enamel, transparency is higher at 1300 nm than at 1600 nm. In general, however, axial resolution decreases with increasing imaging wavelength. On a human tooth in the range 1060 nm–1700 nm, this effect clearly manifested for enamel and dentin, whereas image contrast and imaging depth were enhanced at 1700 nm [36]. It was observed that there was more variation in the wavelength dependence of total signal attenuation in enamel than in dentine. For OCT users, a high axial resolution in particular, plus adequate image contrast and penetration depth are essential parameters. Against this background and considering object-specific characteristics, the center wavelength, bandwidth and optics must be adjusted so that the advantages outweigh the drawbacks.

3. Applications of OCT

3.1. Caries Diagnosis

Today the big challenge in caries diagnosis is the detection of very early stages of demineralization, especially in enamel [12,39]. Effective caries management presupposes that these early lesions (represented by porous areas very different in extension) can be reliably recorded. On smooth tooth surfaces, beginning demineralizations can be detected visually as white spot lesions on dried enamel surfaces (ICDAS II Code 1, Figure 1) [3,40], which are a common complication during treatment with fixed orthodontic appliances [41]. At this stage, the subsurface enamel porosity can be stopped or reversed using appropriate non- or minimally invasive therapies and biofilm control [8]. White spot lesions appear whitish as the incident light is backscattered off porous regions to a considerable extent [29,39]. OCT makes use of the same phenomenon. In contrast to the clinical detection solely at the surface, OCT can image structures up to a depth of 2.5 mm and might therefore be a useful supplement to the visual-tactile assessment of tooth surfaces (ICDAS II) and radiography. More reliable differentiation between the really early lesion signs (ICDAS II Codes 0–2) might become possible, irrespective of existing color changes or surface moisture (Figures 1 and 2). Unlike other diagnostic methods, cross-sectional OCT images can present the axial and lateral extension of different demineralized zones (Figure 5).

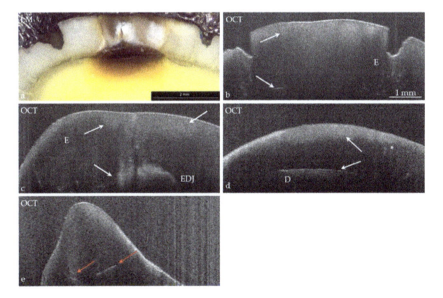

Figure 5. (**a**) Extracted human molar. A carious lesion (ICDAS-II Code 2) without cavitation, but with involvement of dentin (D); (**b–d**) Using different positions or angles, SS-OCT cross-sectional images reveal bright areas (signals) indicating the extension of the lesion (white arrows) and to detect further tissue destructions like enamel cracks ((**e**), red arrows, *), which are primarily invisible in light microscopy. EDJ: enamel–dentin junction.

In vitro studies showed an adequate to strong agreement, when SS- or SD-OCT were compared to histology [29,42,43], confocal microscopy, X-ray microtomography, or transverse microradiography and a diagnostic superiority compared to bitewing radiography [44] (Figures 1, 2 and 5–11 and 14).

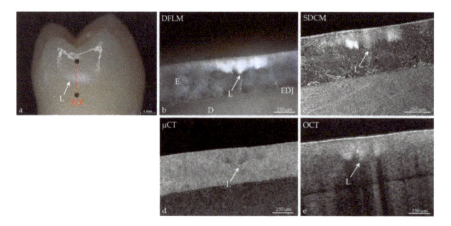

Figure 6. Extracted human premolar with a carious lesion (L) of ICDAS II Code 1. (**a**) ROI; (**b**) The demineralization within the ROI (white arrow) is equally imaged by dark-field light microscopy (DFLM); (**c**) spinning-disc confocal microscopy (SDCM); (**d**) X-ray microtomography (μCT); as well as (**e**) SD-OCT. E: enamel, D: dentin, EDJ: enamel-dentin-junction.

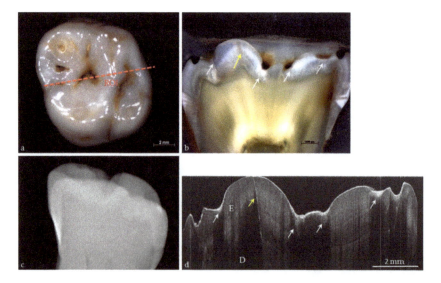

Figure 7. (**a**) Extracted human molar with occlusal caries (ICDAS II Code 2); (**b**) In the histological section through ROI, lesions (white arrows) and enamel cracks (yellow arrows) appear; (**c**) In the conventional digital radiograph no defects or lesions are visible; (**d**) Bright signals in the OCT image of the ROI demonstrate caries lesions (white arrows) and enamel cracking (yellow arrow). E: enamel; D: dentin.

Image contrast and the spatial separation of the lesion body versus sound enamel can be further enhanced using polarization-sensitive OCT [45]. Additionally, variation of the imaging beam angle by using different positions of the OCT probe can improve the detectability of non-cavitated occlusal carious lesions (Figure 8). The sensitivity of the method can be increased among others by increasing the integration time (A-scan rate) or using coupling media [46] to minimize surface reflections, which can also contribute to enhance the image contrast (Figure 9).

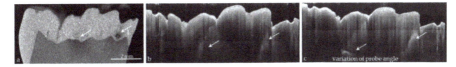

Figure 8. Extracted human molar with occlusal carious lesions without cavitation (L, arrow). (**a**) μCT image of a ROI as reference; (**b**) Bright signals show extended lesion areas (arrows); (**c**) The varying of the OCT probe angle provides increased OCT signals and additional information about dentin involvement.

Figure 9. Extracted human molar with occlusal carious lesions without cavitation (L). (**a**) μCT image of ROI. After application of glycerin gel signal-to-noise ratio in the SD-OCT cross-sectional image (**c**) can be improved towards the image (**b**) (without glycerine, white arrows) leading to an adequate conformance of μCT with OCT.

The International Caries Detection and Assessment System in combination with radiography is insensitive to early caries detection. This has been confirmed by a number of the authors' own OCT evaluations of apparently sound enamel surfaces (ICDAS II Code 0), particularly in cross-sectional OCT images where early stages of decay can often be detected without any clinical findings (Figures 10 and 11).

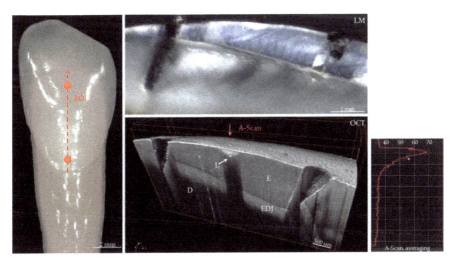

Figure 10. Extracted human canine. There is no conformance of visual assessment with OCT. Visual inspection of the dry smooth surface does not reveal signs of an early-stage carious lesion (L). In contrast, the 3D SD-OCT cross-sectional image of the ROI showed a distinct bright area (L) within the first quarter of enamel (E) with a shadow underneath. The lesion can be confirmed using light microscopy (LM). The A-scan of the position marked in the OCT-image shows the depth profile of the demineralized zone in the enamel surface (A-Scan average 5, additional signal median averaging 6). * grey scale between demineralized and healthy region.

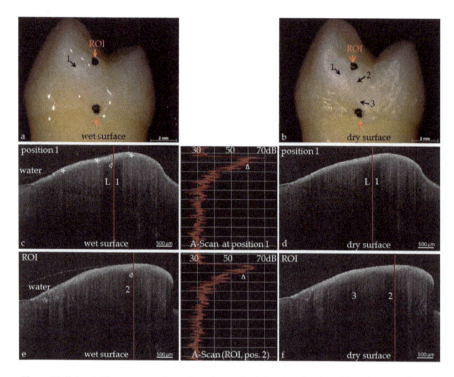

Figure 11. Extracted human premolar with early proximal demineralizations. There is a partial match of the visual and OCT assessment. (**a**) Hardly visible on wet enamel, the early demineralization (1) could also be seen on the dry surface (**b**); (**c–f**) SD-OCT reveals a clinically invisible demineralization in the outer enamel zone (*); (**c,d**) The OCT images reveal a bright signal in position 1 (L, shadow); (**e**) The OCT of the wet surface shows an early demineralization in position 2 of image (**b**); (**f**) After air drying, an additional signal arises in position 3. Visually both lesions are invisible (**a,b**). With the knowledge of the OCT images, at points 2 and 3, slight opacities could also be anticipated on the images (**a,b**). The A-scans (average 5) of the marked positions 1 and 2 in the OCT-images demonstrate the depth profiles of the demineralized zones of the enamel surface. $^\Delta$ grey scales between demineralized and healthy region.

Another highly relevant scenario in need of diagnostic information is the field of restoration assessment. In Figure 12, a clinical case of a discolored restoration margin gap is shown. However, it could be seen that discoloration at margins of composite restorations is not an indicator for deeper interfacial gap formations. OCT allows the identification of weak spots in the composite restorations and the differentiation of the findings associated with discolorations. If a carious lesion is detected, the treatment depends on its longitudinal stability. If progression is identified, treatment will be related to the extent of the lesion and speed of progression (activity).

Today, there is no single method that can be considered the gold standard for imaging the morphology of the hard tooth tissue while simultaneously assessing the longitudinal activity of a carious lesion. Quantitative light-induced fluorescence (QLF) is known as an imaging technique with the option to visualize present bacterial activity in carious lesions [11,47]. The method is based on autofluorescence of tooth hard tissues and quantifies the autofluorescence difference between sound and carious (pathogenic) enamel. This difference results from the decrease of autofluorescence with increasing porosity of progressing subsurface enamel lesions [48]. With the QLF-D Biluminator™ 2 (Inspektor Research Systems bv, Amsterdam, The Netherlands), a tool becomes available that makes it possible to additionally image and quantify the red fluorescence of oral bacteria induced by various porphyrin species as products of anaerobic bacterial metabolism. Thus, detection of low porphyrin concentrations would be an indicator of early stage caries formation [49]. Now, OCT, contrary to histopathological assessment, can provide submicrometer resolved tomographic images of tissues in vivo. We hypothesized that the combination of both imaging techniques would be beneficial as the three-dimensional lesion morphology could be brought together with the microbial and metabolic findings in real-time, as well as of one and the same tooth area. Such an approach should hopefully facilitate more accurate treatment decisions and the assessment of the restoration outcome during follow-ups.

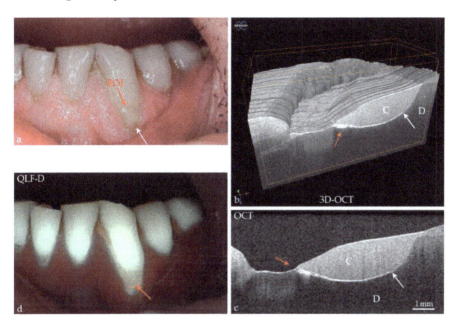

Figure 12. Canine tooth in vivo. (**a**) A cervical composite restoration (C) shows a margin discoloration (white arrow); (**b,c**) In the 3D and 2D OCT images, a margin gap with ingress of material is detected (red arrows). At dentin (D), an extensive interfacial gap has formed (white arrows); (**d**) Using quantitative light-induced fluorescence the image showed red fluorescence (red arrow) at the cervical restoration margin indicating bacterial activity by porphyrin invasion into the gap. There are no signs of "secondary caries" at the restoration margin, but the stability of the situation requires observation over time (monitoring).

3.2. Erosive Tooth Wear

Within the oral cavity, demineralization and remineralization of tooth surfaces (enamel, dentin) are in a state of equilibrium. Even without the participation of bacteria (caries) in a multifactorial process, a shift in the balance can occur towards demineralization, which is caused by different patient-related and nutritional factors, e.g., eating/drinking habits, oral hygiene, medication and saliva conditions. The destructive effect of acid-induced tissue loss of tooth surface and subsurface regions is known as dental erosion (erosive tooth wear) [50].

Clinical diagnosis indices for recording the erosive wear include morphological as well as quantitative criteria, particularly the Basic Erosive Wear Examination (BEWE) [51]. However, the validity of diagnostic criteria and grading/scoring are unsatisfactory or unclear. Thus, there is a need to develop practicable and preferably chairside diagnostic tools [51]. Different assessment methods have been applied for in vitro and in vivo studies to evaluate erosive tooth wear, e.g., profilometry, measuring microscopy techniques, TMR, SEM, atomic force microscopy, nano- and micro-hardness tests, and the iodide permeability test, as well as methods which chemically analyze minerals dissolved from dental hard tissue [52]. In this context, OCT has been described as a suitable tool for the investigation of early erosions of unaffected surfaces [52].

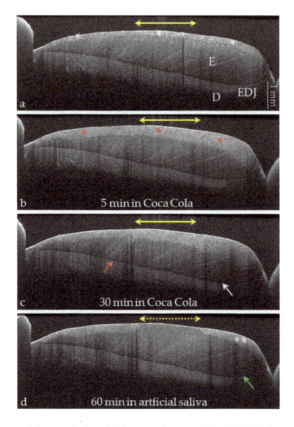

Figure 13. Extracted human incisor. (**a**) Starting situation. The SD-OCT shows a small bright band in the enamel surface (E), which could represent a lower mineralized zone (white asterisks). The yellow double-headed arrow indicates the surface area, which was covered by composite to prevent demineralization of the enamel; (**b**) After 5 min exposure of enamel to Coca Cola (37 °C) a broader bright area (red asterisks) can already clearly be distinguished, which represents a zone of erosion that also extended below the composite coating; (**c**) After 30 min of exposure, the demineralization has made further progress, which is indicated by the reduced contrast at the enamel-dentin junction (EDJ, red arrow) or its disappearance (white arrow); (**d**) After removal of the composite coating the tooth was stored in artificial saliva. Within 60 min, the bright demineralized area previously covered by composite (dotted double-headed arrow) completely disappeared indicating a process of repair ("remineralization"). In uncovered areas, the EDJ appears in a higher contrast compared to (c) and is more extended (green arrow). Particularly in the incisal area, the erosion is clearly reduced (**) compared to (**b,c**).

OCT allows the assessment of the mineralization degree of enamel or dentin and the visualization of very early demineralizations, and thus permits early erosive lesions to be characterized [52]. Until now, erosion could only be assessed clinically through long-term observations of the loss of hard tooth tissue over months to years, well after the erosion has been initiated. With OCT, incipient erosion without loss of enamel height could be recognized and a therapeutic remineralization process could be monitored and assessed [8]. Figure 13 shows an initial enamel erosion before the thickness of hard tissue had changed, and its partial repair. The erosion was produced by the soft drink Coca Cola [53,54]. The method can visualize the partial reversal of demineralization using artificial saliva and "effects", which clearly took place within covered and uncovered areas of the enamel surface. It is clear that the surface area of the partially protected enamel was repaired to a higher degree by artificial saliva.

3.3. Caries Therapy

3.3.1. Assessment of Pit and Fissure Sealing

A major focus of prevention is on pit and fissure sealing as one measure to prevent early caries initiation and progression in occlusal surfaces [40]. In cross-sectional OCT images, fissure depth and caries lesions can be detected due to increased backscattering of light. After sealant application, images reveal its penetration into fissures and its adaptation to the tooth surface (Figure 14). Incompletely filled zones or bubbles can be detected and monitoring over time is possible. Strong reflections at the enamel surface can hamper the detection of caries lesions by covering the light backscattered from lesions. Using polarization-sensitive optical coherence tomography, the contrast can be enhanced by minimizing this effect [55].

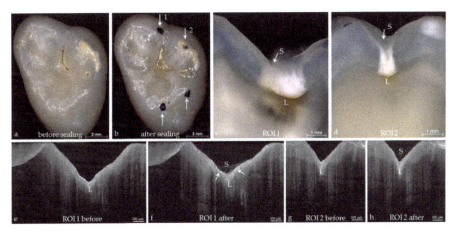

Figure 14. (**a**) Extracted human molar before; and (**b**) after fissure sealing with Helioseal. (**c,d**) In cross-sections through ROI 1 and 2 occlusal caries beneath the sealer can be identified (L). Using SD-OCT, carious lesions (L, shadow) can be localized (**e,g**) before; and (**f,h**) after sealing. The sealer (S) has penetrated into the fissures, is inhomogeneously distributed and adapted to the enamel surface with (**f**, white arrows) or (**h**) without interfacial gap; (**f**) The bright signals reveal an interfacial gap (white arrows) between sealant and enamel. Monitoring will reveal changes over time.

3.3.2. Remaining Dentin during Caries Removal

Opening the pulp chamber during caries removal of deep lesions or tooth preparation is a major risk in the maintenance of the pulp vitality and the tooth, and should therefore be avoided. However, during the treatment of dentin areas close to the pulp, an iatrogenic opening of the pulp complex is sometimes difficult to prevent due to the unknown and invisible location of the pulp chamber.

In contrast to an introduced indirect measuring method based on the determination of the electrical resistance of the residual dentin (prepometer™, Hager and Werken, Duisburg, Germany), the

OCT offers great potential for imaging the outline of the pulp chamber. The remaining dentin thickness and pulp geometry, e.g., pulp horns, can be visualized in vitro during dental procedures using optical coherence tomography [56–59]. We found that the roof of the pulp chamber could be displayed using SD-OCT at a mean distance through dentin of 850 μm, whereas the minimum remaining dentin thickness which could be imaged was 45 μm (Figure 15), and is in accordance with Fonseca et al. [57]. Therefore, the OCT technique could be a valuable diagnostic tool during dental treatment and help prevent accidental pulp exposures by estimating the remaining dentin thickness.

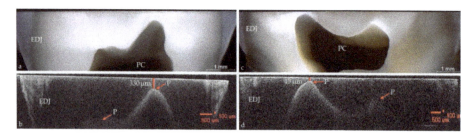

Figure 15. Extracted human molars trimmed and cross-sectioned longitudinally through pulp chamber (PC). Imaging of the pulp borders by (**a,c**) light microscopy, and (**b,d**) SD-OCT, the pulp horns and pulp chamber roof (P) can be clearly visualized in the OCT images. EDJ: enamel-dentin junction.

3.3.3. Composite Restorations

Today, the adhesive bonding of composite materials to enamel and dentin is the predominant restoration technique. The concept is based on both a durable micro retentive and a chemical bond between composite, adhesive and tooth to obtain restorations, which are free of interfacial adhesive defects. As composites shrink during polymerization, the adhesive technique is the crucial prerequisite for minimally invasive treatment. Conversely, this means that the adhesive failure of the bond can be considered as a parameter to assess the quality of restorations [60] (Figure 16).

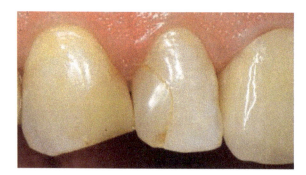

Figure 16. A clinical situation. It is unclear and difficult to determine visually whether the discolored margin of the composite restoration results from an interfacial margin gap.

The morphological correlate for this is the formation of interfacial gaps, or so-called adhesive defects. For the evaluation of composite restorations, clinical trials are still the gold standard, with clinical success assessed by functional, biological, and esthetic criteria [15,61]. Clinically interfacial adhesive defects can only be detected at restoration margins. According to FDI (Fédération Dentaire Internationale) [15], marginal gaps of score 2 (50 μm width) are to be removed by polishing. Gaps of score 3 (50–150 μm), cannot be removed by polishing but are to be monitored. Restorations with marginal gaps of score 4 (width > 250 μm) must be repaired or replaced if these are loose (score 5). However, statements on the clinical performance of a restoration system are possible after the passage of at least three years [61–63]. Furthermore, clinical studies have additional disadvantages, for instance, personnel and financial expenses are relatively high and the filling systems are often no longer on the market at the conclusion of clinical trials after 3–5 years. Consequently, early experimental evaluation of adhesive systems and composites is desirable and a prerequisite for further developing restoration systems. It is also desirable to formulate clear statements regarding the quality of the adhesive bond and the long-term prognosis of restorations immediately after restoration placement, or at early follow-up. On this matter, OCT could provide relevant information. The method enables high-resolution real-time images to be made of composite restorations in 2D and 3D in vitro, ex vivo and in vivo, up to an imaging depth of 2.5 mm. The homogeneity of different restoration materials can be checked, as well as interfacial adhesive defects. Especially the clinically relevant integrity of enamel- and dentin-restoration margins can be reliably evaluated, including (at least partially) the situation beneath the gingiva, as well as the seal quality at these locations (Figure 17). The validity and reproducibility of the method has been demonstrated [64,65].

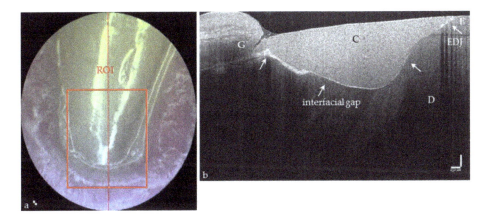

Figure 17. (**a**) Screenshot of the clinical image of a cervical composite restoration (C) on a premolar; (**b**) The cross-sectional OCT image reveals a homogeneous filling showing an extended bond failure at the dentin and enamel (bright signal lines, white arrows). At enamel margin (E), no gap appears, whereas the gingival margin (G) shows a gap which is extended along the composite-dentin interface. In multiple equidistantly distributed cross-sectional images, the length of the interfacial adhesive failures can be measured for dentin (D) and enamel (E). From these, the "percentage gap" of this restoration can be derived separately for enamel and dentin. EDJ: enamel-dentin junction.

Quality Control Using OCT

Thanks to the fast processing of highly resolved images and its non-invasive and harmless approach, OCT has great potential for time-resolved visualization of fast processes [9]. This is also the case in the fields of learning and teaching, and consequently in quality or process control and in monitoring therapy outcomes. The technique allows many opportunities for self-checking. Individual work steps during restoration placement can be demonstrated, better understood and single procedural errors and resulting failures can be clearly seen. This results in great potential in supporting the development of dental materials, and to teach and learn therapeutic procedures (Figures 18 and 19).

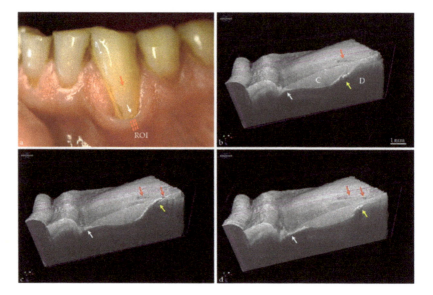

Figure 18. OCT quality control in vivo. 3D OCT imaging of a cervical composite (C) restoration at a premolar after filling the cavity in small increments (**a–d**). Images (**b–d**) show the image stack using a sequence of three OCT B-scans in y-direction (ROI shown in **a**). The bright lines at the tooth-composite interface demonstrate a margin gap at dentin (D) and extended interfacial bond failures (white arrows). Additionally, communicating bubbles are clearly evident (yellow arrows) and reach the restoration surface, forming an entrance for material to penetrate into the restoration (**a–d**, red arrows).

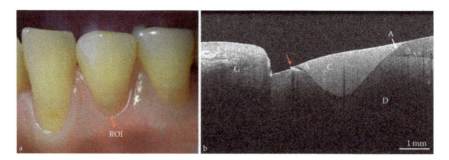

Figure 19. OCT quality control in vivo. (**a**) A cervical composite (C) restoration at a premolar 34 immediately after filling the cavity. The gingiva (G) is slightly retracted. In contrast to Figure 18 the composite has been homogeneously dispersed; (**a**) Using visual-tactile assessment the restoration margins appear smooth; (**b**) However, in the OCT image at dentin (D, ROI) a margin gap in combination with a composite overhang indicates a critical area (bright line, red arrow). In knowledge of the OCT image, the overhang could easily be removed. An adhesive layer (A) could only be partially detected.

Monitoring of Composite Restorations

As OCT is non-invasive, it inevitably follows that it offers the possibility of monitoring the bond failure at the same composite restorations in the same cross-sectional images at any time. In long-term clinical studies conducted by the authors, composite restorations will be imaged parallel to clinical assessment using SD-OCT (vestibular areas, Class V, Figure 4c). In a B-scan by measurement the lengths of adhesive failure and interface the proportion (%) of the interfacial adhesive failure can be calculated separately for enamel, dentin or cement. In 25 equidistantly distributed B-scans of an image stack the mean values of a restoration result with regard to enamel, dentin or cement. From these the values per group can be calculated over the entire period of study. The results of the clinical studies will reveal whether OCT can be an early, cost-effective, and time-saving tool for evaluating the performance of adhesives or composite restoration systems in vivo. In vitro and ex vivo, an exceptional methodological advantage is that identical cross-sectional images will be assessed and compared before and after artificial aging. This results in pairs of measured values. In this case the variance between specimens can be excluded and just the variance within the specimens remains. Compared with unpaired values, at identical ratings of statistical power the number of specimens can be reduced.

The quality assessment of composite restorations is based on the evaluation of functional, biological, and esthetic criteria [61]. It is also recognized that restoration margins should be sealed or intact. The question of what degree of interfacial bond failure is "clinically acceptable" is still open for discussion. From this point of view, we believe that development over time for each individual case must be a focus of consideration as interfacial gap is not just like another. To date, nobody has reliably assessed and estimated these individual dynamics under clinical conditions.

Figures 20 and 21 demonstrate two specific cases of gap formation between composite restoration and tooth during clinical function. Figure 20a–c show a rapid increase of bond failure at dentin over a short period of six months, while the composite-enamel bond has proven itself. In contrast, a much slower dynamic of gap progression over the longer period of 24 months is seen in Figure 21a–e. There are no signs of caries at the defective restoration margins (secondary caries). Figures 22 and 23 provide an example of partial restoration loss in images and the percentage of interfacial gap formation. Within the limits of the method, the concept of restoration monitoring using OCT can be transferred onto other restoration materials.

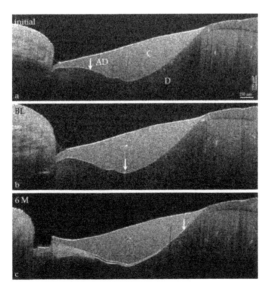

Figure 20. SD-OCT monitoring of tooth-composite bond at a cervical composite restoration (C). (**a**) The restoration immediately after filling the cavity in small composite increments. The composite has been homogeneously dispersed. Initially, the bright line marks an extensive adhesive defect (AD, white arrow) at dentin (D), which progressed rapidly up to the 6-month reevaluation (red arrow, the same cross-section as in image a). (**b**,**c**) Along the interface the gap is at an advanced stage (bright line, white arrow). There are no signs of secondary caries, but of ingressed material (red star, c). No distinct adhesive layer could be detected. * Vertical bar is related to refractive index $n = 1.5$.

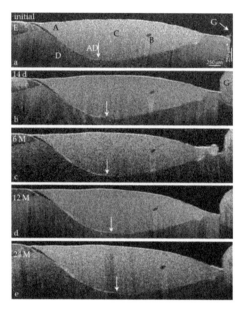

Figure 21. Monitoring of tooth-composite bond using SD-OCT in vivo. (**a**) A cervical composite restoration (C) immediately after filling the cavity in small composite increments. The gingiva (G) is slightly retracted. The composite has been homogeneously dispersed. Initially the bright line marks an extensive adhesive defect (AD) at enamel (E) and dentin (D); (**b–e**) The same cross-section (see bubble in composite, B), imaged using OCT in the period 14 days (**b**) to 24 months (M, **e**); the interfacial gap has progressed slightly (bright line, white arrow). There are no signs of secondary caries. A distinct adhesive layer (A) could only be partially detected. * Vertical bar is related to refractive index n = 1.5.

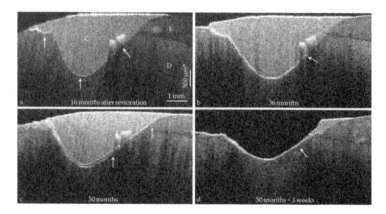

Figure 22. In vivo monitoring of the tooth-composite bond using SD-OCT. (**a**) A cervical composite restoration (C) at premolar 34, sixteen months after restoration. The cavity was filled in increments. The composite contains imperfections (yellow arrow). The bright line marks an extensive adhesive defect (white arrows) at dentin (D). (**b**,**c**) The same cross-section imaged using OCT in the period up until 50 months; the interfacial gap has progressed (bright line, white arrows), but there are no signs of secondary caries. (**d**) Three weeks later, the restoration failed. The imperfection has probably acted as a hinge (yellow arrow). At enamel (E) this saved the composite from loss. * Vertical bar is related to refractive index n = 1.5.

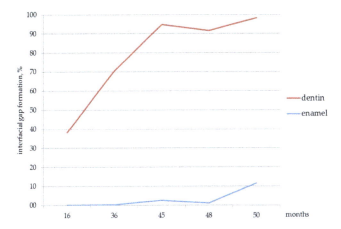

Figure 23. Percentage progression of the interfacial gap formation according to Figure 22, based on OCT cross-sectional images.

4. Conclusions

As a non-invasive imaging technique, optical coherence tomography opens up enormous potential for its application in more effective and reproducible diagnosis and therapy of caries. Caries lesions can be detected significantly earlier and the progress of early dental caries and the effectiveness of restorative treatment options could be monitored longitudinally. This explorative technique could provide additional clinically relevant information, is fast, and thus suitable and cost-effective to use in research. In clinical practice it could be applied equally well if appropriate equipment becomes available.

Acknowledgments: This work was supported by the European Regional Development Fund (ERDF, 100175024; 100175035) and by the German Research Foundation (DFG)/the Saxon State Ministry for Science and the Arts (SMWK) (376/7-1 FUG). We acknowledge support from the German Research Foundation (DFG) and Universität Leipzig within the program of Open Access Publishing. We would like to thank Thorlabs GmbH, Dachau, Germany for provision of the SS-OCT equipment and Timothy Jones (IALT, University of Leipzig, Germany) for editorial assistance.

Author Contributions: Hartmut Schneider and Rainer Haak conceived the ideas. Hartmut Schneider, Kyung-Jin Park, Matthias Häfer, Felix Krause and Rainer Haak designed the experiments. Hartmut Schneider, Kyung-Jin Park, Matthias Häfer, Claudia Rüger, Gerhard Schmalz and Jana Schmidt performed the experiments. Hartmut Schneider, Kyung-Jin Park, Matthias Häfer, Claudia Rüger, Gerhard Schmalz, Felix Krause, Jana Schmidt, Dirk Ziebolz and Rainer Haak processed the data. Kyung-Jin Park, Matthias Häfer, Claudia Rüger, Gerhard Schmalz, Jana Schmidt contributed with the materials and samples. Hartmut Schneider, Felix Krause, Dirk Ziebolz and Rainer Haak wrote the paper.

Conflicts of Interest: The authors declare no conflict of interest. The founding sponsors had no role in the design of the studies; in the collection, analyses, or interpretation of data; in the writing of the manuscript, and in the decision to publish the results.

References

1. Brouwer, F.; Askar, H.; Paris, S.; Schwendicke, F. Detecting Secondary Caries Lesions: A Systematic Review and Meta-analysis. *J. Dent. Res.* **2016**, *95*, 143–151. [CrossRef] [PubMed]

2. Schwendicke, F.; Brouwer, F.; Paris, S.; Stolpe, M. Detecting Proximal Secondary Caries Lesions: A Cost-effectiveness Analysis. *J. Dent. Res.* **2016**, *95*, 152–159. [CrossRef] [PubMed]

3. Ismail, A.I. The International Caries Detection and Assessment System (ICDAS II). Available online: https://www.icdas.org/uploads/Rationale%20and%20Evidence%20ICDAS%20II%20September%2011-1.pdf (accessed on 12 January 2017).

4. Ismail, A.I.; Sohn, W.; Tellez, M.; Amaya, A.; Sen, A.; Hasson, H.; Pitts, N.B. The International Caries Detection and Assessment System (ICDAS): An integrated system for measuring dental caries. *Community Dent. Oral Epidemiol.* **2007**, *35*, 170–178. [CrossRef] [PubMed]
5. Diniz, M.B.; Rodrigues, J.A.; Hug, I.; Cordeiro, R.d.C.L.; Lussi, A. Reproducibility and accuracy of the ICDAS-II for occlusal caries detection. *Community Dent. Oral Epidemiol.* **2009**, *37*, 399–404. [CrossRef] [PubMed]
6. Deligeorgi, V.; Mjör, I.A.; Wilson, N.H.H. An Overview of and Replacement Reasons for the of Restorations Placement. *Prim. Dent. Care* **2001**, *8*, 5–11. [CrossRef] [PubMed]
7. Abdullah, Z.; John, J. Minimally Invasive Treatment of White Spot Lesions-A Systematic Review. *Oral Health Prev. Dent.* **2016**, *14*, 197–205. [PubMed]
8. Mandurah, M.M.; Sadr, A.; Shimada, Y.; Kitasako, Y.; Nakashima, S.; Bakhsh, T.A.; Tagami, J.; Sumi, Y. Monitoring remineralization of enamel subsurface lesions by optical coherence tomography. *J. Biomed. Opt.* **2013**, *18*, 46006. [CrossRef] [PubMed]
9. Schneider, H.; Park, K.-J.; Rueger, C.; Ziebolz, D.; Krause, F.; Haak, R. Imaging resin infiltration into non-cavitated carious lesions by optical coherence tomography. *J. Dent.* **2017**. [CrossRef] [PubMed]
10. Douglas, S.M.; Fried, D.; Darling, C.L. Imaging Natural Occlusal Caries Lesions with Optical Coherence Tomography. *Proc. SPIE. Int. Soc. Opt. Eng.* **2010**, *7549*, 75490N. [PubMed]
11. Alammari, M.R.; Smith, P.W.; de Jong, E.D.J.; Higham, S.M. Quantitative light-induced fluorescence (QLF): A tool for early occlusal dental caries detection and supporting decision making in vivo. *J. Dent.* **2013**, *41*, 127–132. [CrossRef] [PubMed]
12. Pretty, I.A. Caries detection and diagnosis: Novel technologies. *J. Dent.* **2006**, *34*, 727–739. [CrossRef] [PubMed]
13. Gimenez, T.; Piovesan, C.; Braga, M.M.; Raggio, D.P.; Deery, C.; Ricketts, D.N.; Ekstrand, K.R.; Mendes, F.M. Visual Inspection for Caries Detection: A Systematic Review and Meta-analysis. *J. Dent. Res.* **2015**, *94*, 895–904. [CrossRef] [PubMed]
14. Swets, J. Measuring the accuracy of diagnostic systems. *Science* **1988**, *240*, 1285–1293. [CrossRef] [PubMed]
15. Hickel, R.; Peschke, A.; Tyas, M.; Mjor, I.; Bayne, S.; Peters, M.; Hiller, K.-A.; Randall, R.; Vanherle, G.; Heintze, S.D. FDI World Dental Federation—Clinical criteria for the evaluation of direct and indirect restorations. Update and clinical examples. *J. Adhes. Dent.* **2010**, *12*, 259–272. [CrossRef] [PubMed]
16. American Dental Association. American Dental Association acceptance program guidelines: Composite resins for posterior restorations. *JADA* **2003**, *134*, 510–512.
17. Roulet, J.F.; Reich, T.; Blunck, U.; Noack, M. Quantitative margin analysis in the scanning electron microscope. *Scanning Microsc.* **1989**, *3*, 147–159. [PubMed]
18. Huang, D.; Swanson, E.A.; Lin, C.P.; Schuman, J.S.; Stinson, W.G.; Chang, W.; Hee, M.R.; Flotte, T.; Gregory, K.; Puliafito, C.A.; et al. Optical Coherence Tomography. *Science* **1991**, *22*, 1178–1181. [CrossRef]
19. Wylegala, A.; Teper, S.; Dobrowolski, D.; Wylegala, E. Optical coherence angiography: A review. *Medicine* **2016**, *95*, e4907. [CrossRef] [PubMed]
20. Adhi, M.; Duker, J.S. Optical coherence tomography–current and future applications. *Curr. Opin. Ophthalmol.* **2013**, *24*, 213–221. [CrossRef] [PubMed]
21. Hangai, M.; Ojima, Y.; Gotoh, N.; Inoue, R.; Yasuno, Y.; Makita, S.; Yamanari, M.; Yatagai, T.; Kita, M.; Yoshimura, N. Three-dimensional imaging of macular holes with high-speed optical coherence tomography. *Ophthalmology* **2007**, *114*, 763–773. [CrossRef] [PubMed]
22. Bernardes, R.; Cunha-Vaz, J. *Optical Coherence Tomography: A Clinical and Technical Update*; Springer: Berlin/Heidelberg, Germany, 2012.
23. Ulrich, M. Optical coherence tomography for diagnosis of basal cell carcinoma: Essentials and perspectives. *Br. J. Dermatol.* **2016**, *175*, 1145–1146. [CrossRef] [PubMed]
24. Cheng, H.M.; Lo, S.; Scolyer, R.; Meekings, A.; Carlos, G.; Guitera, P. Accuracy of optical coherence tomography for the diagnosis of superficial basal cell carcinoma: A prospective, consecutive, cohort study of 168 cases. *Br. J. Dermatol.* **2016**, *175*, 1290–1300. [CrossRef] [PubMed]
25. Vignali, L.; Solinas, E.; Emanuele, E. Research and Clinical Applications of Optical Coherence Tomography in Invasive Cardiology: A Review. *Curr. Cardiol. Rev.* **2014**, *10*, 369–376. [CrossRef] [PubMed]
26. Colston, B.; Sathyam, U.; Dasilva, L.; Everett, M.; Stroeve, P.; Otis, L. Dental OCT. *Opt. Express* **1998**, *14*, 230–238. [CrossRef]

27. Colston, B.W.; Everett, M.; Sathyam, U.S.; Dasilva, L.; Otis, L.L. Imaging of the oral cavity using optical coherence tomography. *Monogr. Oral. Sci.* **2000**, *17*, 32–55. [PubMed]
28. Feldchtein, F.I.; Gelikonov, G.V.; Gelikonov, V.M.; Iksanov, R.R.; Kuranov, R.V.; Sergeev, A.M.; Gladkova, N.D.; Ourutina, M.N.; Warren, J.A.; Reitze, D.H. In vivo OCT imaging of hard and soft tissue of the oral cavity. *Opt. Express* **1998**, *3*, 239–250. [CrossRef] [PubMed]
29. Shimada, Y.; Sadr, A.; Sumi, Y.; Tagami, J. Application of Optical Coherence Tomography (OCT) for Diagnosis of Caries, Cracks, and Defects of Restorations. *Curr. Oral Health Rep.* **2015**, *2*, 73–80. [CrossRef] [PubMed]
30. Min, J.H.; Inaba, D.; Kwon, H.K.; Chung, J.H.; Kim, B.I. Evaluation of penetration effect of resin infiltrant using optical coherence tomography. *J. Dent.* **2015**, *43*, 720–725. [CrossRef] [PubMed]
31. Drexler, W.; Liu, M.; Kumar, A.; Kamali, T.; Unterhuber, A.; Leitgeb, R.A. Optical coherence tomography today: Speed, contrast, and multimodality. *J. Biomed. Opt.* **2014**, *19*, 71412. [CrossRef] [PubMed]
32. Hsieh, Y.-S.; Ho, Y.-C.; Lee, S.-Y.; Chuang, C.-C.; Tsai, J.-C.; Lin, K.-F.; Sun, C.-W. Dental optical coherence tomography. *Sensors* **2013**, *13*, 8928–8949. [CrossRef] [PubMed]
33. Wijesinghe, R.E.; Cho, N.H.; Park, K.; Jeon, M.; Kim, J. Bio-Photonic Detection and Quantitative Evaluation Method for the Progression of Dental Caries Using Optical Frequency-Domain Imaging Method. *Sensors* **2016**, *16*. [CrossRef] [PubMed]
34. Wu, T.; Wang, Q.; Liu, Y.; Wang, J.; He, C.; Gu, X. Extending the Effective Ranging Depth of Spectral Domain Optical Coherence Tomography by Spatial Frequency Domain Multiplexing. *Appl. Sci.* **2016**, *6*, 360. [CrossRef]
35. Nakagawa, H.; Sadr, A.; Shimada, Y.; Tagami, J.; Sumi, Y. Validation of swept source optical coherence tomography (SS-OCT) for the diagnosis of smooth surface caries in vitro. *J. Dent.* **2013**, *41*, 80–89. [CrossRef] [PubMed]
36. Ishida, S.; Nishizawa, N. Quantitative comparison of contrast and imaging depth of ultrahigh-resolution optical coherence tomography images in 800–1700 nm wavelength region. *Biomed. Opt. Express* **2012**, *3*, 282–294. [CrossRef] [PubMed]
37. Sharma, U.; Chang, E.W.; Yun, S.H. Long-wavelength optical coherence tomography at 17 μm for enhanced imaging depth. *Opt. Express* **2008**, *16*, 19712. [CrossRef] [PubMed]
38. Kodach, V.M.; Kalkman, J.; Faber, D.J.; van Leeuwen, T.G. Quantitative comparison of the OCT imaging depth at 1300 nm and 1600 nm. *Biomed. Opt. Express* **2010**, *1*, 176–185. [CrossRef] [PubMed]
39. Darling, C.L.; Huynh, G.D.; Fried, D. Light scattering properties of natural and artificially demineralized dental enamel at 1310 nm. *J. Biomed. Opt.* **2006**, *11*, 34023. [CrossRef] [PubMed]
40. Ito, S.; Shimada, Y.; Sadr, A.; Nakajima, Y.; Miyashin, M.; Tagami, J.; Sumi, Y. Assessment of occlusal fissure depth and sealant penetration using optical coherence tomography. *Dent. Mater. J.* **2016**, *35*, 432–439. [CrossRef] [PubMed]
41. Srivastava, K.; Tikku, T.; Khanna, R.; Sachan, K. Risk factors and management of white spot lesions in orthodontics. *J. Orthod. Sci.* **2013**, *2*, 43–49. [CrossRef] [PubMed]
42. Shimada, Y.; Sadr, A.; Burrow, M.F.; Tagami, J.; Ozawa, N.; Sumi, Y. Validation of swept-source optical coherence tomography (SS-OCT) for the diagnosis of occlusal caries. *J. Dent.* **2010**, *38*, 655–665. [CrossRef] [PubMed]
43. Schneider, H.; Gottwald, R.; Meißner, T.; Krause, F.; Becker, K.; Attin, T.; Haak, R. Assessment of uncavitated carious enamel lesions by optical coherence tomography and X-ray microtomography. *Caries Res.* **2016**, *50*, 239–240.
44. Shimada, Y.; Nakagawa, H.; Sadr, A.; Wada, I.; Nakajima, M.; Nikaido, T.; Otsuki, M.; Tagami, J.; Sumi, Y. Noninvasive cross-sectional imaging of proximal caries using swept-source optical coherence tomography (SS-OCT) in vivo. *J. Biophotonics* **2014**, *7*, 506–513. [CrossRef] [PubMed]
45. Fried, D.; Xie, J.; Shafi, S.; Featherstone, J.D.B.; Breunig, T.M.; Le, C. Imaging caries lesions and lesion progression with polarization sensitive optical coherence tomography. *J. Biomed. Opt.* **2002**, *7*, 618–627. [CrossRef] [PubMed]
46. Kang, H.; Darling, C.L.; Fried, D. Enhancing the detection of hidden occlusal caries lesions with OCT using high index liquids. *Proc. SPIE Int. Soc. Opt. Eng.* **2014**, *8929*, 89290O. [PubMed]
47. Volgenant, C.M.C.; Fernandez Y Mostajo, M.; Rosema, N.A.M.; van der Weijden, F.A.; Cate, J.M.; van der Veen, M.H. Comparison of red autofluorescing plaque and disclosed plaque-a cross-sectional study. *Clin. Oral Investig.* **2016**, *20*, 2551–2558. [CrossRef] [PubMed]

48. Karlsson, L. Caries Detection Methods Based on Changes in Optical Properties between Healthy and Carious Tissue. *Int. J. Dent.* **2010**. [CrossRef] [PubMed]
49. Timoshchuk, M.-A.I.; Ridge, J.S.; Rugg, A.L.; Nelson, L.Y.; Kim, A.S.; Seibel, E.J. Real-Time Porphyrin Detection in Plaque and Caries: A Case Study. *Proc. SPIE* **2015**. [CrossRef]
50. Lussi, A.; Carvalho, T.S. Erosive tooth wear: A multifactorial condition of growing concern and increasing knowledge. *Monogr. Oral Sci.* **2014**, *25*, 1–15. [PubMed]
51. Ganss, C.; Lussi, A. Diagnosis of erosive tooth wear. *Monogr. Oral Sci.* **2014**, *25*, 22–31. [PubMed]
52. Attin, T.; Wegehaupt, F.J. Methods for assessment of dental erosion. *Monogr. Oral Sci.* **2014**, *25*, 123–142. [PubMed]
53. Poggio, C.; Lombardini, M.; Colombo, M.; Bianchi, S. Impact of two toothpastes on repairing enamel erosion produced by a soft drink: An AFM in vitro study. *J. Dent.* **2010**, *38*, 868–874. [CrossRef] [PubMed]
54. Carvalho Sales-Peres, S.H.; Magalhães, A.C.; Andrade Moreira Machado, M.A.; Buzalaf, M.A.R. Evaluation of The Erosive Potential of Soft Drinks. *Eur. J. Dent.* **2007**, *1*, 10–13. [PubMed]
55. Jones, R.S.; Staninec, M.; Fried, D. Imaging artificial caries under composite sealants and restorations. *J. Biomed. Opt.* **2004**, *9*, 1297–1304. [CrossRef] [PubMed]
56. Majkut, P.; Sadr, A.; Shimada, Y.; Sumi, Y.; Tagami, J. Validation of Optical Coherence Tomography against Micro-computed Tomography for Evaluation of Remaining Coronal Dentin Thickness. *J. Endod.* **2015**, *41*, 1349–1352. [CrossRef] [PubMed]
57. Fonsêca, D.D.D.; Kyotoku, B.B.C.; Maia, A.M.A.; Gomes, A.S.L. In vitro imaging of remaining dentin and pulp chamber by optical coherence tomography: Comparison between 850 and 1280 nm. *J. Biomed. Opt.* **2009**, *14*, 24009. [CrossRef] [PubMed]
58. Fujita, R.; Komada, W.; Nozaki, K.; Miura, H. Measurement of the remaining dentin thickness using optical coherence tomography for crown preparation. *Dent. Mater.* **2014**, *33*, 355–362. [CrossRef]
59. Sinescu, C.; Negrutiu, M.L.; Bradu, A.; Duma, V.-F.; Podoleanu, A.G. Noninvasive Quantitative Evaluation of the Dentin Layer during Dental Procedures Using Optical Coherence Tomography. *Comput. Math. Methods Med.* **2015**. [CrossRef] [PubMed]
60. Heintze, S.D. Clinical relevance of tests on bond strength, microleakage and marginal adaptation. *Dent. Mater.* **2013**, *29*, 59–84. [CrossRef] [PubMed]
61. Hickel, R.; Roulet, J.-F.; Bayne, S.; Heintze, S.D.; Mjor, I.A.; Peters, M.; Rousson, V.; Randall, R.; Schmalz, G.; Tyas, M.; et al. Recommendations for conducting controlled clinical studies of dental restorative materials. *Clin. Oral Investig.* **2007**, *11*, 5–33. [CrossRef] [PubMed]
62. Häfer, M.; Jentsch, H.; Haak, R.; Schneider, H. A three-year clinical evaluation of a one-step self-etch and a two-step etch-and-rinse adhesive in non-carious cervical lesions. *J. Dent.* **2015**, *43*, 350–361. [CrossRef] [PubMed]
63. Häfer, M.; Schneider, H.; Rupf, S.; Busch, I.; Fuchß, A.; Merte, I.; Jentsch, H.; Haak, R.; Merte, K. Experimental and clinical evaluation of a self-etching and an etch-and-rinse adhesive system. *J. Adhes. Dent.* **2013**, *15*, 275–286. [PubMed]
64. Park, K.-J.; Schneider, H.; Haak, R. Assessment of defects at tooth/self-adhering flowable composite interface using swept-source optical coherence tomography (SS-OCT). *Dent. Mater.* **2015**, *31*, 534–541. [CrossRef] [PubMed]
65. Park, K.J.; Schneider, H.; Haak, R. Assessment of interfacial defects at composite restorations by swept source optical coherence tomography. *J. Biomed. Opt.* **2013**, *18*. [CrossRef] [PubMed]

Article

Improved Imaging of Magnetically Labeled Cells Using Rotational Magnetomotive Optical Coherence Tomography

Peter Cimalla [1,2], Julia Walther [1,3], Claudia Mueller [4], Seba Almedawar [2], Bernd Rellinghaus [5], Dierk Wittig [4], Marius Ader [2], Mike O. Karl [2,6], Richard H. W. Funk [4], Michael Brand [2] and Edmund Koch [1,*]

1   Anesthesiology and Intensive Care Medicine, Clinical Sensoring and Monitoring,
    Faculty of Medicine Carl Gustav Carus, Technische Universität Dresden, 01307 Dresden, Germany;
    peter_cimalla@yahoo.de (P.C.); julia.walther@tu-dresden.de (J.W.)
2   DFG-Center for Regenerative Therapies Dresden (CRTD), Technische Universität Dresden,
    01307 Dresden, Germany; seba.almedawar@crt-dresden.de (S.A.); marius.ader@crt-dresden.de (M.A.);
    mike.karl@crt-dresden.de (M.O.K.); michael.brand@biotec.tu-dresden.de (M.B.)
3   Department of Medical Physics and Biomedical Engineering, Faculty of Medicine Carl Gustav Carus,
    Technische Universität Dresden, 01307 Dresden, Germany
4   Institute of Anatomy, Faculty of Medicine Carl Gustav Carus, Technische Universität Dresden,
    01307 Dresden, Germany; cmueller605@gmail.com (C.M.); diewit@googlemail.com (D.W.);
    richard.funk@tu-dresden.de (R.H.W.F.)
5   Institute for Metallic Materials, Leibniz Institute for Solid State and Materials Research,
    01069 Dresden, Germany; B.Rellinghaus@ifw-dresden.de
6   German Center for Neurodegenerative Diseases (DZNE), 01307 Dresden, Germany
*   Correspondence: edmund.koch@tu-dresden.de; Tel.: +49-351-458-6131

Academic Editor: Michael Pircher
Received: 20 January 2017; Accepted: 14 April 2017; Published: 27 April 2017

Abstract: In this paper, we present a reliable and robust method for magnetomotive optical coherence tomography (MM-OCT) imaging of single cells labeled with iron oxide particles. This method employs modulated longitudinal and transverse magnetic fields to evoke alignment and rotation of anisotropic magnetic structures in the sample volume. Experimental evidence suggests that magnetic particles assemble themselves in elongated chains when exposed to a permanent magnetic field. Magnetomotion in the intracellular space was detected and visualized by means of 3D OCT as well as laser speckle reflectometry as a 2D reference imaging method. Our experiments on mesenchymal stem cells embedded in agar scaffolds show that the magnetomotive signal in rotational MM-OCT is significantly increased by a factor of ~3 compared to previous pulsed MM-OCT, although the solenoid's power consumption was 16 times lower. Finally, we use our novel method to image ARPE-19 cells, a human retinal pigment epithelium cell line. Our results permit magnetomotive imaging with higher sensitivity and the use of low power magnetic fields or larger working distances for future three-dimensional cell tracking in target tissues and organs.

Keywords: optical coherence tomography; cell imaging; dynamic contrast agents; magnetic particles; laser speckle; fluorescence microscopy

1. Introduction

Magnetomotive optical coherence tomography (MM-OCT) is a promising imaging method for noninvasive three-dimensional (3D) tracking of magnetically labeled cells in target tissues or organs [1–3]. In this situation, selected cells are labeled with magnetic micro- or nanoparticles as dynamic contrast agents, which are excited into motion by a modulated external magnetic field. By analysis of OCT signal alterations, this motion can be detected and visualized for enhanced contrast of labeled cells.

In conventional MM-OCT, the external magnetic field is generated by a solenoid close to the sample. Ideally, magnetomotion is characterized by particle displacement towards the solenoid when the current is switched on, and return of the particle to its initial position when the current is switched off. Hence, besides the magnetic force, also the restoring force of the elastic tissue environment is essential for the retrieval of appropriate magnetomotive signals. To expand MM-OCT to applications and environments where no elastic restoring force is present, such as in liquids, dual-coil configurations were introduced, in which the particles are attracted alternately by a primary and an opposite secondary coil on a common axis [4]. Such a setup would also be convenient for cell imaging, since cells are characterized by viscoelastic properties rather than a true elastic behavior [5]. However, because of the limited space between the two solenoids, this method is restricted to small samples only. Therefore, we introduce an alternative method called rotational magnetomotive (rMM)-OCT which uses two magnets on the same side of the sample but with two different axes. This off-axis configuration allows the generation of modulated longitudinal and transverse magnetic fields, which cause magnetomotion by the alignment and rotation of anisotropic magnetic particles in the sample volume.

For our experiments, we used a multimodal imaging system presented recently by our group [6]. This system allows MM-OCT imaging in combination with light microscopy and laser speckle reflectometry for structural and motion analysis of single cells. For the implementation of rMM-OCT, we added a small permanent magnet to the existing setup to generate the transverse magnetic field. After analysis of magnetic field components and particle behavior, we provide proof-of-principle evidence for rMM-OCT imaging of single mesenchymal stem cells (MSCs) embedded in agar scaffolds and demonstrate the magnetomotive signal enhancement compared to previous pulsed MM-OCT. Finally, we use our method to image ARPE-19 cells [7], a human retinal pigment epithelium (RPE) cell line, in order to demonstrate the potential of rMM-OCT to track therapeutically relevant cell types, such as RPE cells transplanted into the retina [8,9].

2. Materials and Methods

2.1. Experimental Setup for MM-OCT

All experiments were carried out with a multimodal imaging system for MM-OCT, laser speckle reflectometry and light microscopy, which was presented recently by our group [6]. In brief, the setup shown in Figure 1 consists of a conventional epi-fluorescence microscope (Leica DMRB, Leica Microsystems GmbH, Wetzlar, Germany) for bright-field transmission and multi-color fluorescence microscopy, an integrated MM-OCT probe including an electromagnet for magnetic field generation, and a dark-field laser illumination for laser speckle imaging. The setup allows parallel visualization of the same cells by means of two-dimensional (2D) light microscopy and 3D OCT, as well as investigation of sample motion by means of laser speckle variance analysis. Laser illumination of the sample is generated by a fiber-coupled laser diode at 638 nm (Lasiris PTL-500-635-1-2.0, Coherent Inc., Santa Clara, CA, USA) with an output power of 1 mW. Light microscopy and laser speckle images are detected by a video camera (1.3 megapixel monochrome CMOS, Sumix SMX-M71M, Sumix, Oceanside, CA, USA) mounted on the microscope. The microscope-integrated OCT probe is fiber-coupled to a self-developed spectral domain OCT system operating at 880 nm [10]. This system is equipped with a superluminescent diode (Exalos EXS8810-2411, Exalos AG, Schlieren, Switzerland) with a center wavelength of 876 nm and a spectral bandwidth of 64 nm (FWHM), and a linear-in-wavenumber spectrometer. In summary, the system allows OCT imaging at an axial depth scan (A-scan) rate of 11.9 kHz with an axial resolution of 6.7 μm in air, an optical imaging depth of 3 mm and a sensitivity of −102 dB.

The electromagnet for magnetic field generation (Intertec Components GmbH, ITS-MS 5030 12 VDC, Hallbergmoos, Germany) is attached to the distal end of the OCT probe with parts of the OCT imaging optics incorporated into a center bore of the solenoid's ferromagnetic core. With a working distance of 6 mm and an imaging numerical aperture (NA) of 0.1 yielding a theoretical lateral resolution of 3.3 µm, this MM-OCT probe allows observation of single cells in a three-dimensional tissue-like environment. The field of view of the OCT probe in the focal plane is 1.5 mm in diameter.

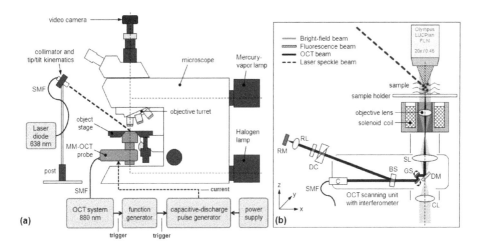

Figure 1. Multimodal imaging setup for magnetomotive OCT (MM-OCT), laser speckle reflectometry and light microscopy. (**a**) Light microscope with integrated MM-OCT probe and dark-field laser illumination for laser speckle imaging. The microscope is equipped with a halogen lamp for bright-field transmission imaging and a mercury-vapor lamp for epi-fluorescence imaging; (**b**) the inset shows a detailed view of the sample region and the OCT scanning unit with attached electromagnet. The coordinate system (x, y, z) indicates the orientation of the MM-OCT probe. The x-direction is the fast scanning direction of the OCT probe. Hence, cross-sectional OCT images (tomographic B-scans) are created in the x–z plane Abbreviations: BS—beam splitter, C—collimator, CL—condensing lens, DC—dispersion compensation, DM—Vis/NIR dichroic mirror, GS—galvanometer scanners, RL—reference lens, RM—reference mirror, SL—scanning lens, SMF—single-mode fiber (color online).

The solenoid coil (18.5 Ω) is usually driven in pulsed operation mode by a self-developed capacitive-discharge pulse generator synchronized to the OCT system. Typically, the magnetic pulse length is set to 80 ms, which corresponds to the acquisition time of two tomographic images (B-scans), and the pulse repetition rate is set to 1 Hz. In the current setup, the maximum charging voltage of the capacitor is 120 V yielding a peak magnetic flux density of 0.2 T and an axial field gradient of 18 T/m in the sample region.

2.2. Additional Experimental Setup for Rotational MM-OCT

For the implementation of rMM-OCT, we added a permanent ring magnet (Conrad Electronic SE, 506014-62, Hirschau, Germany) with an internal remanent field of 1.35 T and a thickness of 2 mm on top of the electromagnet. This ring magnet is a highly textured NdFeB sinter magnet whose easy axis of magnetization is parallel to its rotational symmetry axis. As indicated in Figure 2, both magnets are not concentric but have a lateral axis offset of approximately 15 mm. Hence, as long as the solenoid is switched off (Figure 1b), the laterally displaced permanent magnet creates a magnetic field with a significant horizontal component in the sample region. This causes the anisotropic magnetic particles to align horizontally with their long axis in parallel to the field lines (such as a compass needle). However, when the solenoid is switched on (Figure 2c), the field of the permanent magnet is superimposed by the solenoid's magnetic field, which has a significant vertical component in the sample region. Thus, the particles will rotate and align vertically. As a consequence, periodic switching of the solenoid, such as in pulsed operation mode, will create a nod-like motion of the particles, which should be detectable by OCT and laser speckle reflectometry.

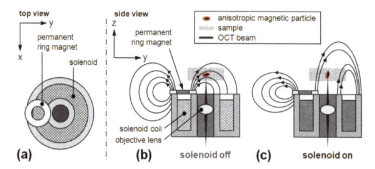

Figure 2. Principle of rotational magnetomotive OCT (rMM-OCT) using an electromagnet (solenoid) and a laterally displaced permanent magnet. (**a**) Top view and (**b**,**c**) side view of the dual-magnet configuration. When the solenoid is turned off (**b**), anisotropic magnetic particles in a soft sample are aligned horizontally (*y*-direction) towards the field of the permanent magnet; when the solenoid is turned on (**c**), particles are aligned vertically (*z*-direction) towards the field of the solenoid. Drawn magnetic field lines are a guide to the eye only. The indicated coordinate system (x, y, z) is identical to Figure 1b. The coordinate origin is at the optical axis of the OCT probe ($x = y = 0$) and the surface of the solenoid ($z = 0$) (color online).

3. Results

3.1. Validation of the Rotational Magnetomotive Concept

To validate the hypothesis mentioned above, we first analyzed the transverse and longitudinal magnetic field components of our rMM-OCT setup during the off- and on-states of the solenoid. For this experiment, we operated the solenoid in DC (direct current) mode and successively measured the field components (Figure 3a,b) using a conventional magnetometer (KOSHAVA 5, Wuntronic GmbH, München, Germany). In order to prevent the permanent magnet from being pushed off the solenoid during the on-state, we limited the supply voltage to 30 V. As it can be seen in Figure 3c, switching off the solenoid results in a varying magnetic field vector in the sample region, i.e., the focus region of the OCT beam. When the solenoid is switched on, the tilt angle of the field vector changes from 32° to 83°, thus, there is a dominant vertical field during the on-state compared to a rather horizontal field during the off-state. The magnitude of the field vector amounts to 17 mT in the off-state and 101 mT in the on-state. This significant difference between the two states is most likely due to the empirical and non-optimized character of our dual-magnet configuration. Ideally, one would assume that the best results will be obtained with equal field strengths in both states. However, the setup presented in this work is considered more a proof-of-principle rather than an optimized system. Related, the corresponding field gradient (in the direction of the respective tilt angle) is approximately 4 T/m and 9 T/m in the off- and on-states, respectively. However, we believe that the field gradient plays at least a minor role in this context, as the particles are supposed to rotate rather than move via a magnetic gradient force-induced displacement.

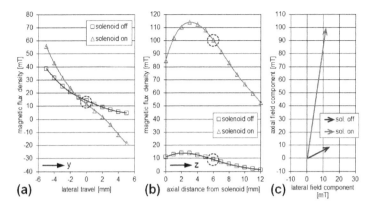

Figure 3. Magnetic field in the rMM-OCT setup. (**a**) Transverse magnetic field; and (**b**) longitudinal magnetic field measured with the solenoid turned off and on (30 V). The measurements were performed in the y–z plane with $x = 0$, as indicated in Figure 2b,c. The coordinate origin is again at the optical axis of the OCT probe ($x = y = 0$ mm) and the surface of the solenoid ($z = 0$ mm). The transverse field in (**a**) was measured at the working distance of the OCT probe ($z = 6$ mm); The longitudinal field in (**b**) was measured along the optical axis of the OCT probe ($x = y = 0$ mm). The focus position of the OCT beam along the optical axis ($x = y = 0$ mm; $z = 6$ mm) is indicated by the dashed circles; (**c**) Resulting magnetic field vectors in the central OCT focus position ($x = y = 0$ mm; $z = 6$ mm) during off- and on-states of the solenoid. The field vectors were calculated from the respective transverse and longitudinal magnetic field components indicated by the dashed circles in (**a,b**).

In a second step, we proved magnetic particle alignment and rotation in the switchable field. We dispersed magnetic particles in glycerin (0.5 mg/mL) and investigated them using the bright-field microscopy channel of our multimodal imaging setup in combination with a high-magnification objective (Leica C Plan L $40\times$/0.50). The particles were identical to the ones used for cell experiments in our previous study [6]; they are sub-micron particles with a nominal size of 200 nm (nano-screen MAG/R-DXS, Chemicell GmbH, Berlin, Germany) consisting of a magnetite core, a fluorescent shell emitting in the red visible spectral range, and a coating made of dextran-sulfate for biomolecule coupling.

As shown in Figure 4a, the magnetic particles are rather homogeneously distributed in the freshly prepared sample. However, after 30 min exposure to the permanent magnetic field, the particles have formed long chains that are transversally aligned towards the permanent magnet (y-direction, Figure 4b). Finally, when the electromagnet is switched on, these chains rotate and align longitudinally towards the solenoid (z-direction, Figure 4c). A corresponding time-lapse video of the particle movement is shown in Video S1. This type of chaining, alignment and rotation behavior is also reported in the literature for magnetic particles inside cells. Wilhelm, C. et al. [11] describe that magnetic particles internalized by cells are accommodated in "magnetic endosomes," i.e., vesicles containing several magnetic particles. These magnetic endosomes cluster to elongated chains within single cells when exposed to a permanent magnetic field, and these chains also rotate when the direction of the magnetic field is changed.

Interestingly, as Figure 4b appears brighter than Figure 4c, the rotation of the magnetic particle chains also seems to influence the transparency of the sample.

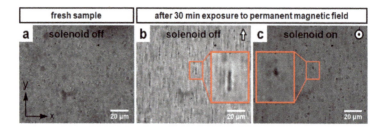

Figure 4. Magnetic particle behavior in the switchable magnetic field. (**a**) Bright-field transmission microscopy of magnetic nanoparticles dispersed in glycerin (0.5 mg/mL) in a freshly prepared sample; (**b**) after 30 min exposure to the permanent magnetic field with the solenoid switched off; and (**c**) shortly after with the solenoid switched on (30 V). The dominant direction of the magnetic field is indicated by the symbols in the upper left corner of the images (↑—transverse, ⊙—longitudinal). The indicated coordinate system (*x*, *y*) is identical to Figure 2a. As illustrated by the magnified views (red rectangles), the particles in (**b**) have formed long chains that are aligned parallel to the image plane; when the solenoid is switched on in (**c**), these chains rotate and align perpendicular to the image plane. This rotation effect is reversible across many on/off cycles of the solenoid (see Scheme 1 and Video S1) (color online).

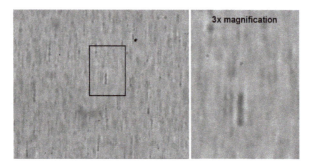

Scheme 1. Time-lapse video of the magnetic particle movement when the solenoid is turned on and off. (**Left**) Bright-field transmission image as shown in Figure 4; (**Right**) Three-fold magnified view of a single particle chain. The position of the magnified view is indicated by the dark rectangle. Due to the fluidic properties of glycerin, there is a certain drift of the particles caused by the gradient of the magnetic field.

3.2. Cell Imaging

Based on these findings, we applied rMM-OCT to cell imaging. For this, we magnetically labeled mesenchymal stem cells (MSCs) with the fluorescent magnetic particles, as mentioned in Section 3.1, and embedded them in agar scaffolds resembling the elastic properties of soft tissue, as described in [6]. Concerning cellular particle load and its effect on cell function, recent literature indicates that MSCs incubated with 200-nm magnetic particles coated with dextran-sulfate take up ~10 pg of iron oxide per cell, and show a cell viability of ~70% after 24 h [12]. After preparation, we placed the samples on the object stage of our multimodal imaging setup containing the permanent magnet and kept them under resting conditions for 30 min in order to provoke the formation and alignment of magnetic chains inside the cells. After that, we analyzed single cells using rMM-OCT in combination with light microscopy and laser speckle reflectometry. In this context, we applied bright-field transmission and fluorescence microscopy in order to identify single cells inside the sample volume and evaluated cell labeling using the fluorescent properties of the magnetic particles. Additionally, we applied laser speckle reflectometry as a 2D reference imaging method to evaluate the sample motion. This time, we operated the solenoid in pulsed mode (pulse duration 80 ms; repetition rate 1 Hz) in order to induce repetitive magnetomotion. According to our previous experiments, we set the capacitor charging voltage to 30 V, expecting the magnitude of the pulsed magnetic field to be similar to the values measured in DC mode (Figure 3a–c). Finally, we retrieved motion contrast from both OCT and laser speckle (LS) images ($f_{OCT} = f_{LS} = 25$ Hz) by calculating the temporal variance of the OCT signal amplitude and the laser speckle intensity over an interval of five B-scans and five video frames, respectively.

Exemplary imaging results for five labeled and two unlabeled control cells are shown in Figure 5. It can be seen that labeled cells are characterized by an intense red fluorescence signal in the microscopy channel originating from the fluorescent magnetic particles, as well as significant color-coded motion signals in the OCT and laser speckle channels resulting from magnetomotion. In comparison, unlabeled cells show no fluorescence and no motion contrast.

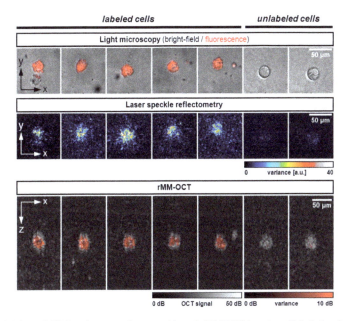

Figure 5. Multimodal light microscopy, laser speckle and rMM-OCT imaging of labeled and unlabeled mesenchymal stem cells (MSCs) embedded in agar scaffolds. The labeled cells are characterized by an intense red fluorescence signal in the microscopy images and a significantly increased variance in both the laser speckle images and the OCT cross-sections, which is caused by magnetically induced motion. In contrast, the unlabeled control cells show no fluorescence and no motion signal (color online).

Hypothetically, the pulse length of 80 ms is not long enough to allow the sub-micron particles to fully respond to the magnetic field. However, we choose this pulse duration to make the results comparable to previous work [6].

In this context, we measured the magnetomotive signal of five different cells imaged with rMM-OCT and compared the results to our previous pulsed MM-OCT method. For measurement, we selected a region-of-interest (ROI) representing the cell, and calculated the normalized magnetomotive signal as the quotient of the ROI's mean variance and the average amplitude of the OCT signal across the magnetic pulse. Similarly, we measured the magnetomotive signal in laser speckle reflectometry by calculating the mean laser speckle variance divided by the average laser speckle intensity. The results in Figure 6 show that the magnetomotive signal in rMM-OCT is significantly increased in both OCT and laser speckle analysis by a factor of 2.8 and 2.5, respectively. We note that the only differences of the rMM-OCT setup compared to the MM-OCT setup are the addition of the permanent magnetic ring and the lower charging voltage of the pulse generator capacitor (30 V compared to 120 V, respectively). However, rMM-OCT provides a significantly improved magnetomotive contrast compared to MM-OCT, although the charging voltage was four times smaller and, thus, the solenoid's power consumption was 16 times lower.

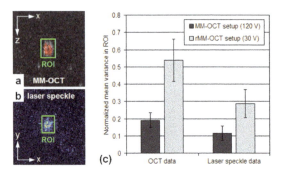

Figure 6. Quantification of the cellular magnetomotive signal using OCT and laser speckle analysis. In each OCT (**a**) and laser speckle image (**b**) a region of interest (ROI) representing the cell is chosen and the mean magnetomotive signal is determined; (**c**) Magnetomotive signals retrieved from OCT data (left column) and laser speckle data (right column) for five different cells imaged with our previous pulsed MM-OCT setup (dark bars) and the novel rMM-OCT setup (bright bars) using a capacitor charging voltage of 120 V and 30 V, respectively. The magnetomotive signals represent the normalized mean variance in the ROI, i.e., the mean pixel variance divided by the average pixel intensity across the magnetic pulse. Data represents mean values ± standard deviation (color online).

In a final step, we tested our rMM-OCT imaging method on alternative cells next to MSCs that are used for retinal transplantation studies concerning the RPE, the ARPE-19 cell line [13–15]. These cells were labeled and embedded in agar scaffolds identical to MSCs, as described before. Representative OCT imaging results for labeled ARPE-19 cells are shown in Figure 7. It can be seen that the magnetomotive contrast is very similar to the MSCs shown in Figure 5.

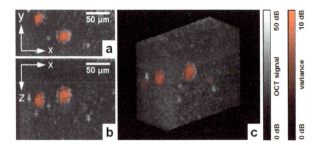

Figure 7. Volumetric rMM-OCT imaging of labeled ARPE-19 cells embedded in an agar scaffold. (**a**) Top view and (**b**) side view of a recorded 3D OCT image stack showing two cells next to each other. (**c**) 3D view of the two cells. All views represent maximum intensity projections of the 3D image data (color online).

4. Discussion

In conclusion, we have shown that rMM-OCT provides a higher magnetomotive signal at a lower power consumption compared to previous pulsed MM-OCT. This improvement was mostly achieved

Appl. Sci. **2017**, *7*, 444

by adding a small permanent magnet in an off-axis configuration to the MM-OCT setup. The thickness of the permanent magnet is only 2 mm; hence, the reduction of working distance from 6 mm (relative to the surface of the solenoid) in the original setup to 4 mm (relative to the surface of the permanent magnet) in the new setup is still acceptable for retinal imaging in small animal models used in cell transplantation studies, such as mice (eye length 3 mm). Further, the dual-magnet configuration of rMM-OCT is a single-sided setup that circumvents the restrictions of previous dual-coil configurations to small samples.

In this context, we have also shown that rMM-OCT allows robust and reliable contrast-enhancement in different cell types, such as MSCs and transplantable ARPE-19 cells. We note that magnetomotion in cells can also be induced by other MM-OCT methods, such as pulsed MM-OCT using our previous setup. However, our results shown in Figure 6 imply that the magnetomotive signal of rMM-OCT is significantly higher compared to the previous pulsed MM-OCT. Besides particle chaining and rotation, this signal increase might be additionally supported by particle clumping induced by the permanent magnet leading to enhanced magnetic susceptibility of the contrast agent and, thus, improved displacement in response to the applied magnetic field. As a consequence, rMM-OCT permits higher sensitivity, the use of low power magnetic fields or corresponding larger working distances, and single-sided access to biological specimens. This makes it a promising method for future 3D cell tracking in target tissues and organs, such as the retina in small animal models, under in vivo conditions.

However, further investigation is required to adress several aspects of the method presented herein. Most important, as the induced magnetic particle changes to elongated chains (as shown in Figure 4) end up being very long in glycerin (~10 μm) and approach the diameters of cells, cell viability experiments are necessary to investigate whether particle chain formation and rotation during rMM-OCT imaging may be destructive to cells. Additionally, it is not yet clear what the size these chains in cells may reach, how controllable this is in conjunction with the magnetization time (currently 30 min), and what the minimum detectable concentration of magnetic particles is or the minimum detectable size of particle aggregates. In this context, the current magnetization time of 30 min is relatively long, especially with regards to prospective in vivo applications. Therefore, we plan further experiments in which we investigate the time-dependency of magnetic chain formation and the correlation with magnetomotive signal enhancement in order to find the minimum magnetization time appropriate for in vivo imaging.

As an alternative to further enhance the sensitivity of rotational magnetomotive measurements, particles with potentially increased anisotropy such as magnetic nanostars [16,17] could be applied. Moreover, for future in vivo application, faster imaging is desired. In this study, the magnetomotive imaging speed was limited by the fixed pulse repetition rate of 1 Hz in order to make the results comparable to previous work. However, the lower power consumption of the solenoid in this work facilitates the use of higher duty cycles, and thus faster pulse repetition rates. In this context, an adequate pulse generator for faster pulse series is the subject of current, ongoing investigation.

Supplementary Materials: The following are available online at http://www.mdpi.com/2076-3417/7/5/444/s1, Video S1: rMM-OCT magnetic particle movement.avi.

Acknowledgments: This research was supported by the TU Dresden CRTD (Center for Regenerative Therapies Dresden) Seed Grant Program and grants from the MeDDrive program of the Faculty of Medicine Carl Gustav Carus of the TU Dresden.

Author Contributions: All authors contributed to this work. Peter Cimalla conceived the method of rotational MM-OCT, designed the experimental setup and performed the imaging experiments; Claudia Mueller and Seba Almedawar performed the cell labeling experiments of the MSC and ARPE-19 cells; Peter Cimalla, Julia Walther and Edmund Koch processed and analyzed the data; Bernd Rellinghaus characterized the magnetic particles and analyzed the magnetic field; Dierk Wittig, Marius Ader, Mike O. Karl, Richard H. W. Funk and Michael Brand analyzed the cell labeling and the imaging data; Peter Cimalla and Julia Walther wrote the paper.

Conflicts of Interest: The authors declare no conflict of interest.

References

1. Oldenburg, A.L.; Gunther, J.R.; Boppart, S.A. Imaging magnetically labeled cells with magnetomotive optical coherence tomography. *Opt. Lett.* **2005**, *30*, 747–749. [CrossRef] [PubMed]
2. Oldenburg, A.L.; Gallippi, C.M.; Tsui, F.; Nichols, T.C.; Beicker, K.N.; Chhetri, R.K.; Spivak, D.; Richardson, A.; Fischer, T.H. Magnetic and contrast properties of labeled platelets for magnetomotive optical coherence tomography. *Biophys. J.* **2010**, *99*, 2374–2383. [CrossRef] [PubMed]
3. Oldenburg, A.L.; Wu, G.; Spivak, D.; Tsui, F.; Wolberg, A.S.; Fischer, T.H. Imaging and elastometry of blood clots using magnetomotive optical coherence tomography and labeled platelets. *IEEE J. Sel. Top. Quantum Electron.* **2012**, *18*, 1100–1109. [CrossRef] [PubMed]
4. Kim, J.; Ahmad, A.; Boppart, S.A. Dual-coil magnetomotive optical coherence tomography for contrast enhancement in liquids. *Opt. Express* **2013**, *21*, 7139–7147. [CrossRef] [PubMed]
5. Nawaz, S.; Sanchez, P.; Bodensiek, K.; Li, S.; Simons, M.; Schaap, I.A.T. Cellvisco-elasticity measured with AFM and optical trapping at sub-micrometer deformations. *PLoS ONE* **2012**, *7*, e45297. [CrossRef] [PubMed]
6. Cimalla, P.; Werner, T.; Winkler, K.; Mueller, C.; Wicht, S.; Gaertern, M.; Mehner, M.; Walther, J.; Rellinghaus, B.; Wittig, D.; et al. Imaging of nanoparticle-labeled stem cells using magnetomotive optical coherence tomography, laser speckle reflectometry, and light microscopy. *J. Biomed. Opt.* **2015**, *20*, 036018. [CrossRef] [PubMed]
7. Dunn, K.C.; Aotaki-Keen, A.E.; Putkey, F.R.; Hjemeland, L.M. ARPE-19, a human retinal pigment epithelial cell line with differentiated properties. *Exp. Eye Res.* **1996**, *62*, 155–170. [CrossRef] [PubMed]
8. Zhu, Y.; Carido, M.; Meinhardt, A.; Kurth, T.; Karl, M.O.; Ader, M.; Tanaka, E.M. Three-dimensional neuroepithelial culture from human embryonic stem cells and its use for quantitative conversion to retinal pigment epithelium. *PLoS ONE* **2013**, *8*, e54552. [CrossRef] [PubMed]
9. Carido, M.; Zhu, Y.; Postel, K.; Benkner, B.; Cimalla, P.; Karl, M.O.; Kurth, T.; Paquet-Durand, F.; Koch, E.; Münch, T.A.; et al. Characterization of a mouse model with complete RPE loss and its use for RPE cell transplantation. *Investig. Ophthalmol. Vis. Sci.* **2014**, *55*, 5431–5444. [CrossRef] [PubMed]
10. Burkhardt, A.; Walther, J.; Cimalla, P.; Mehner, M.; Koch, E. Endoscopic optical coherence tomography device for forward imaging with broad field of view. *J. Biomed. Opt.* **2012**, *17*, 071302. [CrossRef] [PubMed]
11. Wilhelm, C.; Gazeau, F.; Bacri, J.C. Rotational magnetic endosome microrheology: Viscoelastic architecture inside living cells. *Phys. Rev. E* **2003**, *67*, 061908. [CrossRef] [PubMed]
12. Yanai, A.; Häfeli, U.O.; Metcalfe, A.L.; Soema, P.; Addo, L.; Gregory-Evans, C.Y.; Po, K.; Shan, X.; Moritz, O.L.; Gregory-Evans, K. Focused magnetic stem cell targeting to the retina using superparamagnetic iron oxide nanoparticles. *Cell Transplant.* **2012**, *21*, 1137–1148. [CrossRef] [PubMed]
13. Coffey, P.J.; Girman, S.; Wang, S.M.; Hetherington, L.; Adamson, P.; Greenwood, J.; Lund, R.D. Long-term preservation of cortically dependent visual function in RCS rats by transplantation. *Nat. Neurosci.* **2002**, *5*, 53–56. [CrossRef] [PubMed]
14. Wang, S.; Lu, B.; Lund, R.D. Morphological changes in the Royal College of Surgeons rat retina during photoreceptor degeneration and after cell-based therapy. *J. Comp. Neurol.* **2005**, *491*, 400–417. [CrossRef] [PubMed]
15. Pinilla, I.; Cuenca, N.; Sauvé, Y.; Wang, S.; Lund, R.D. Preservation of outer retina and its synaptic connectivity following subretinal injections of human RPE cells in the Royal College of Surgeons rat. *Exp. Eye Res.* **2007**, *85*, 381–392. [CrossRef] [PubMed]
16. Wei, Q.; Song, H.M.; Leonov, A.P.; Hale, J.A.; Oh, D.; Ong, Q.K.; Ritchie, K.; Wei, A. Gyromagnetic imaging: Dynamic optical contrast using goldnanostars with magnetic cores. *J. Am. Chem. Soc.* **2009**, *131*, 9728–9734. [CrossRef] [PubMed]
17. Song, H.M.; Wei, Q.; Ong, Q.K.; Wei, A. Plasmon-resonant nanoparticles and nanostars with magnetic cores: Synthesis and magnetomotive imaging. *ACS Nano* **2010**, *4*, 5163–5173. [CrossRef] [PubMed]

Article

Full-Field Optical Coherence Tomography as a Diagnosis Tool: Recent Progress with Multimodal Imaging

Olivier Thouvenin [1], Clement Apelian [1,2], Amir Nahas [1], Mathias Fink [1] and Claude Boccara [1,2,*]

1   Institut Langevin, ESPCI Paris, PSL Research University, 1 rue Jussieu, Paris 75005, France; olivierthouvenin3@gmail.com (O.T.); clement.apelian@espci.fr (C.A.); amir.nahas@espci.fr (A.N.); mathias.fink@espci.fr (M.F.)
2   LLTech SAS, Pépinière Paris Santé Cochin 29 rue du Faubourg Saint Jacques, Paris 75014, France
*   Correspondence: claude.boccara@espci.fr; Tel.: +331-80-96-30-44

Academic Editor: Michael Pircher
Received: 20 January 2017; Accepted: 21 February 2017; Published: 2 March 2017

Abstract: Full-field optical coherence tomography (FF-OCT) is a variant of OCT that is able to register 2D *en face* views of scattering samples at a given depth. Thanks to its superior resolution, it can quickly reveal information similar to histology without the need to physically section the sample. Sensitivity and specificity levels of diagnosis performed with FF-OCT are 80% to 95% of the equivalent histological diagnosis performances and could therefore benefit from improvement. Therefore, multimodal systems have been designed to increase the diagnostic performance of FF-OCT. In this paper, we will discuss which contrasts can be measured with such multimodal systems in the context of ex vivo biological tissue examination. We will particularly emphasize three multimodal combinations to measure the tissue mechanics, dynamics, and molecular content respectively.

Keywords: full-field optical coherence tomography; multimodality; biomechanics; mechanical properties; dynamics; cellular metabolism; fluorescence microscopy

1. Introduction

Over the past 25 years, optical coherence tomography (OCT) has rapidly developed and spread into many fields of biomedical research thanks to its ability to capture label-free structural contrast deep inside a tissue [1]. Over the years, OCT has been developed into numerous different modalities that emphasize the 3D interplay between tissue microstructures and various specific complementary contrasts [2–5].

Full-field optical coherence tomography (FF-OCT) is a less familiar OCT-derived technology [6,7]. FF-OCT differs from standard OCT by acquiring 2D *en face* images in a widefield configuration at a given depth in a tissue. To record a 3D tomogram, the sample is further scanned in depth, instead of acquiring series of A lines as in regular OCT. Such a configuration offers the possibility to use high numerical aperture (NA) microscope objectives, without limiting the exploration depth, and can therefore access submicrometric and subcellular resolution.

FF-OCT has been successfully applied for non-invasive label-free ex vivo pathology assessment in various tissues [8–15]. Thanks to its isotropic micrometric resolution, FF-OCT can often be compared to histology slices. In contrast to histology, FF-OCT does not require time for sample fixation, slicing, or labeling, and an histology-like image can be produced within a few minutes after tissue extraction. It also helps to reduce fixation and slicing artifacts, as FF-OCT does not necessarily require physical slicing, and can image in depth, eventually far from the incision edges for a biopsy.

However, FF-OCT may suffer from its lack of specificity, and sometimes fails to equal the diagnostic capacity of gold standard histology. Yet, FF-OCT detection scores are above 80% for most pathologies in both sensitivity and specificity [14,16], and Manu Jain and colleagues could even report a 100% correct identification of tumors in ex vivo kidneys [17], obtaining similar results to histology, even though the tumor subtyping is still not 100%. In order to increase diagnosis scores, FF-OCT has therefore expanded to multimodal systems combining FF-OCT structural contrast with other pathology biomarkers inside a single microscope. The interest of such multimodal systems is to reduce the ambiguity in structure identification that can arise from sequential imaging with two different microscopes, and to reduce operation time. Some FF-OCT multimodal systems have been reviewed in the recently published *Handbook of Full-Field Optical Coherence Microscopy*, edited by Arnaud Dubois [16]. We can non-exhaustively mention the multimodal combinations of FF-OCT with fluorescence [18], photothermal imaging [19], elastography [20], polarization [21], multispectral and hyperspectral imaging [22–24]. Most of these multimodal systems have been inspired by similar systems developed in standard OCT, but they needed to be adapted and redesigned to fit FF-OCT's particular configuration.

In this paper, our goal was not to exhaustively describe these multimodal systems, or to systematically review all the potential fields in which multimodal FF-OCT systems can be of interest. Instead, we will start from the need to increase the FF-OCT's diagnostic ability up to equal that of gold standard histology to describe three different multimodal systems that we are currently developing or improving in our lab. After a brief description of FF-OCT, and the optical systems that we commonly use, we describe and emphasize the differences between FF-OCT and histology contrasts, to better apprehend what contrasts would be optimal to add to FF-OCT. Then, we will describe our recent development related to multimodal systems combining FF-OCT with elastography, with cellular metabolism, and with fluorescence, and how these multimodal systems could increase the diagnosis performance of FF-OCT.

2. Scanning OCT versus Full-Field OCT and Technological Description of the Setups Description

The main difference between classical OCT systems and FF-OCT lies in the absence of transverse scanning. In OCT, or in optical coherence microscopy (OCM), an axial line (A-line) is captured with a single point [1,25,26] or with a line detection [27,28], and further scanned in the transverse plane, to successively obtain 2D planes (B-scan) and 3D tomograms (C-scan). On the contrary, the full-field configuration captures a 2D image at a single depth with a 2D camera, and an axial scan enables the recording of a 3D tomogram. In both configurations, the optical section is mostly controlled by the temporal coherence length, given by the inverse of the spectral bandwidth. Scanning in one dimension compared to two dimensions has several major consequences:

- Because it looks at a single depth at a time, FF-OCT can be performed with a small depth of field, and therefore with high numerical aperture (NA) objectives. This enables high transverse resolution to be obtained, below 1 μm for visible light. Other elegant OCT systems have greatly increased their transverse resolution [1,29–32], however these techniques remain more complex and often have to sacrifice their temporal resolution.
- Moreover, scanning in two directions imposes a trade-off between exposure time and the field of view. In scanning OCT, each pixel is only illuminated for a few microseconds, while in FF-OCT, the entire field of view is usually evenly illuminated during a few milliseconds. It can make a difference in the light dose delivered to the sample, and the most rapid speckle fluctuations are often averaged out [33].
- The time required to get information from the sample is much longer in FF-OCT, since it is limited by the camera frame rate, which is hardly below 1 ms, while some scanning OCT systems can now acquire lines at more than 100,000 lines per second [34]. The consequence is that FF-OCT is much more sensitive to sample motion, and is difficult to perform in vivo, other than on anesthetized animal or by pressing the sample against an imaging window [35].

- A last important difference when looking at *en face* images either performed in scanning OCT or FF-OCT is that all the pixels have been acquired at different times in OCT, while they are acquired at the same time in FF-OCT (if the camera is operated in global shutter mode), which can introduce some artifacts when looking at moving objects. On the other hand, all axial pixels are acquired simultaneously in OCT, while they are largely separated in time for FF-OCT. An important consequence is that the optical phase can be directly retrieved in OCT by subtracting two adjacent axial pixels [36] while FF-OCT requires multiple phase-shifted measurements to retrieve the optical phase.

In practice, we mainly use two different FF-OCT layouts presented in Figure 1. The first setup is the LLTech commercial system [37], based on a Linnik interferometer using two identical 10X, 0.3 NA microscope objectives, and a thermal source to produce an isotropic resolution of 1 µm. We also developed custom-built versions of this system, exhibiting higher mechanical stability, or different cameras for example, but the general layout stays identical. We used these systems to add elastography [20,38] and motility contrasts [39,40].

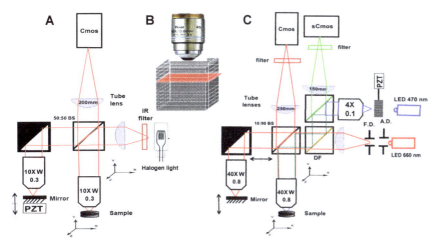

Figure 1. FF-OCT systems layouts. (**A**) Standard layout, as in the LLTech commercial system. It is based on a Linnik interferometer, with identical microscope objectives in both arms. The system can be illuminated by a simple thermal source, or a standard LED, and the image is recorded with a CMOS camera; (**B**) Principle of 3D acquisition in FF-OCT. A 2D plane can be recorded in a single camera acquisition, then the field is scanned in the axial direction; (**C**) Another FF-OCT layout developed in the lab, offering the possibility to work with higher NA objectives, and to add a fluorescence channel simultaneously to the FF-OCT image.

The second setup is a multimodal setup combining FF-OCT and structured illumination fluorescence microscopy (SIM) [41], in a relatively similar layout to the one developed by H.Makhlouf and colleagues [42]. To enhance the photon collection, we used higher NA objectives, either 40X, 0.8 NA in water, or 30X, 1.05 NA in silicone oil, achieving transverse resolution below 500 nm, and depth of focus between 1 and 2 µm. Interestingly, in this system, the optical sectioning ability comes from the lateral spatial coherence [43,44] controlled by the depth of focus, and therefore the objective NA rather than the temporal coherence, meaning that simple LEDs can be used without sacrificing the axial resolution. Both systems are illuminated by spatially incoherent sources, which eliminate the cross talks [45], and preserve the transverse resolution from degradation due to low order aberrations [46]. The illumination sources are mounted in a Kohler illumination to provide an illumination as homogeneous as possible.

Both systems are now operated with a new CMOS camera developed by Adimec during an ERC project (CAReIOCA) led by LLTech [47]. This camera can handle a Full Well Capacity of about $2Me^-$ on 2 million pixels at frame rates up to 700 Hz, which typically enhanced our signal-to-noise ratio by a three-fold factor, and acquisition speed by a five-fold factor, with respect to previously used cameras (PhotonFocus, MV1-D1024E-160CL, Lachen, Switzerland).

3. FF-OCT and Histology Contrasts

In order to further compare histology and FF-OCT, we describe in this section histology contrast and procedures.

It is commonly said that Marcello Malpighi is the precursor of modern histology, as he was the first scientist to carefully describe micro-anatomic features of plants and animal tissues by the end of the 17th century with the first microscopes [48,49]. Since then, histology has continuously benefited from advances in chemical labeling and fixatives, and has benefited from a better understanding of tissue nanostructures and organization thanks to electron microscopy [50]. It now benefits from an extensive literature, and from digital atlases [51–53] enabling direct comparisons with normal or pathological slices of most tissues. However, the basic microscopy principles have mainly remained unchanged since the beginning of histology. They consist of imaging a thin slice of a sample with a simple transmission optical microscope. Since light absorption and scattering by biological structures exhibit a low contrast, histology commonly uses chemical agents to reveal specific contrasts of interest [54–56]. The most common staining uses two associated stains, Hematoxylin and Eosin (H & E), that respectively label chromatin and chromosomes in blue, and cytoplasm in red. Hemoglobin usually appears orange, and collagen fibers appear pink. This staining is particularly versatile, as it can be used with a variety of fixatives. A wide range of different staining can be applied to a tissue slice. We can non-exhaustively evoke Golgi staining that labels 1% to 2% of neurons with silver that absorbs light to appear dark, Nissl staining that labels nucleic acids, or tetraoxide osmium that labels unsaturated lipids and lipoproteins. If many subcellular structures can be selectively labeled, it is often difficult to combine different stains in a single slide. However, using specific labels imposes a prior knowledge of the tissue composition, and of the pathology, and some unlabeled biomarkers could be missed.

Moreover, histology requires tissue fixation, performed either chemically or by freezing the sample, and physical slicing of the tissue, usually performed by a microtome. The fixation first goal is to harden the sample to be able to cut it in thin sections. However, fixation can introduce strong artifacts, as it often removes lipidic structures, and can increase the tissue volume. The physical slicing can induce physical damage, notably in neuronal tissues, in which many axial neurites are usually cut, leading to neuronal death. Finally, these two steps are usually time consuming, and can lead to the physical destruction of the samples if not adequately performed.

In the case of histopathology, the tissue is often fixated by replacing water by paraffin, and is then sliced in 4 to 5 µm slices. For each slice, the dyes have then to diffuse through the paraffin scaffold. In total, these processes can last for a few days.

In the case of intra-operative histology, the time allowed to the pathologist is usually not more than 15 to 30 min. To meet these time requirements, the sample is frozen, and directly sectioned and colored. However, many tissues, such as fat, brain, or retina, are not well suited for this process, and the number of available labeling is limited.

On the other hand, OCT was only invented 25 years ago, and its full-field version is 18 years old, so comparatively little literature and atlases can be found [57]. However, OCT and FF-OCT provide a label-free, and optically-sectioned imaging. This means that imaging can be performed right after tissue extraction, which can provide a very fast diagnosis, and could even ideally provide an intra-operative diagnosis to efficiently define a tumor margin for example.

The main advantage of OCT is its virtual sectioning ability, enabling imaging at a few hundred microns below the tissue surface, hence reasonably far from the sectioned surface no matter which section is required [58]. To completely remove slicing artifacts and reach large depths, OCT systems have successfully been coupled to a vibratome, in a procedure where an entire volume is imaged before a physical slice smaller than the imaged volume is performed [59].

Additionally, OCT systems capture the light naturally backscattered by the sample, without using any label. Working label-free can be advantageous, as staining artifacts can be avoided, and as it can importantly reduce the time prior to imaging. On the other hand, the recorded signal often lacks specificity. The obtained contrast is complicated to predict, as it is described by multi-scale complex theories, notably combining Rayleigh and Mie theories, depending on the scatterer size [60]. However, for a given tissue, the obtained contrast is highly reproducible, so the diagnosis performance could be increased only by simply better characterizing the typical contrast obtained on each tissue, and each condition.

Quite generally, we can nevertheless emphasize the strong ability of FF-OCT to detect high refractive index (high lipid or high fat content) fibers, such as collagen fibers [8], myelin fibers [61], and even some large unmyelinated axons in the eye [62]. Moreover, in terms of contrast, looking at the backscattered light can differ substantially from transmission microscopes in thick samples. However, one can assume that for thin slices, the contrast in reflection is mainly the inverted contrast for scattering structures and the same contrast from absorbing structures compared to a transmission microscope.

The label-free ability of OCT has a final practical advantage, since the imaging can be performed prior to any tissue modification. It means that the same tissue can be further used for another procedure, including a standard histology procedure [8], or fluorescence labeling. It is usually preferred to perform the OCT imaging before any sample modification, as OCT signal alteration can occur after fixation, depending on the tissue content [62] (especially since fixatives often remove lipidic structures of the cells). Keeping this in mind, it is nevertheless possible to perform histology-like measurements in fixed tissues [58].

Scanning OCT has also been used successfully to perform histology-like measurements [58,63–65], and to detect tumor regions [66,67] through changes in the collagen content [68], the tissue microangiogram and backscattering [69], or in the tissue fractal dimension [70]. However, we believe that the high resolution of FF-OCT can significantly improve the diagnostic abilities of OCT [8,17]. With higher numerical apertures, the light collection is improved which helps to reveal more structures and increase the signal-to-noise ratio for the calculation of backscaterring coefficients or fractal dimensions [40,71]. Signal from cell contours and from intracellular structures can also be detected. This will be described in Section 5 on Dynamic FF-OCT. Finally, the *en face* view of the sample is quite natural for histologists, as it looks similar to histology slides. FF-OCT is also faster than scanning OCT to record a single *en face* image.

Figure 2 shows a comparison between FF-OCT and a regular H & E histology in a breast sample, the organization of which is altered by a ductal carcinoma in situ (DCIS). FF-OCT notably reveals normal adipocytes, normal and abnormal collagen fibers (based on their scattering strength), and enlarged ducts. These indicators are characteristics of DCIS, that could be identified with 90% sensitivity and 75 % specificity in this case [14].

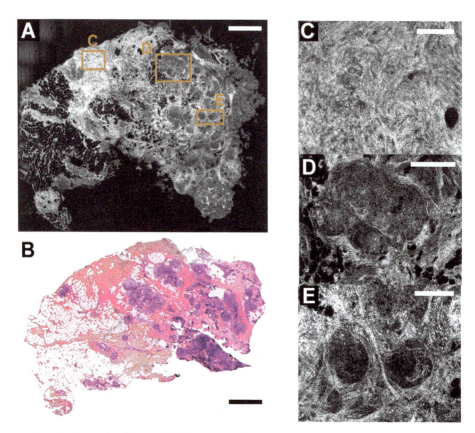

Figure 2. Comparison between FF-OCT and Histology. (**A,B**) Large field image of a breast sample affected by ductal carcinoma in situ (DCIS) acquired with FF-OCT (**A**) and histology (H & E staining; (**B–E**) Zoom in from boxes represented in panel A, respectively showing a normal fibrous tissue, enlarged ducts, and the collagen organization around the carcinomatous cells in a tumorous region [14]. These images have been acquired by LLTech, and are displayed here with their authorization. Scale bars in A and B represent 1 mm, and 200 μm in C to E.

4. FF-OCT and Biomechanics

Since the early beginning of medicine, palpation has been used to qualitatively assess abnormal tissue stiffness associated with several diseases. Palpation-based diagnosis of heart, abdomen, and wounds diseases or infections has been reported in many ancient cultures, such as Greek, Egyptian, or Chinese medicine [72,73]. The mechanical properties of tissues and cells are mostly related to their structure and function, and changes in a tissue's biomechanical properties can reveal its pathological state [74]. Notably, biomechanical changes can include the stiffness increase due to an extracellular matrix accumulation, including collagen accumulation, associated with liver fibrosis, and ultimately cirrhosis [75,76]. Moreover, tumor development significantly and progressively impacts the tissue biomechanics [77]. More importantly, the tumor stiffness actively increases its progression, and thus its malignancy, partly through a mechanically-induced molecular circuit, leading to the reduction in the number of tumor suppressors [78]. Therefore, not only tissue stiffness, but also stiffness anisotropy can be used as relevant diagnosis parameters.

In the 1980s, the thorough and quantitative measurement of tissue elasticity was developed, giving access to stiffness in the approximation of purely elastic deformation of tissues. Elastography has been introduced in biomedical imaging with ultrasound and MRI elastography [79,80], and has been proven to be a precious diagnosis biomarker for physicians since then [81,82]. In this paper, we will only discuss tissue biomechanics in comparison to palpation-based diagnosis, and therefore at macroscopic and mesoscopic scales. However, we can add that cellular and molecular biomechanics can also be significant biomarkers of diseases, but they stay difficult to measure in tissues. To our knowledge, the most promising technique to study microscopic biomechanics inside a tissue is fluorescent FRET sensors [83].

In OCT, mechanical contrast has been introduced by Schmitt in 1998 [84]. Since this pioneering work, many methods have been developed with success such as 2D, 3D, static or dynamic elastography [84–88]. In FF-OCT, our group adapted two similar approaches to measure mechanical contrast. The first approach is a static method based on a finite element 3D correlation algorithm that gives access to the 3D local strain tensor inside the samples [20], giving access to the local shear modulus, which is itself proportional to the tissue elasticity, or stiffness for soft tissues. The second approach is a quantitative dynamic method based on shear wave speed imaging, inspired by similar experiments performed in ultrasounds developed in our lab. Using a fast FF-OCT system working at up to 30,000 frames per second, we could measure the local shear wave speed which is directly related to the local shear modulus [38].

In this section, we present a brief description of both methods and some results obtained on biological samples.

4.1. Description of the Different Mechanical Parameters and Approximations

Before describing how it is possible to optically measure tissue biomechanics, we briefly summarize here the numerous different mechanical parameters that can be found in the extensive literature on tissue biomechanics. We will emphasize the relationship between these parameters, and their relevance for diagnosis.

Most studies in MRI, as well as in ultrasound or optics, focus on the elastic properties of tissues, meaning the reversible linear deformation $\bar{\bar{\varepsilon}}$ caused by a small mechanical stress $\bar{\bar{\sigma}}$. In the elastic regime, and neglecting viscosity, the generalized Hooke's law gives:

$$\bar{\bar{\sigma}} = \bar{\bar{\bar{\bar{C}}}}.\bar{\bar{\varepsilon}} \tag{1}$$

where $\bar{\bar{\bar{\bar{C}}}}$ is a fourth order tensor, named stiffness tensor, that depends on the mechanical parameters of the sample. All the other parameters that we will describe are elements of this fourth order tensor, and are usually obtained when the dimensions of the generalized problem are reduced. For example, Young's modulus E corresponds to the linear coefficient that relates a purely axial stress, and its associated deformation along the stress axis. In this case, the problem is reduced to a single dimension:

$$E = \frac{\varepsilon}{\sigma} \tag{2}$$

We can add that Young's modulus is qualitatively the main measurement that is performed with manual palpation. Basically, the first information that one can get with manual palpation is how the tissue resists to an axial deformation or stress, which is given by the Young's modulus.

From Equation (1), it is easy to realize that a given sample is not only deformed in the direction of the stress, but can also be deformed in the transverse direction, with a preferential axis for anisotropic material. We will describe in Section 4.2 an example metric to quantify the local anisotropy of a sample.

In addition to these parameters, many mechanical measurements rely on shear wave speed measurement that we will describe more carefully in Section 4.3. The mechanical parameters on which the propagation of mechanical waves depend are simply defined as a result of the wave equation in a medium of a given stiffness. Inside biological media, two waves can propagate [89]: A longitudinal wave that corresponds to compression and traction of the medium with a mechanical displacement parallel to the direction of propagation. Additionally, shear waves can propagate, which correspond to perpendicular mechanical displacements with respect to the direction of propagation. The two waves propagate at speeds c_L and c_{shear} respectively that are controlled by the mechanical parameters of the medium. We can therefore define the bulk modulus K, and the shear modulus μ with:

$$c_L = \sqrt{\frac{K}{\rho}} \tag{3}$$

$$c_{shear} = \sqrt{\frac{\mu}{\rho}} \tag{4}$$

where ρ is the medium local density. These parameters are related to Young's modulus [89]:

$$E = \mu \frac{3K}{K + \frac{1}{3}.\mu} \tag{5}$$

Finally, for soft tissues, which represent most biological tissues, except bones, the bulk modulus is much larger than the shear modulus. Bulk moduli of biological samples range from 1 GPa to 10 GPa, while shear moduli range from 0.1 kPa in the brain to 0.1 GPa in epidermis and cartilage, and up to 10 GPa in bones [90]. When considering Equation (5) for soft tissues, the Young's modulus becomes dependent on the shear modulus only:

$$E \simeq 3.\mu \simeq 3\rho c_{shear}^2 \tag{6}$$

With such approximations, we have shown that it was possible to locally measure the Young's modulus by measuring the propagation speed of shear waves.

4.2. Static Elastography Based on Digital Volume Correlation (DVC)

The principle of the static approach is to register a volumetric image before and after mechanical solicitation of the sample [20]. From these two sets of images, we can estimate the 3D local displacement field, $\vec{u}$, inside the sample using a finite element 3D digital image correlation algorithm developed by S. Roux and his team [91,92]. From the optically measured 3D displacement field, we can calculate the 3D strain tensor $\bar{\bar{\varepsilon}}$:

$$\bar{\bar{\varepsilon}} = \frac{1}{2}(\vec{grad}(\vec{u}) + {}^t\vec{grad}(\vec{u})) \tag{7}$$

However, biological samples can be very complex and mechanically heterogeneous. This complexity usually prevents direct access to the local stress tensor and so, in most cases, methods based on digital image correlation are limited to strain contrast and do not provide quantitative stiffness information. This is the main limitation of this type of approach but already the strain information recreates a sort of "virtual local palpation" that can highlight mechanical heterogeneity at the micrometer scale given by FF-OCT or OCT . From the local strain tensor, it is possible to extract different parameters that give access to mechanical information such as strain magnitude or strain anisotropy.

- **Strain magnitude**

  In order to quantify the strain magnitude, we chose to use an equivalent Von Mises strain defined as:

$$\forall \vec{x}, \ \varepsilon_{eq}(\vec{x}) = \sqrt{\frac{2}{3}tr(\bar{\bar{\varepsilon}}(\vec{x})^2)} \tag{8}$$

This parameter is independent of the base in which the strain tensor is expressed and allows quantification of the strain amplitude in the case of incompressible media such as biological samples. To illustrate this, we present strain amplitude maps calculated for a human breast tissue in Figure 3A,B, which highlight mechanical differences.

- **Strain anisotropy**

    The full 3D strain tensor also allows calculation of eigenvalues and eigenvectors of the strain tensor that gives access to information on the mechanical anisotropy of the sample. One should keep in mind that, since the compression of the sample is generally performed along the optical axis, the local stress field is inherently anisotropic. In order to display and to quantify strain anisotropy, we mainly use two parameters. The first parameter is the principal strain direction. For each pixel, the eigenvector corresponding to the largest eigenvalue of the strain tensor is calculated. This vector gives the principal strain direction. For example, Figure 3C is a cross-sectional slice from the volumetric FF-OCT image of a rat heart. In the FF-OCT image, it is difficult to evaluate the fiber direction, and the two different fiber orientations are not visible. However, if the projection angle of the principal strain direction vector on the plane perpendicular to the compression (see Figure 3D) is plotted, it is clear that there are two different fiber orientations.

    Another parameter used to quantify the strain anisotropy is the fractional anisotropy *FA*. In a similar manner to classical diffusion tensor imaging (DTI) in MRI [93,94] or ultrasound [95], we define the fractional anisotropy as:

$$FA = \sqrt{\frac{3}{2}} \frac{\sqrt{(\varepsilon_1 - \bar{\varepsilon})^2 + (\varepsilon_2 - \bar{\varepsilon})^2 + (\varepsilon_3 - \bar{\varepsilon})^2}}{\sqrt{\varepsilon_1^2 + \varepsilon_2^2 + \varepsilon_3^2}} \tag{9}$$

    where ε_1, ε_2 and ε_3 are the strain tensor eigenvalues and $\bar{\varepsilon}$ is the mean of the absolute value of the strain tensor eigenvalues.

    Using this parameter, it is possible to differentiate isotropic from anisotropic strain. In Figure 3E,G, we show that computing the fractional anisotropy in isotropic and anisotropic polyvinyl alcohol (PVA) polymer gels [96] enables discrimination of the two samples, in a case where FF-OCT fails to detect a difference.

In parallel to our work on 3D digital image correlation, we adapt the *"tactile"* approaches developed by Kennedy et al [97] for standard OCT. The principle is to use a thin layer of polymer deposited on the imaging windows to mimic manual palpation at the micrometer scale [98]. Knowing the geometry and the mechanical properties of the polymer layer, it is possible to directly measure the stress at the surface of the sample by measuring the displacement at the surface of the polymer layer. With FF-OCT, the 2D *en face* displacement can be measured by taking the difference between two phase images taken at the surface of the polymer before and after the compression. This method allows a fast stress measurement at the surface of the sample simultaneously with FF-OCT imaging. This method is particularly interesting in the case of micro-biopsy where different tissues are usually juxtaposed. Figure 3I,J show a measurement on a biopsy of bovine kidney. At the micrometer scale, the stress map shows different mechanical behaviors for the muscle fibers and the kidney at the same stress level. The higher pressure at the muscle fiber level shows that this part of the sample is stiffer than the kidney area because the polymer is more deformed by this tissue. According to an experiment that consists of the observation of the tissue behavior during the compression, the pressure is negative at the kidney level because the tissue tends to move sideways. The negative pressure where there is no tissue can be explained by the fact that the silicone is incompressible and needs to expand somewhere to conserve its total volume. Finally, this picture shows the high resolution of this method because each muscle fiber with 20 μm diameter can be distinguished by comparing the FF-OCT image and the stress map.

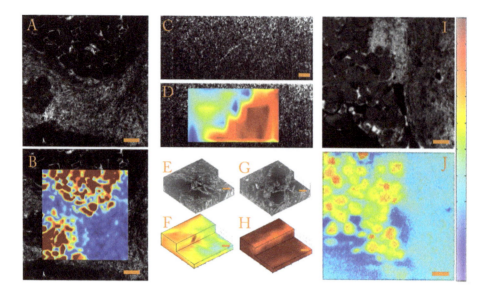

Figure 3. FF-OCT and biomechanical contrast in various tissues obtained with static elastography. **Panels A and B** show a standard four-phases FF-OCT of an ex vivo human breast tissue and the corresponding strain map (equivalent von Mises strain) after sample compression. As expected, a greater deformation can be observed around the adipose inclusions (black domains in panel A) when at the surrounding fibrous breast tissue. The color bar corresponds to a strain in percent from 0 for blue to 10 for red. **Panel C** corresponds to a FF-OCT cross-sectional image of an ex vivo rat heart. In **Panel D**, we show the corresponding angle of the principal strain direction. The color bar corresponds to the angle in degree from 0 for blue to 90 for red. **Panels E and G** respectively correspond to 3D volumetric FF-OCT images of a mechanically isotropic and anisotropic polyvinyl alcohol (PVA) polymers, and **panels F and H** correspond to their respective fractional anisotropy maps. The color bar corresponds to the fractional anisotropy from 0 for blue to 0.7 for red. **Panels I and J** finally show a standard FF-OCT image of a biopsy of bovine kidney, and its associated stress map obtained with the "tactile" approach. The color coded stress map varies from −0.5 kPa in blue to 0.4 kPa in red. All the orange scale bars correspond to 100 μm.

FF-OCT static elastography measurements can be performed with a standard FF-OCT system, such as the LLTech commercial system [37], or can be associated with other multimodal FF-OCT systems to diversify the types of contrast available. However, this technique can be computationally intensive, and requires the calculation of two 3D stacks and a physical compression of the sample. In total, the acquisition time can last for one to a few minutes, and the processing time for several hours. FF-OCT static elastography measurements only require a static FF-OCT signal to correlate, so penetration depth is only limited by FF-OCT capabilities.

4.3. Transient Elastography Based on Shear Wave Imaging

Transient elastography based on shear wave imaging is the second approach used to combine elastography with FF-OCT [38]. Transient elastography based on shear wave imaging was introduced in OCT by Razani et al. [99]. As shown in Equation (6), a direct measurement of the local shear wave speed gives direct access to a quantitative value of Young's modulus. Dynamic elastography methods involve two steps: first, a shear wave is locally sent onto the sample; then, the shear wave propagation is recorded. We were inspired by a technique named supersonic shear wave imaging, partly developed in our lab, which uses the ultrasound radiation force to create short shear impulses and ultrafast ultrasound imaging to image the shear wave propagation [100,101]. We adapted this method to FF-OCT by generating shear waves using an ultrasound system and by recording the generated wave using a FF-OCT system operated with a high speed camera that can run at up to 30 kHz for a 512 × 512 pixel field of view [38]. Shear waves were induced using a programmable ultrasound scanner (Aixplorer®, Supersonic Imaging, Aix-en-Provence, France), and a 15-MHz ultrasound linear array (128 elements, pitch 100 µm, elevation focus at 12 mm, Vermon, France), as described in A. Nahas et al. [38]. This system focuses ultrasonic beams for tens of microseconds in the sample, which create a local radiation force, localized at the ultrasound focal spot and oriented along the beam axis. In response to this force, the tissue relaxation generates a shear wave, polarized along the ultrasonic beam and propagating transversely [90,102]. To increase the amplitude of the radiated shear wave, an original solution consisting of successively focusing the ultrasonic beams at different depths along the ultrasonic path is performed [38,102]. If the shear source is moved at a higher speed than the speed of the radiated shear wave, the source induces constructive interference along a Mach cone between all the generated shear waves, and the resulting wave can propagate with minimized diffraction over a large area. For short ultrasonic bursts (shorter than one millisecond), the spatial distribution of large tissue displacements initially corresponds to the acoustic intensity distribution, i.e., a cylinder with a lateral extension of 200 µm and an axial extension of about 1 mm in this configuration. The shear wave location can further be swept transversely by electronically focusing the ultrasound beam at different lateral locations, without any mechanical translation involved. The pressure levels involved are within the limits set by the Food and Drug Administration (FDA) for the application of diagnostic ultrasound in several organs.

By taking advantage of the interferometric sensitivity of FF-OCT to small axial displacement, we can further detect the propagation of the shear wave with high sensitivity. Unfortunately, high speed cameras do not offer large full well capacities (and it would be hard to generate hundreds of thousands of photoelectrons per pixel at 30 KHz), so our axial sensitivity is about 10 times lower than in our standard systems. Nevertheless, we could successfully follow shear waves in a sample and calculate their local speeds, by computing the temporal cross-correlation between neighboring pixels. One such propagation in an ex vivo brain tissue is shown in Figure 4. To better capture the propagation in time, we did not modulate the reference piezo, but only recorded direct images. However, we get rid of the incoherent intensity captured by the camera by subtracting the direct images to a few initial frames captured before the generation of the shear wave. The signal of each pixel is therefore directly proportional to the cosine of the phase:

$$I_{diff} = I_0.a_\delta[cos(\frac{2\pi\delta}{\lambda_0} + \varphi(t)) - cos(\frac{2\pi\delta}{\lambda_0} + \varphi_{ini})] \tag{10}$$

where a_δ is linked to the backscattered wave amplitude interfering with the reference signal; δ is the path length difference at the considered pixel; λ_0 the central wavelength; $\varphi(t)$ the relative phase that depends on time; and φ_{ini} the initial phase. In Figure 4, we can see the wavefront of the shear wave crossing the field of view. With this method, we measured a shear wave speed of 0, 64 m·s^{-1}, corresponding to a Young's modulus of 1.25 kPa, which agrees well with the literature [103,104].

We should add that although this dynamic measurement seems quite promising, it is performed here in an ideal case, since the brain is one of the softest tissues of an organism, and is quite reflective. In other cases, dynamic elastography FF-OCT can be challenging mainly due to current technological limitations. Indeed, we could not find any high sensitivity and high speed camera with parameters which would be sufficient to accurately follow shear waves in most tissues. However, since the shear wave source is controlled here, it should be possible to average several measurements to increase the sensitivity, or to perform a stroboscopic measurement by shifting the time delay between the shear wave generation and therefore measure the wave propagation with lower speed cameras. Ultimately, a shear speed can be calculated from only three measurements, so mechanical maps can be computed almost at video rate with dynamic elastography FF-OCT, and has been demonstrated at several thousand frames per second here. Here again, the penetration depth is only limited by FF-OCT characteristics. However, due to the low sensitivity provided by fast cameras, the penetration depth is, in practice, limited to 50 μm.

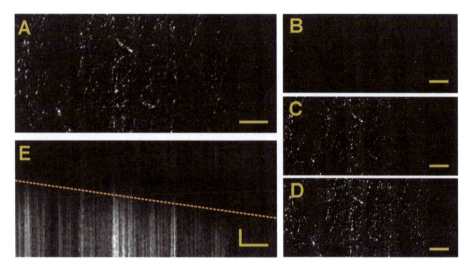

Figure 4. Dynamic elastography performed with FF-OCT by measuring shear wave propagation. This figure displays different FF-OCT images of an ex vivo rat brain recorded at 30,000 Hz with a field of view of 471 × 235 μm^2, and their evolution during shear wave propagation. **Panel A** shows a standard four-phase FF-OCT amplitude image of the sample. **Panels B to D** show the evolution of direct images after subtraction of the incoherent intensity, during shear wave propagation. Panels B to D have been respectively acquired at $t = 0$ ms, $t = 4$ ms, and $t = 5$ ms after the shear wave emission. Finally, **Panel E** shows a plot of the temporal evolution of the average image intensity along the y axis. The horizontal scale bars represent 50 μm and the vertical scale bar 1 ms.

To conclude this section, we have shown that FF-OCT can be coupled with elastography to measure the local mechanical parameters of a tissue. In terms of coupling with biomechanical contrast, FF-OCT and standard OCT measurements are not very different, since they rely on similar techniques to generate the mechanical contrast. Although FF-OCT-based mechanical measurements should theoretically have an isotropic 1 μm resolution, we could never measure any relevant differences at this scale. Additionally, the mechanical models used to link local deformations with mechanical stiffness are mesoscopic models, whose validity can become questionable when decreasing the resolution below 10 μm.

We should add that as the deformation is applied along the optical axis, it seems more natural to look at its propagation in the transverse plane, especially when looking at shear waves. Therefore, FF-OCT, or any other *en face* OCT techniques should be well adapted for such dynamic mechanical measurements. However, mainly due to current technological limitations, these measurements are still difficult to perform in FF-OCT.

Finally, a promising technique, named passive elastography [105,106], allows one to measure dynamic mechanical parameters in the case where the sampling rate that is lower than the shear wave propagation has recently been demonstrated in spectral domain OCT in our lab [107]. Passive elastography is well adapted to FF-OCT measurements, and is currently under development in our group.

5. Dynamic FF-OCT

In this section, we will briefly review D-FF-OCT, including details of available contrasts and structures that can be revealed, and we present a few potential applications.

We developed dynamic full-field OCT (D-FF-OCT), inspired by the work on quasi elastic dynamic light scattering and other dynamic measurements to measure biological activity [33,108–113]. We therefore wondered whether the intracellular movements and vibrations can be important enough to modulate the optical path length of our interferometer by themselves, even without reference piezo modulation. Interestingly, we found out that in fresh biological samples, calculating the standard deviation of the interferometric signal over time gives a substantially different contrast in comparison with the regular FF-OCT signal. We discovered later that this dynamic signal was related to cellular metabolism [39]. Interestingly, D-FF-OCT can access motility contrast in 3D, taking advantage of the FF-OCT low coherence to only select the phase variations at a given plane.

Now, what we generally name D-FF-OCT is a set of mathematical techniques to analyze the temporal fluctuations of the FF-OCT signal. Trying to understand and quantify these fluctuations, we were inspired and influenced by similar measurements performed with more standard OCT [4,114–116], other interferometric techniques [33,111], or general dynamic techniques [117].

The new contrast revealed by D-FF-OCT originates from a different interpretation of the temporal fluctuations of the signal. In regular FF-OCT, the temporal fluctuations are averaged out to increase the signal-to-noise ratio of strongly backscattering structures. On the contrary, in the first versions of D-FF-OCT [39], we emphasized the amplitude of fluctuations of the signal by calculating the standard deviation, as described previously. This amplitude of fluctuations not only depends on the backscattering strength, but also on the scatterers dynamics inside a given voxel. An interesting layout of D-FF-OCT is to remove the piezo modulation, since the incoherent light varies less, and more slowly than the interference term. In such a configuration, the backscattering amplitude and the phase fluctuation amplitude are coupled in the final image. However, we found out that D-FF-OCT images calculated from four phase signals are often less contrasted, so we often stop the piezo movement and only acquire direct images.

To extract more information from intracellular dynamics, we now try to emphasize not only the amplitude of fluctuations, but also the timescales of variations. We measured the interferometric signal autocorrelation to extract two characteristic times, and could measure the metabolic-related effects of a drug that depolymerizes F-actin [40]. Similarly, we also emphasize fluctuation timescales by calculating the Fourier transform of the time signal at a given pixel, or set of pixels. Then, we integrate the signals between different frequency bands to measure the dynamic signal originating from different timescales. We displayed the values by using an HSV color space, with the central frequency as the hue, the spectral width as the saturation and the amplitude of fluctuations as the value for each pixel of an image. This method is illustrated in Figure 5B,E, and enables segregation of cell populations based not only on their shapes but also on their fluctuation dynamics. We can distinguish cells with highly dynamic membranes, such as red blood cells [118] or lymphocytes [119] from slowly varying cells.

Finally, both our group and H. Ammari et al. [120] also used singular value decomposition (SVD) to better separate large structures that are slowly varying from smaller structures that are varying more rapidly. SVD can also eliminate slow movements from all the samples, similarly to what was obtained in ultrasound [121].

Although the location of cell bodies can often be inferred from the FF-OCT image in regions with low signal, D-FF-OCT can distinguish between a cell and a local oedema (which appears black in D-FF-OCT) for example, and gives a better precision of the cell contour. For example, in Figure 6 and in the cortex region of coronal brain slices, the position of neuron somas can be inferred from the FF-OCT image, but the dynamic image can better reveal individual neuron shapes and sizes, from which the neuron type might be inferred (pyramidal or stellate neurons). Ultimately, the identification of cortical layers can be performed in situ.

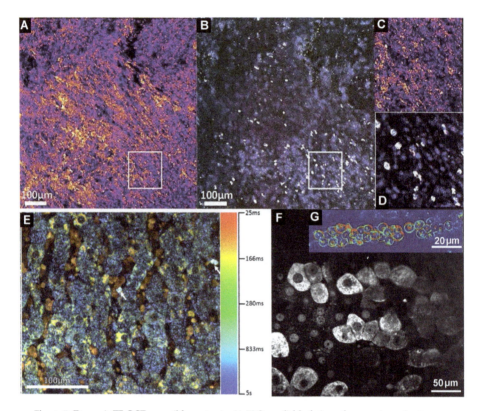

Figure 5. Dynamic FF-OCT accessible contrasts. (**A,B**) Same field of view of an ex vivo subcutaneous pancreatic tumor respectively with FF-OCT and D-FF-OCT. Panel A shows a fibrous environment dense in collagen fibers without specific cues of cancerous cells. On the other hand, panel B identifies slowly varying cell bodies (in blue), and rapidly varying immune cells (bright white cells); (**C,D**) Zoom corresponding to the white areas of panels A and B respectively; (**E**) D-FF-OCT image of a fresh rat liver with 10X 0.3 NA objectives, based on a frequency analysis to compute the color. D-FF-OCT can reveal hepatocyte cytoplasms, and nuclei (black region at the center of cells), organized along stationary capillaries, characterized by the presence of rapidly varying red blood cells trapped inside; (**F**) D-FF-OCT image of a similar fresh rat liver with 40X 0.8 NA objectives. The increased resolution allows to better capture intracellular dynamics, and to capture some dynamic signal inside the nucleus; (**G**) High resolution image of red blood cells trapped in a capillary of a fresh rat retina, acquired with the 40X 0.8 objectives. The better resolution allows to capture the enhanced dynamics localized at the red blood cells membrane.

In addition, D-FF-OCT may play an important role as a diagnostic tool as it can detect nuclei in large cells (Figures 5E,F and 6). Nuclear size and shape is an important biomarker of cancer, with higher nuclear to cytoplasmic ratios in tumor cells [122,123]. The nuclei mainly appear black in D-FF-OCT, even though the inner dynamics of the nuclei are significant and usually faster than cytoplasm dynamics [110]. Our hypothesis for the dark appearance of the nuclei is that nuclei are more densely packed than cytoplasm, and appear somehow homogeneous at a 1 μm resolution. Inside a homogenous nucleus, there should be no backscattering except on its contour and no dynamics to image, which can explain why nuclei appear black in both FF-OCT and D-FF-OCT. Interestingly, imaging cells with higher NA objectives, and with a transverse resolution below 500 nm, enables us to recover some dynamic signal inside nuclei (See Figure 5F).

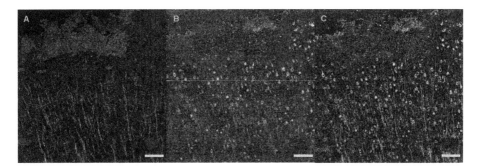

Figure 6. D-FF-OCT as a histology tool for the brain. This figure presents a combination of FF-OCT (**Panel A**) and D-FF-OCT (**Panel B**) in a thick coronal section of fresh ex vivo rat brain. This image has been acquired at a few dozen microns below the section surface and presents the first cellular layers of the cortex. It reveals both myelinated axons in FF-OCT, as well as the neuronal cell bodies in D-FF-OCT. **Panel C** shows a frequency analysis performed in D-FF-OCT, which reveals highly dynamic nuclei in red, neuronal cytoplasms with intermediate dynamics (mostly in green) and mostly static myelin fibers in blue. The scale bar represents 60 μm.

Another interesting D-FF-OCT feature for diagnosis is its ability to detect red blood cells and immune cells (See Figure 5B,D,E). Indeed, in ex vivo tissues, the red blood cells are trapped inside the capillaries, and can be imaged even at low frame rates. Due to their strong membrane fluctuations [118,124], they produce an intense and fast dynamic signal (See Figure 5E,G) on their edges. Similarly, immune cells are known for their intense motility as they are constantly exploring their environment and changing their shapes, as their protrusions are looking for abnormal markers to suppress [119]. Interestingly, the nature of the immune cell might be inferred from their motility coefficient [119]. Firstly, D-FF-OCT can therefore detect immune cells infiltration inside the tumors, which seems to be correlated with the tumor grade [125,126]. Secondly, FF-OCT can usually detect capillaries, but cannot make the difference between a blood vessel and a lymphatic vessel. Thanks to the additional information from D-FF-OCT, and by detecting either red blood cells, or lymphocytes in a vessel, a more accurate angiogram and lymphogram could be established in cancerous tissues. Interestingly, correlations between an increase of the blood vessel density and tumor grades have been well established for a long time ([127,128] among many others). Similar results with the lymphangiogram have been obtained [129]. Quite generally, denser angiograms and lymphangiograms increase the risk of metastasis.

D-FF-OCT is promising, when compared with digital holography techniques [33], dynamic light scattering techniques [110], or quantitative phase imaging [124,130,131], as it provides optical sectioning, and can access a motility-based contrast in 3D in scattering tissues with a subcellular resolution. Additionally, compared with dynamic OCT techniques [116,132], we take advantage of the higher spatial resolution to capture subcellular variations, which further enable the cells to be revealed. To our knowledge, the contrast revealed by D-FF-OCT is therefore quite unique. Similar dynamic measurements in OCT mainly focus on the blood flow velocity [116,133,134], even with high transverse resolution [135] similar to ours, but, to our knowledge, they never directly show dynamic signals coming from intracellular dynamics. We are wondering whether it comes from a loss in the signal collection due to the use of lower NA illumination, from a difference in the algorithms used (if only blood flow was expected), or from a difference in the exposure time per pixel.

In terms of multimodality, D-FF-OCT is attractive, as it does not require any modification from a regular FF-OCT. It can be associated with any other dual modality system to provide three different contrasts [41], as will be pointed out in the next section. D-FF-OCT is highly complementary to standard FF-OCT especially in fibrous environments, as fibers are highly backscattering structures and often hide cells entangled inside the matrix. However, fibers are usually still, while cells have intracellular activity that produces a highly dynamic signal. Hence, D-FF-OCT can easily reveal these cells, and suppress the stationary signal from the fibers. Nevertheless, one should keep in mind that D-FF-OCT significantly lowers the acquisition rates of FF-OCT microscopy. Indeed, even though each image is acquired at the camera frame rate, significant cellular dynamics occur in the second timescale, which is not very compressible because it is mainly sample dependent. In practice, our fastest D-FF-OCT measurements were at one to two seconds per plane that therefore expand to several minutes for a 3D volume. As for penetration depth, D-FF-OCT shows similar features to FF-OCT, except that the D-FF-OCT signal is often lost before the FF-OCT signal when going in depth, mainly because D-FF-OCT is generally weaker. Nevertheless, we succeeded in imaging D-FF-OCT signals at 100 μm in brain slices, or even through the entire retina, going through up to 200 μm.

6. Full-Field OCT and Fluorescence

In principle, usual histology staining and fluorescence techniques are quite similar, except that fluorescence techniques use fluorescence probes instead of absorbing probes in histology. Some fluorescence techniques have been part of the histology toolkit for decades [136], such as immunofluorescence, which aims at detecting the interaction of a labeled antibody with a structure of interest [137], or fluorescence in situ hybridization (FISH) ([138,139], Chapter 6), which aims at detecting the interaction of a labeled nucleotide sequence with its specific complementary sequence in a cell. More recently, confocal microscopy [140] and two photon microscopy [141] have been used in order to detect fluorescent probes in depth in tissues. Interestingly, thanks to the optical sectioning provided by these techniques, it has become possible to follow some fluorescent structures in ex vivo tissues, and in vivo [142,143].

However, fluorescence techniques still require fluorescent labels, that need to be injected in the tissue. It can take a long time before the label diffuses in depth inside a tissue. Moreover, and similarly to classic histology stains, these labels only tag a few structures of interest, and it might be hard to understand the organization of these structures inside their microenvironment. Multiple labeling can be used to tag several structures, but it increases the complexity, and the cost of such experiments.

We can also add that usually, confocal and two photon microscopy can only access shallow depth inside a tissue, down to a few hundred μm at best. OCT techniques on the other hand can access deeper regions down to one millimeter, since the signal is often enhanced by the reference arm. Recently, clearing techniques [144,145] with low tissue distortions have been developed in order to increase tissue transparency, and to image through the entire tissue. OCT can also benefit from these approaches [146], even if we would expect the intrinsic signal to be reduced in OCT, since clearing tends to remove lipid structures and to average the refractive index fluctuations. However, it can help to detect exogenous OCT contrast agent at important depths, such as metallic nanoparticles. Here again, clearing processes are often long, and can induce tissue distortions.

Fluorescence and OCT techniques are quite complementary, and multimodal systems combining the two can help with diagnosis, as we will illustrate throughout this section. Many such multimodal setups have been developed with both scanning OCT approaches [5,147] and full-field approaches, using either a flip mirror to switch from one modality to the other [18,148], or a simultaneous approach [41,42]. If all these multimodal approaches share a common interest, scanning OCT approaches are more easily combined with scanning fluorescence approaches, such as confocal or two photon microscopies, whereas FF-OCT systems have been developed in combination with widefield fluorescence techniques, such as structured illumination microscopy (SIM). SIM acquisition speed intrinsically depends on the fluorescence camera frame rate divided by the number of grid modulation steps. The SIM acquisition rate can go up to 100 frames per second in our measurements, even though the usual frame rate limit comes from the fluorophores' brightness, and from the minimal exposure time required to produce a decent image. Interestingly, simultaneous combination of FF-OCT and SIM [41,42] enable the parallel acquisition of both modalities, so images with both modalities can be acquired at more than 100 frames per second. However, SIM is not very robust to aberrations and multiple scattering, and can hardly penetrate inside scattering samples after 30 to 50 μm. Nevertheless, in mostly transparent samples, such as cornea or retina, with only a few structures labeled, we could perform multimodal FF-OCT and SIM images through the entire sample.

The first interest of combining OCT with fluorescence is to identify structures that are captured in OCT, in order to better understand and characterize OCT contrast [62]. The second interest is to use fluorescence to reveal some structures, proteins, or ultimately any structure that cannot be captured by OCT, therefore combining fluorescence molecular contrast with structural contrast from OCT. It can help to understand how a structure of interest integrates and evolves inside its micro environment, such as neurons in the intestine [149], or to find the location of stem cells in the cornea [12]. Figure 7 shows the example of imaging a tumorous intestine tissue, in which cell nuclei are labeled with fluorescence. This image shows the complementary contrast that can be obtained with both modalities. Additionally, with both sources of information, we can localize the cells on the contours of organized structures, named intestine villi, indicating that this field of view has been acquired in a healthy region of the cancerous tissue.

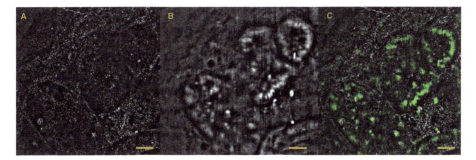

Figure 7. Revealing cells in tumors with FF-OCT/SIM. A fixed tumorous tissue of intestine is imaged in FF-OCT (**A**); in fluorescence (**B**); and the overlay is displayed in **Panel C**. Fluorescence reveals cell nuclei in the tissue. These images show that the cells are located within a well defined structure captured by FF-OCT, indicating that we are imaging in a healthy part of the tissue. The scale bars represent 40 μm.

A final interest, that is often neglected, especially in the context of histology, is that fluorescence enables dynamic measurements that can probe the tissue biochemistry and its evolution, using specific probes. For examples, fluorescence measurements can follow the electrical activity of cells by quantitatively measuring their intracellular calcium concentration [150,151], or directly measure the membrane potential [152]. It can be useful to assess the viability of a cortical or retinal explant from its spontaneous activity, or can assess the endocrine pancreatic function, as insulin production is triggered by calcium uptake. Fluorescence can also directly measure cell function by measuring the intracellular ATP concentration [153], and can specifically detect apoptotic cells, with the ClickIt TUNEL assay (ThermoFisher). This can be of major importance to study the efficiency of pharmaceuticals on a cancerous tissue, as the cellular recovery of apoptosis within the tumor is usually a promising sign of a treatment's success.

Figure 8 shows the result of the ClickIt TUNEL assay (from ThermoFisher Scientific [154]) on a fresh liver tissue. The tissue is first purposely excised with a corner shape to elicit and follow the cell death propagation from the margin to the center of the tissue. After several minutes, the tissue is imaged with FF-OCT and D-FF-OCT, and an inactive region (dark region from the tissue margin) can be detected with D-FF-OCT (panels A to C). Directly after the imaging, the tissue is fixed, in order to start the TUNEL assay and detect the apoptotic regions of this tissue. Unfortunately, this assay requires fixed tissues, so we cannot overlay a D-FF-OCT, and a fluorescence image. However, we could detect that almost all cells in a 200 μm region from the tissue border were in apoptosis (panels D to F). The fluorescence signal started decreasing as we went further from the tissue margin, and is almost equal to zero at the center of the tissue. Figure 8 tends to show that apoptotic cells do not produce a dynamic contrast, at least for cells in apoptosis for a long time. Surprisingly, this indirect result is different from what was obtained with OCT measurements [155], where the authors showed a decrease of the speckle decorrelation time (i.e., an increase of the intracellular activity) in apoptotic cells. Explanations for this result may be the following: In contrast to G.Farhat et al. [155], we used a *brute force* method to generate apoptosis, as we physically cut the sample, and imaged close to the tissue edge. It is likely that the cells in the dark D-FF-OCT region are in late stages of apoptosis. Moreover, since cells in apoptosis tend to dilute and homogenize their intracellular material [156], it should lower the backscattered signal from the cell to a level that may be beneath our system detection limit, thus creating a dark D-FF-OCT image, even if the dynamics have increased.

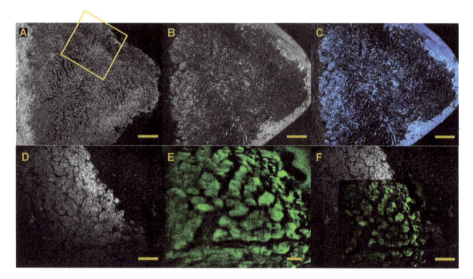

Figure 8. Imaging apoptosis with multimodal FF-OCT systems. This figure presents the imaging of the edge of an ex vivo liver. Here, the liver has been excised with a triangular edge to illustrate the propagation of cellular death from the excised edge. First, a large field FF-OCT and D-FF-OCT images have been acquired with the commercial LLTech system (**A**, **B** and overlay in **C**) 30 min after the excision. Then, the tissue has been fixed, and labeled with ClickIt Plus TUNEL Alexa 488 assay (ThermoFisher) to reveal apoptotic cells; **Panels D**, **E** and **F** respectively show the FF-OCT and fluorescence signals (and the overlay) acquired from the part of the tissue in the yellow box in panel A. In this region close to the excised edge, almost all cells are apoptotic, as suggested by the low level of dynamic signal in panel B. Scale bars for panels A, B, and C represent 125 μm, and scale bars in D to F represent 40 μm.

7. Comparison with Other Popular Novel Microscopies

As previously pointed out, although fixation and labeling protocols have been greatly improved these past years in histology, this field could still benefit from the important technological developments that optical imaging has known over the past twenty years. So far, we compared FF-OCT and its multimodal versions to standard histology procedures. However, to be fair, we should also compare them to other recently developed optical techniques, such as multiphoton microscopies, and coherent anti-Stokes Raman scattering (CARS) microscopy, that also are label-free and enable optical sectioning. Similarly to OCT, these techniques rely on the intrinsic properties of tissues to generate contrast. They both can provide optical sectioning, since they rely on the combination of two or three photons, that can statistically only happen at the focal plane where the power density is high enough. Second harmonic generation (SHG) is measured in dense non-centrosymmetric structures, such as microtubule bundles, and collagen I fibers [157]. Third harmonic generation (THG) is measured at interfaces displaying high index mismatch, such as cell or nuclear contours, and lipid droplets [158]. Both techniques, especially when combined, are therefore useful to study 3D tissue organization, e.g., in the cornea [159], or in developing embryos [160]. The association of the two techniques gives a contrast similar to FF-OCT, but non-linear microscopes are a little more sensitive and specific than the FF-OCT signal, and can be naturally combined with two-photon fluorescence microscopy [161]. The transverse resolution is often higher than FF-OCT, as high NA objectives are required to generate a high power density at the focal point. We can add that SHG has already been extensively used to assess tissue disorganization, especially in the context of liver fibrosis [162]. A commercial SHG microscope for histology is already available at HistoIndex [163].

CARS [164], and its hyperspectral version [165], are other promising non-linear techniques for histology, which reveal a high density of either a single or multiple molecular vibrations inside a sample. Interestingly, CARS is one of the only techniques that is both label-free, and specific at the molecular level, as vibration spectra of given structures can be measured, and computationally extracted [166]. It is therefore one of the most promising label-free techniques today for quantification of tissue structure, and molecular content. Unfortunately, to our knowledge, there is no commercial CARS system available, and CARS systems often require careful and time consuming alignment steps.

Finally, we can add that all these techniques are based on non-linear processes that require similar kinds of sources, i.e., femtosecond lasers, and can be ingeniously combined all together from a single laser source [167]. However, such non-linear processes are quite weak. They typically display a low signal-to-noise ratio, especially in dense tissues, and often require long exposure times to generate a measurable signal. Combined with high NA objectives, the measurable field of view within one hour of imaging is fairly limited. Additionally, femtosecond lasers are expensive, especially compared with FF-OCT sources that are simple thermal light or LEDs. Finally, as non-linear processes, SHG, THG, and CARS signals are mainly forward scattered, so they often have to be used within thin tissue sections (a few hundred microns to 1 mm). We can nevertheless note that multiple scattering happening in highly scattering tissues allows the observation of non-linear processes in an epi configuration [164].

FF-OCT appears better suited than competing techniques for fast imaging of large tissue volumes in the operating room, to complement frozen-section histology procedures.

8. Conclusions and Perspectives

Full-field optical coherence tomography is a modified version of OCT that can access submicrometric resolution deep inside a sample. Similarly to OCT, it can provide label-free 3D imaging of ex vivo tissues, and can perform measurements similar to histology in terms of contrast and resolution. The interest of FF-OCT lies in its ability to obtain 3D images similar to histology with almost no tissue preparation, which can reduce slicing, staining, and fixation artifacts. FF-OCT can also significantly shorten the biopsy-to-diagnosis time, down to a few minutes. Unfortunately, FF-OCT diagnosis scores are not yet 100% when compared with histology. In this review, we highlighted three promising recent multimodal systems combining FF-OCT structural contrast with other specific contrasts such as mechanical, motility, and molecular contrasts, providing hope that FF-OCT diagnosis scores will increase in the future, and might even ultimately perform better than histology. We need to add that despite our choice here, there are other promising multimodal systems that could be of interest for diagnosis improvement. For example, polarization FF-OCT [21] can be used to increase the sensitivity to fibers, and should be able to detect fiber disorganizations, similarly to OCT measurements, e.g., that could detect polarization changes caused by age-related macular degeneration (AMD) or glaucoma [168]. Multicolor FF-OCT [22] might also help the detection of subtle wavelength-dependent scattering changes, related to tumor malignancy [69].

Despite all these exciting developments, and the possibility to combine many different modalities in a single optical system, we have to emphasize that a compromise will have to be found between speed, processing time, compactness, penetration depth and the number of contrasts available, especially when FF-OCT is used during surgery. We believe that a better characterization of FF-OCT (and D-FF-OCT) contrasts that would be performed specifically for each tissue would help to increase FF-OCT diagnosis ability, without even changing the optical design of the microscope.

Beyond the scope of histology, we hope that the development of such multimodal systems will increase the potential interest in FF-OCT, not only as an ex vivo diagnosis tool, but also as a tool to answer complex biological questions. We showed in this article that FF-OCT reveals a label-free original contrast, and can be used either in ex vivo tissues, or in vivo in small animals. If scanning OCT techniques were now to be extensively used in vivo and in clinics, FF-OCT would suffer from the slow speed of 2D detectors that do not allow a dynamic correction of the sample position, which could cancel the sample motion; in vivo implementation is therefore challenging. However, a few in vivo experiments have been successfully performed in paralyzed or anesthetized small animals [35,61], and in human skins pressed against an imaging window. We recently successfully performed a simultaneous FF-OCT, D-FF-OCT, and fluorescence measurement in living zebrafish larvae. We illustrate our ability to obtain a motility contrast in vivo, even in the presence of blood flow in Figure 9. The dynamic signal is saturated in the blood vessels, and below, but cells can be detected in between the vessels. Additionally, an important effort is being made to enable FF-OCT measurements in patients, with the recent development of two adaptive optics FF-OCT systems [169,170] for retinal imaging, a FF-OCT system coupled with an OCT system to dynamically compensate for eye motion for cornea and retina assessment, and a dark field FF-OCT system [171] to record internal fingerprints. Additionally, our group recently developed an *en face* matrix based approach of OCT, which allows to increase the penetration depth by a factor of two [172], and could be advantageous for in vivo experiments.

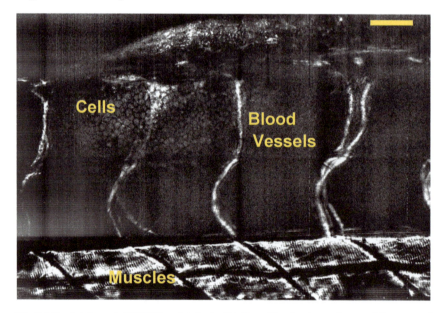

Figure 9. In vivo dynamic FF-OCT in a 2 dpf zebrafish larva. Despite strong phase variations caused by the blood flow, which generates a strong dynamic signal, some cells can still be revealed in between the capillaries. The scale bar represents 40 μm.

Finally, we are now interested in the development of functional imaging in FF-OCT, in fresh tissues, or in living animals, either by measuring changes in the blood flow, similarly to what has been achieved in OCT [132,173], or by directly measuring changes in the tissue biomechanics, or cellular activity. We believe that diagnosis capabilities will further benefit from this research, since it might soon be possible not only to detect static structural changes in a tissue, but also to detect how it reacts to different stimuli, or different drugs.

Acknowledgments: The authors want to thank Kate Grieve for her expertise on cornea and retina experiments, and for revising the manuscript. We acknowledge the important help from Thu Mai Nguyen on tissue biomechanics, and passive elastography. We thank Peng Xiao for his help to understand adaptive optics FF-OCT and the role of incoherent illumination for FF-OCT. We are also grateful for the help of Jena Sternberg and Claire Wyart for discussion and preparation of the live zebrafish larvae experiments. We thank Ralitza Stavena for preparations and discussions about tumorous intestine samples. We finally express our gratitude to the team of Diana Zala, for fresh tissue preparation, and the numerous discussions. This work was supported by grants from the European Research Council SYNERGY Grant scheme (HELMHOLTZ, ERC Grant Agreement # 610110; European Research Council, Europe) and LABEX WIFI (Laboratory of Excellence ANR-10-LABX-24) as part of the French Program 'Investments for the Future' under reference ANR-10-IDEX-0001-02 PSL.

Conflicts of Interest: This article has been written independently from LLTech. However, Claude Boccara is the CSO and co-founder of LLTech, and Clément Apelian has a PhD scholarship funded by LLTech.

References

1. Drexler, W.; Fujimoto, J.G. *Optical Coherence Tomography: Technology and Applications*, 2nd ed.; Springer: Gewerbestrasse, Switzerland, 2015.
2. Zaitsev, V.Y.; Vitkin, I.A.; Matveev, L.A.; Gelikonov, V.M.; Matveyev, A.L.; Gelikonov, G.V. Recent Trends in Multimodal Optical Coherence Tomography. II. The Correlation-Stability Approach in OCT Elastography and Methods for Visualization of Microcirculation. *Radiophys. Quant. Electron.* **2014**, *57*, 210–225.
3. Zaitsev, V.Y.; Gelikonov, V.M.; Matveev, L.A.; Gelikonov, G.V.; Matveyev, A.L.; Shilyagin, P.A.; Vitkin, I.A. Recent trends in multimodal optical coherence tomography. I. Polarization-sensitive oct and conventional approaches to OCT elastography. *Radiophys. Quant. Electron.* **2014**, *57*, 52–66.
4. Oldenburg, A.L.; Chhetri, R.K.; Cooper, J.M.; Wu, W.C.; Troester, M.A.; Tracy, J.B. Motility-, autocorrelation-, and polarization-sensitive optical coherence tomography discriminates cells and gold nanorods within 3D tissue cultures. *Opt. Lett.* **2013**, *38*, 2923–2926.
5. Beaurepaire, E.; Moreaux, L.; Amblard, F.; Mertz, J. Combined scanning optical coherence and two-photon-excited fluorescence microscopy. *Opt. Lett.* **1999**, *24*, 969–971.
6. Beaurepaire, E.; Boccara, A.C.; Lebec, M.; Blanchot, L.; Saint-Jalmes, H. Full-field optical coherence microscopy. *Opt. Lett.* **1998**, *23*, 244–246.
7. Dubois, A.; Vabre, L.; Boccara, A.C.; Beaurepaire, E. High-resolution full-field optical coherence tomography with a Linnik microscope. *Appl. Opt.* **2002**, *41*, 805–812.
8. Jain, M.; Shukla, N.; Manzoor, M.; Nadolny, S.; Mukherjee, S. Modified full-field optical coherence tomography: A novel tool for rapid histology of tissues. *J. Pathol. Inf.* **2011**, *2*, 28.
9. Grieve, K.; Mouslim, K.; Assayag, O.; Dalimier, E.; Harms, F.; Bruhat, A.; Boccara, C.; Antoine, M. Assessment of Sentinel Node Biopsies With Full-Field Optical Coherence Tomography. *Technol. Cancer Res. Treat.* **2016**, *15*, 266–274.
10. De Leeuw, F.; Casiraghi, O.; Lakhdar, A.B.; Abbaci, M.; Laplace-Builhé, C. Full-field OCT for fast diagnostic of head and neck cancer. In Proceedings of the SPIE BiOS, San Francisco, CA, USA, 7–12 February 2015; p. 93031Z.
11. Yang, C.; Ricco, R.; Sisk, A.; Duc, A.; Sibony, M.; Beuvon, F.; Dalimier, E.; Delongchamps, N.B. High efficiency for prostate biopsy qualification with full-field OCT after training. In Proceedings of the SPIE BiOS, San Francisco, CA, USA, 13–18 February 2016; p. 96891J.
12. Grieve, K.; Georgeon, C.; Andreiuolo, F.; Borderie, M.; Ghoubay, D.; Rault, J.; Borderie, V.M. Imaging Microscopic Features of Keratoconic Corneal Morphology. *Cornea* **2016**, *35*, 1621.
13. Assayag, O.; Grieve, K.; Devaux, B.; Harms, F.; Pallud, J.; Chretien, F.; Boccara, C.; Varlet, P. Imaging of non-tumorous and tumorous human brain tissues with full-field optical coherence tomography. *NeuroImage Clin.* **2013**, *2*, 549–557.

14. Assayag, O.; Antoine, M.; Sigal-Zafrani, B.; Riben, M.; Harms, F.; Burcheri, A.; Grieve, K.; Dalimier, E.; de Poly, B.L.C.; Boccara, C. Large field, high resolution full-field optical coherence tomography: A pre-clinical study of human breast tissue and cancer assessment. *Technol. Cancer Res. Treat.* **2014**, *13*, 455–468.

15. Peters, I.T.A.; Stegehuis, P.L.; Peek, R.; Boer, F.L.; Zwet, E.W.V.; Eggermont, J.; Westphal, J.R.; Kuppen, P.J.K.; Trimbos, J.B.M.Z.; Hilders, C.G.J.M.; et al. Non-invasive detection of metastases and follicle density in ovarian tissue using full-field optical coherence tomography. *Clin. Cancer Res.* **2016**, doi:clincanres.0288.2016.

16. *Handbook of Full-Field Optical Coherence Microscopy: Technology and Applications*; Pan Stanford Publishing Pte. Ltd.: Singapore, 2016.

17. Jain, M.; Robinson, B.D.; Salamoon, B.; Thouvenin, O.; Boccara, C.; Mukherjee, S. Rapid evaluation of fresh ex vivo kidney tissue with full-field optical coherence tomography. *J. Pathol. Inf.* **2015**, *6*, doi:10.4103/2153-3539.166014.

18. Auksorius, E.; Bromberg, Y.; Motiejūnaitė, R.; Pieretti, A.; Liu, L.; Coron, E.; Aranda, J.; Goldstein, A.M.; Bouma, B.E.; Kazlauskas, A.; et al. Dual-modality fluorescence and full-field optical coherence microscopy for biomedical imaging applications. *Biomed. Opt. Express* **2012**, *3*, 661–666.

19. Nahas, A.; Varna, M.; Fort, E.; Boccara, A.C. Detection of plasmonic nanoparticles with full field-OCT: Optical and photothermal detection. *Biomed. Opt. Express* **2014**, *5*, 3541–3546.

20. Nahas, A.; Bauer, M.; Roux, S.; Boccara, A.C. 3D static elastography at the micrometer scale using Full Field OCT. *Biomed. Opt. Express* **2013**, *4*, 2138–2149.

21. Moneron, G.; Boccara, A.C.; Dubois, A. Polarization-sensitive full-field optical coherence tomography. *Opt. Lett.* **2007**, *32*, 2058–2060.

22. Yu, L.; Kim, M. Full-color three-dimensional microscopy by wide-field optical coherence tomography. *Opt. Express* **2004**, *12*, 6632–6641.

23. Federici, A.; Dubois, A. Three-band, 1.9-μm axial resolution full-field optical coherence microscopy over a 530–1700 nm wavelength range using a single camera. *Opt. Lett.* **2014**, *39*, 1374–1377.

24. Dubois, A.; Moreau, J.; Boccara, C. Spectroscopic ultrahigh-resolution full-field optical coherence microscopy. *Opt. Express* **2008**, *16*, 17082–17091.

25. Huang, D.; Swanson, E.A.; Lin, C.P.; Schuman, J.S.; Stinson, W.G.; Chang, W.; Hee, M.R.; Flotte, T.; Gregory, K.; Puliafito, C.A.; et al. Optical coherence tomography. *Science* **1991**, *254*, 1178.

26. Izatt, J.A.; Swanson, E.A.; Fujimoto, J.G.; Hee, M.R.; Owen, G.M. Optical coherence microscopy in scattering media. *Opt. Lett.* **1994**, *19*, 590–592.

27. Fercher, A.F.; Hitzenberger, C.K.; Kamp, G.; El-Zaiat, S.Y. Measurement of intraocular distances by backscattering spectral interferometry. *Opt. Commun.* **1995**, *117*, 43–48.

28. De Boer, J.F.; Cense, B.; Park, B.H.; Pierce, M.C.; Tearney, G.J.; Bouma, B.E. Improved signal-to-noise ratio in spectral-domain compared with time-domain optical coherence tomography. *Opt. Lett.* **2003**, *28*, 2067–2069.

29. Pircher, M.; Götzinger, E.; Hitzenberger, C.K. Dynamic focus in optical coherence tomography for retinal imaging. *J. Biomed. Opt.* **2006**, *11*, 054013.

30. Holmes, J.; Hattersley, S.; Stone, N.; Bazant-Hegemark, F.; Barr, H. Multi-channel Fourier domain OCT system with superior lateral resolution for biomedical applications. In Proceedings of the Biomedical Optics (BiOS), San Jose, CA, USA, 19–24 January 2008; p. 68470O.

31. Ding, Z.; Ren, H.; Zhao, Y.; Nelson, J.S.; Chen, Z. High-resolution optical coherence tomography over a large depth range with an axicon lens. *Opt. Lett.* **2002**, *27*, 243–245.

32. Ralston, T.S.; Marks, D.L.; Carney, P.S.; Boppart, S.A. Interferometric synthetic aperture microscopy. *Nat. Phys.* **2007**, *3*, 129–134.

33. Kazmi, S.S.; Wu, R.K.; Dunn, A.K. Evaluating multi-exposure speckle imaging estimates of absolute autocorrelation times. *Opt. Lett.* **2015**, *40*, 3643–3646.

34. Potsaid, B.; Gorczynska, I.; Srinivasan, V.J.; Chen, Y.; Jiang, J.; Cable, A.; Fujimoto, J.G. Ultrahigh speed spectral/Fourier domain OCT ophthalmic imaging at 70,000 to 312,500 axial scans per second. *Opt. Express* **2008**, *16*, 15149–15169.

35. Grieve, K.; Dubois, A.; Simonutti, M.; Paques, M.; Sahel, J.; Le Gargasson, J.F.; Boccara, C. In vivo anterior segment imaging in the rat eye with high speed white light full-field optical coherence tomography. *Opt. Express* **2005**, *13*, 6286–6295.

36. Song, S.; Huang, Z.; Nguyen, T.M.; Wong, E.Y.; Arnal, B.; O'Donnell, M.; Wang, R.K. Shear modulus imaging by direct visualization of propagating shear waves with phase-sensitive optical coherence tomography. *J. Biomed. Opt.* **2013**, *18*, 121509.

37. Commercial System LLTech. Avaiable online: http://www.lltechimaging.com/products-applications/products/ (accessed on 20 January 2017).

38. Nahas, A.; Tanter, M.; Nguyen, T.M.; Chassot, J.M.; Fink, M.; Boccara, A.C. From supersonic shear wave imaging to full-field optical coherence shear wave elastography. *J. Biomed. Opt.* **2013**, *18*, 121514.

39. Apelian, C.; Harms, F.; Thouvenin, O.; Boccara, A.C. Dynamic full field optical coherence tomography: Subcellular metabolic contrast revealed in tissues by temporal analysis of interferometric signals. *arXiv* **2016**, arXiv:1601.01208.

40. Leroux, C.E.; Bertillot, F.; Thouvenin, O.; Boccara, A.C. Intracellular dynamics measurements with full field optical coherence tomography suggest hindering effect of actomyosin contractility on organelle transport. *Biomed. Opt. Express* **2016**, *7*, 4501–4513.

41. Thouvenin, O.; Fink, M.; Boccara, C. Dynamic multimodal full-field optical coherence tomography and fluorescence structured illumination microscopy. *J. Biomed. Opt.* **2017**, *22*, 026004.

42. Makhlouf, H.; Perronet, K.; Dupuis, G.; Lévêque-Fort, S.; Dubois, A. Simultaneous optically sectioned fluorescence and optical coherence microscopy with full-field illumination. *Opt. Lett.* **2012**, *37*, 1613–1615.

43. Abdulhalim, I. Spatial and temporal coherence effects in interference microscopy and full-field optical coherence tomography. *Ann. Phys.* **2012**, *524*, 787–804.

44. Dubois, A.; Selb, J.; Vabre, L.; Boccara, A.C. Phase measurements with wide-aperture interferometers. *Appl. Opt.* **2000**, *39*, 2326–2331.

45. Karamata, B.; Lambelet, P.; Laubscher, M.; Salathé, R.; Lasser, T. Spatially incoherent illumination as a mechanism for cross-talk suppression in wide-field optical coherence tomography. *Opt. Lett.* **2004**, *29*, 736–738.

46. Xiao, P.; Fink, M.; Boccara, A.C. Full-field spatially incoherent illumination interferometry: A spatial resolution almost insensitive to aberrations. *Opt. Lett.* **2016**, *41*, 3920–3923.

47. CMOSIS New Camera by Adimec. Avaiable online: http://info.adimec.com/blogposts/careioca-project-results-in-several-new-products-including-cmosis-csi2100-adimec-q-2a750-and-lltech-ffoct-microscope-and-endoscope (accessed on 20 January 2017).

48. Mandrioli, P.; Ariatti, A. Marcello Malpighi, a pioneer of the experimental research in biology. *Aerobiologia* **1991**, *7*, 3–9.

49. Reverón, R.R. Marcello Malpighi (1628–1694), founder of microanatomy. *Int. J. Morphol.* **2011**, *29*, 399–402.

50. Cohen, A.L.; Hayat, M.A. *Principles and Techniques of Scanning Electron Microscopy. Biological Applications. Volume 1*; Van Nostrand Reinhold Company: Cincinatti, OH, USA, 1974.

51. Digital Microscope. Avaiable online: http://www.histology.be/digital_microscope_histology_.html (accessed on 20 January 2017).

52. Histology Guide: Virtual Histology Lab. Avaiable online: http://histologyguide.org/index.html (accessed on 20 January 2017).

53. Amunts, K.; Lepage, C.; Borgeat, L.; Mohlberg, H.; Dickscheid, T.; Rousseau, M.É.; Bludau, S.; Bazin, P.L.; Lewis, L.B.; Oros-Peusquens, A.M.; et al. BigBrain: An ultrahigh-resolution 3D human brain model. *Science* **2013**, *340*, 1472–1475.

54. Leslie P. Gartner, L.P. *Color Atlas and Text of Histology*, 6th ed.; Lippincott Williams and Wilkins: Philadelphia, PA, USA, 2013.

55. Auth, A.C. *A Text-Book of Histology. Descriptive and Practical. For the Use of Students*; John Wright and Co.: Bristol, UK, 1986.

56. Spitalnik, P.F. *Histology Lab Manual*; Columbia University: New York, NY, USA, 2015.

57. Clinical Atlas LLTech. Avaiable online: http://www.lltechimaging.com/image-gallery/atlas-of-images/ (accessed on 20 January 2017).

58. Magnain, C.; Augustinack, J.C.; Konukoglu, E.; Frosch, M.P.; Sakadžić, S.; Varjabedian, A.; Garcia, N.; Wedeen, V.J.; Boas, D.A.; Fischl, B. Optical coherence tomography visualizes neurons in human entorhinal cortex. *Neurophotonics* **2015**, *2*, 015004.

59. Wang, H.; Zhu, J.; Akkin, T. Serial optical coherence scanner for large-scale brain imaging at microscopic resolution. *Neuroimage* **2014**, *84*, 1007–1017.

60. Quinten, M. *Optical Properties of Nanoparticle Systems: Mie and Beyond*; John Wiley & Sons: Hoboken, NJ, USA, 2010.

61. Arous, J.B.; Binding, J.; Léger, J.F.; Casado, M.; Topilko, P.; Gigan, S.; Boccara, A.C.; Bourdieu, L. Single myelin fiber imaging in living rodents without labeling by deep optical coherence microscopy. *J. Biomed. Opt.* **2011**, *16*, 116012.

62. Grieve, K.; Thouvenin, O.; Sengupta, A.; Borderie, V.M.; Paques, M. Appearance of the Retina With Full-Field Optical Coherence Tomography. *Investig. Opthalmol. Vis. Sci.* **2016**, *57*, OCT96.

63. Wang, S.; Liu, C.H.; Zakharov, V.P.; Lazar, A.J.; Pollock, R.E.; Larin, K.V. Three-dimensional computational analysis of optical coherence tomography images for the detection of soft tissue sarcomas. *J. Biomed. Opt.* **2014**, *19*, 0211022.

64. Boppart, S.A.; Luo, W.; Marks, D.L.; Singletary, K.W. Optical coherence tomography: Feasibility for basic research and image-guided surgery of breast cancer. *Breast Cancer Res. Treat.* **2004**, *84*, 85–97.

65. McLaughlin, R.A.; Scolaro, L.; Robbins, P.; Hamza, S.; Saunders, C.; Sampson, D.D. Imaging of human lymph nodes using optical coherence tomography: Potential for staging cancer. *Cancer Res.* **2010**, *70*, 2579–2584.

66. Vakoc, B.J.; Fukumura, D.; Jain, R.K.; Bouma, B.E. Cancer imaging by optical coherence tomography: Preclinical progress and clinical potential. *Nat. Rev. Cancer* **2012**, *12*, 363–368.

67. Sharma, M.; Verma, Y.; Rao, K.; Nair, R.; Gupta, P. Imaging growth dynamics of tumour spheroids using optical coherence tomography. *Biotechnol. Lett.* **2007**, *29*, 273–278.

68. Yang, Y.; Wang, T.; Biswal, N.C.; Wang, X.; Sanders, M.; Brewer, M.; Zhu, Q. Optical scattering coefficient estimated by optical coherence tomography correlates with collagen content in ovarian tissue. *J. Biomed. Opt.* **2011**, *16*, 090504.

69. Vakoc, B.J.; Lanning, R.M.; Tyrrell, J.A.; Padera, T.P.; Bartlett, L.A.; Stylianopoulos, T.; Munn, L.L.; Tearney, G.J.; Fukumura, D.; Jain, R.K.; et al. Three-dimensional microscopy of the tumor microenvironment in vivo using optical frequency domain imaging. *Nat. Med.* **2009**, *15*, 1219–1223.

70. Sullivan, A.C.; Hunt, J.P.; Oldenburg, A.L. Fractal analysis for classification of breast carcinoma in optical coherence tomography. *J. Biomed. Opt.* **2011**, *16*, 066010.

71. Gao, W.; Zhu, Y. Fractal analysis of en face tomographic images obtained with full field optical coherence tomography. *Ann. Phys.* **2016**, *93*, doi:10.1002/andp.201600216.

72. Nicolson, M. The art of diagnosis: Medicine and the five senses. In *Companion Encyclopedia of the History of Medicine, Volume 2*; Bynum, W., Porter, R., Eds.; Routledge: London, UK, 1997; pp. 801–825.

73. Medical Diagnosis in Egyptian World. Available online: http://www.arabworldbooks.com/articles8.htm (accessed on 20 January 2017).

74. Sarvazyan, A. Shear acoustic properties of soft biological tissues in medical diagnostics. *J. Acoust. Soc. Am.* **1993**, *93*, 2329–2330.

75. Bataller, R.; Brenner, D.A. Liver fibrosis. *J. Clin. Investig.* **2005**, *115*, 209–218.

76. Mueller, S.; Sandrin, L. Liver stiffness: A novel parameter for the diagnosis of liver disease. *Hepat. Med.* **2010**, *2*, 49–67.

77. Kumar, S.; Weaver, V.M. Mechanics, malignancy, and metastasis: The force journey of a tumor cell. *Cancer Metastasis Rev.* **2009**, *28*, 113–127.

78. Mouw, J.K.; Yui, Y.; Damiano, L.; Bainer, R.O.; Lakins, J.N.; Acerbi, I.; Ou, G.; Wijekoon, A.C.; Levental, K.R.; Gilbert, P.M.; et al. Tissue mechanics modulate microRNA-dependent PTEN expression to regulate malignant progression. *Nat. Med.* **2014**, *20*, 360.

79. Ophir, J.; Cespedes, I.; Ponnekanti, H.; Yazdi, Y.; Li, X. Elastography: A quantitative method for imaging the elasticity of biological tissues. *Ultrason. Imaging* **1991**, *13*, 111–134.

80. Muthupillai, R.; Lomas, D.; Rossman, P.; Greenleaf, J.F.; Manduca, A.; Ehman, R.L. Magnetic resonance elastography by direct visualization of propagating acoustic strain waves. *Science* **1995**, *269*, 1854.

81. Mariappan, Y.K.; Glaser, K.J.; Ehman, R.L. Magnetic resonance elastography: A review. *Clin. Anat.* **2010**, *23*, 497–511.

82. Tanter, M.; Bercoff, J.; Athanasiou, A.; Deffieux, T.; Gennisson, J.L.; Montaldo, G.; Muller, M.; Tardivon, A.; Fink, M. Quantitative assessment of breast lesion viscoelasticity: Initial clinical results using supersonic shear imaging. *Ultras. Med. Biol.* **2008**, *34*, 1373–1386.

83. Gayrard, C.; Borghi, N. FRET-based molecular tension microscopy. *Methods* **2016**, *94*, 33–42.

84. Schmitt, J.M. OCT elastography: Imaging microscopic deformation and strain of tissue. *Opt. Express* **1998**, *3*, 199–211.

85. Wang, R.K.; Ma, Z.; Kirkpatrick, S.J. Tissue Doppler optical coherence elastography for real time strain rate and strain mapping of soft tissue. *Appl. Phys. Lett.* **2006**, *89*, 144103.

86. Kennedy, B.F.; Kennedy, K.M.; Sampson, D.D. A review of optical coherence elastography: Fundamentals, techniques and prospects. *IEEE J. Sel. Top. Quant. Electron.* **2014**, *20*, 272–288.

87. Wang, S.; Larin, K.V. Optical coherence elastography for tissue characterization: A review. *J. Biophoton.* **2015**, *8*, 279–302.

88. Mulligan, J.A.; Untracht, G.R.; Chandrasekaran, S.N.; Brown, C.N.; Adie, S.G. Emerging Approaches for High-Resolution Imaging of Tissue Biomechanics With Optical Coherence Elastography. *IEEE J. Sel. Top. Quant. Electron.* **2016**, *22*, 1–20.

89. Royer, D.; Dieulesaint, E. *Elastic Waves in Solids I: Free and Guided Propagation*; Morgan, D.P., Translator; Springer: New York, NY, USA, 2000.

90. Sarvazyan, A.P.; Rudenko, O.V.; Swanson, S.D.; Fowlkes, J.B.; Emelianov, S.Y. Shear wave elasticity imaging: A new ultrasonic technology of medical diagnostics. *Ultras. Med. Boil.* **1998**, *24*, 1419–1435.

91. Roux, S.; Hild, F.; Viot, P.; Bernard, D. Three-dimensional image correlation from X-ray computed tomography of solid foam. *Compos. A Appl. Sci. Manuf.* **2008**, *39*, 1253–1265.

92. Leclerc, H.; Périé, J.N.; Hild, F.; Roux, S. Digital volume correlation: What are the limits to the spatial resolution? *Mech. Ind.* **2012**, *13*, 361–371.

93. Basser, P.J.; Mattiello, J.; LeBihan, D. MR diffusion tensor spectroscopy and imaging. *Biophys. J.* **1994**, *66*, 259.

94. Le Bihan, D.; Mangin, J.F.; Poupon, C.; Clark, C.A.; Pappata, S.; Molko, N.; Chabriat, H. Diffusion tensor imaging: Concepts and applications. *J. Magn. Reson. Imaging* **2001**, *13*, 534–546.

95. Papadacci, C.; Tanter, M.; Pernot, M.; Fink, M. Ultrasound backscatter tensor imaging (BTI): Analysis of the spatial coherence of ultrasonic speckle in anisotropic soft tissues. *IEEE Trans. Ultrason. Ferroelectr. Freq. Control* **2014**, *61*, 986–996.

96. Chatelin, S.; Bernal, M.; Deffieux, T.; Papadacci, C.; Flaud, P.; Nahas, A.; Boccara, C.; Gennisson, J.L.; Tanter, M.; Pernot, M. Anisotropic polyvinyl alcohol hydrogel phantom for shear wave elastography in fibrous biological soft tissue: A multimodality characterization. *Phys. Med. Biol.* **2014**, *59*, 6923.

97. Kennedy, K.M.; Es'haghian, S.; Chin, L.; McLaughlin, R.A.; Sampson, D.D.; Kennedy, B.F. Optical palpation: Optical coherence tomography-based tactile imaging using a compliant sensor. *Opt. Lett.* **2014**, *39*, 3014–3017.

98. Wellman, P.S.; Howe, R.D.; Dewagan, N.; Cundari, M.A.; Dalton, E.; Kern, K.A. Tactile imaging: A method for documenting breast masses. In Proceedings of the First Joint IEEE Engineering in Medicine and Biology, 21st Annual Conference and the 1999 Annual Fall Meetring of the Biomedical Engineering Society, Atlanta, GA, USA, 13–16 October 1999; Volume 2, p. 1131.

99. Razani, M.; Mariampillai, A.; Sun, C.; Luk, T.W.; Yang, V.X.; Kolios, M.C. Feasibility of optical coherence elastography measurements of shear wave propagation in homogeneous tissue equivalent phantoms. *Biomed. Opt. Express* **2012**, *3*, 972–980.

100. Catheline, S.; Thomas, J.L.; Wu, F.; Fink, M.A. Diffraction field of a low frequency vibrator in soft tissues using transient elastography. *IEEE Trans. Ultrason. Ferroelectr. Freq. Control* **1999**, *46*, 1013–1019.

101. Sandrin, L.; Catheline, S.; Tanter, M.; Hennequin, X.; Fink, M. Time-resolved pulsed elastography with ultrafast ultrasonic imaging. *Ultrason. Imaging* **1999**, *21*, 259–272.

102. Bercoff, J.; Tanter, M.; Fink, M. Supersonic shear imaging: A new technique for soft tissue elasticity mapping. *IEEE Trans. Ultrason. Ferroelectr. Freq. Control* **2004**, *51*, 396–409.

103. Chatelin, S.; Constantinesco, A.; Willinger, R. Fifty years of brain tissue mechanical testing: From in vitro to in vivo investigations. *Biorheology* **2010**, *47*, 255–276.

104. Macé, E.; Cohen, I.; Montaldo, G.; Miles, R.; Fink, M.; Tanter, M. In vivo mapping of brain elasticity in small animals using shear wave imaging. *IEEE Trans. Med. Imaging* **2011**, *30*, 550–558.

105. Catheline, S.; Souchon, R.; Rupin, M.; Brum, J.; Dinh, A.; Chapelon, J.Y. Tomography from diffuse waves: Passive shear wave imaging using low frame rate scanners. *Appl. Phys. Lett.* **2013**, *103*, 014101.

106. Zorgani, A.; Souchon, R.; Dinh, A.H.; Chapelon, J.Y.; Ménager, J.M.; Lounis, S.; Rouvière, O.; Catheline, S. Brain palpation from physiological vibrations using MRI. *Proc. Natl. Acad. Sci. USA* **2015**, *112*, 12917–12921.

107. Nguyen, T.M.; Zorgani, A.; Lescanne, M.; Boccara, C.; Fink, M.; Catheline, S. Diffuse shear wave imaging: Toward passive elastography using low-frame rate spectral-domain optical coherence tomography. *J. Biomed. Opt.* **2016**, *21*, 126013.

108. Pecora, R. *Dynamic Light Scattering: Applications of Photon Correlation Spectroscopy*; Springer Science & Business Media: Heidelberg, Germany, 2013.

109. Suissa, M.; Place, C.; Goillot, E.; Berge, B.; Freyssingeas, E. Dynamic light scattering as an investigating tool to study the global internal dynamics of a living cell nucleus. *EPL Europhys. Lett.* **2007**, *78*, 38005.

110. Suissa, M.; Place, C.; Goillot, E.; Freyssingeas, E. Internal dynamics of a living cell nucleus investigated by dynamic light scattering. *Eur. Phys. J. E* **2008**, *26*, 435–448.

111. Jeong, K.; Turek, J.J. Volumetric motility-contrast imaging of tissue response to cytoskeletal anti-cancer drugs. *Opt. Express* **2007**, *15*, 14057–14064.

112. Nolte, D.D.; An, R.; Turek, J.; Jeong, K. Tissue dynamics spectroscopy for phenotypic profiling of drug effects in three-dimensional culture. *Biomed. Opt. Express* **2012**, *3*, 2825–2841.

113. An, R.; Wang, C.; Turek, J.; Machaty, Z.; Nolte, D.D. Biodynamic imaging of live porcine oocytes, zygotes and blastocysts for viability assessment in assisted reproductive technologies. *Biomed. Opt. Express* **2015**, *6*, 963–976.

114. Tan, W.; Oldenburg, A.L.; Norman, J.J.; Desai, T.A.; Boppart, S.A. Optical coherence tomography of cell dynamics in three-dimensional tissue models. *Opt. Express* **2006**, *14*, 7159–7171.

115. Oldenburg, A.L.; Yu, X.; Gilliss, T.; Alabi, O.; Taylor, R.M.; Troester, M.A. Inverse-power-law behavior of cellular motility reveals stromal–epithelial cell interactions in 3D co-culture by OCT fluctuation spectroscopy. *Optica* **2015**, *2*, 877.

116. Lee, J.; Wu, W.; Jiang, J.Y.; Zhu, B.; Boas, D.A. Dynamic light scattering optical coherence tomography. *Opt. Express* **2012**, *20*, 22262–22277.

117. Chen, H.; Farkas, E.R.; Webb, W.W. In vivo applications of fluorescence correlation spectroscopy. *Methods Cell Biol.* **2008**, *89*, 3–35.

118. Tishler, R.B.; Carlson, F.D. A study of the dynamic properties of the human red blood cell membrane using quasi-elastic light-scattering spectroscopy. *Biophys. J.* **1993**, *65*, 2586.

119. Miller, M.J.; Wei, S.H.; Parker, I.; Cahalan, M.D. Two-photon imaging of lymphocyte motility and antigen response in intact lymph node. *Science* **2002**, *296*, 1869–1873.

120. Ammari, H.; Romero, F.; Shi, C. A signal separation technique for sub-cellular imaging using dynamic optical coherence tomography. *arXiv* **2016**, arXiv:1608.04382.

121. Demené, C.; Deffieux, T.; Pernot, M.; Osmanski, B.F.; Biran, V.; Gennisson, J.L.; Sieu, L.A.; Bergel, A.; Franqui, S.; Correas, J.M.; et al. Spatiotemporal clutter filtering of ultrafast ultrasound data highly increases Doppler and fUltrasound sensitivity. *IEEE Trans. Med. Imaging* **2015**, *34*, 2271–2285.

122. Zink, D.; Fischer, A.H.; Nickerson, J.A. Nuclear structure in cancer cells. *Nat. Rev. Cancer* **2004**, *4*, 677–687.

123. Slater, D.; Rice, S.; Stewart, R.; Melling, S.; Hewer, E.; Smith, J. Proposed Sheffield quantitative criteria in cervical cytology to assist the grading of squamous cell dyskaryosis, as the British Society for Clinical Cytology definitions require amendment. *Cytopathology* **2005**, *16*, 179–192.

124. Popescu, G.; Park, Y.; Choi, W.; Dasari, R.R.; Feld, M.S.; Badizadegan, K. Imaging red blood cell dynamics by quantitative phase microscopy. *Blood Cells Mol. Dis.* **2008**, *41*, 10–16.

125. Sato, E.; Olson, S.H.; Ahn, J.; Bundy, B.; Nishikawa, H.; Qian, F.; Jungbluth, A.A.; Frosina, D.; Gnjatic, S.; Ambrosone, C.; et al. Intraepithelial CD8+ tumor-infiltrating lymphocytes and a high CD8+/regulatory T cell ratio are associated with favorable prognosis in ovarian cancer. *Proc. Natl. Acad. Sci. USA* **2005**, *102*, 18538–18543.

126. Mahmoud, S.M.; Paish, E.C.; Powe, D.G.; Macmillan, R.D.; Grainge, M.J.; Lee, A.H.; Ellis, I.O.; Green, A.R. Tumor-infiltrating CD8+ lymphocytes predict clinical outcome in breast cancer. *J. Clin. Oncol.* **2011**, *29*, 1949–1955.

127. Weidner, N.; Carroll, P.; Flax, J.; Blumenfeld, W.; Folkman, J. Tumor angiogenesis correlates with metastasis in invasive prostate carcinoma. *Am. J. Pathol.* **1993**, *143*, 401.

128. Weidner, N. Current pathologic methods for measuring intratumoral microvessel density within breast carcinoma and other solid tumors. *Breast Cancer Res. Treat.* **1995**, *36*, 169–180.

129. Maier, J.G.; Schamber, D.T. The role of lymphangiography in the diagnosis and treatment of malignant testicular tumors. *Am. J. Roentgenol.* **1972**, *114*, 482–491.

130. Shaked, N.T.; Rinehart, M.T.; Wax, A. Quantitative phase microscopy of biological cell dynamics by wide-field digital interferometry. In *Coherent Light Microscopy*; Springer: Heidelberg, Germany, 2011; pp. 169–198.

131. Ma, L.; Rajshekhar, G.; Wang, R.; Bhaduri, B.; Sridharan, S.; Mir, M.; Chakraborty, A.; Iyer, R.; Prasanth, S.; Millet, L.; et al. Phase correlation imaging of unlabeled cell dynamics. *Sci. Rep.* **2016**, *6*.

132. Berclaz, C.; Szlag, D.; Nguyen, D.; Extermann, J.; Bouwens, A.; Marchand, P.J.; Nilsson, J.; Schmidt-Christensen, A.; Holmberg, D.; Grapin-Botton, A.; et al. Label-free fast 3D coherent imaging reveals pancreatic islet micro-vascularization and dynamic blood flow. *Biomed. Opt. Express* **2016**, *7*, 4569–4580.

133. Bouwens, A.; Szlag, D.; Szkulmowski, M.; Bolmont, T.; Wojtkowski, M.; Lasser, T. Quantitative lateral and axial flow imaging with optical coherence microscopy and tomography. *Opt. Express* **2013**, *21*, 17711–17729.

134. Leitgeb, R.A.; Werkmeister, R.M.; Blatter, C.; Schmetterer, L. Doppler optical coherence tomography. *Progr. Retin. Eye Res.* **2014**, *41*, 26–43.

135. Bouwens, A.; Bolmont, T.; Szlag, D.; Berclaz, C.; Lasser, T. Quantitative cerebral blood flow imaging with extended-focus optical coherence microscopy. *Opt. Lett.* **2014**, *39*, 37–40.

136. Hicks, J.; Matthaei, E. Fluorescence in histology. *J. Pathol. Bacteriol.* **1955**, *70*, 1–12.

137. John, D.; Bancroft, M.G. *Theory and Practice Of Histological Techniques*, 5th ed.; Churchill Livingstone: London, UK, 2002.

138. Volpi, E.V.; Bridger, J.M. FISH glossary: An overview of the fluorescence in situ hybridization technique. *Biotechniques* **2008**, *45*, 385–386.

139. Roulston, J.E.; Bartlett, J.M. *Molecular Diagnosis Of Cancer: Methods and Protocols*; Springer Science & Business Media: New York, NY, USA, 2004; Volume 97.

140. Transidico, P.; Bianchi, M.; Capra, M.; Pelicci, P.G.; Faretta, M. From cells to tissues: Fluorescence confocal microscopy in the study of histological samples. *Microsc. Res. Tech.* **2004**, *64*, 89–95.

141. Yuste, R. Fluorescence microscopy today. *Nat. Methods* **2005**, *2*, 902–904.

142. Jung, J.C.; Mehta, A.D.; Aksay, E.; Stepnoski, R.; Schnitzer, M.J. In vivo mammalian brain imaging using one-and two-photon fluorescence microendoscopy. *J. Neurophysiol.* **2004**, *92*, 3121–3133.

143. Flusberg, B.A.; Cocker, E.D.; Piyawattanametha, W.; Jung, J.C.; Cheung, E.L.; Schnitzer, M.J. Fiber-optic fluorescence imaging. *Nat. Methods* **2005**, *2*, 941–950.

144. Tuchin, V.V. Optical clearing of tissues and blood using the immersion method. *J. Phys. D Appl. Phys.* **2005**, *38*, 2497.

145. Chung, K.; Deisseroth, K. CLARITY for mapping the nervous system. *Nat. Methods* **2013**, *10*, 508–513.

146. Larin, K.V.; Ghosn, M.G.; Bashkatov, A.N.; Genina, E.A.; Trunina, N.A.; Tuchin, V.V. Optical clearing for OCT image enhancement and in-depth monitoring of molecular diffusion. *IEEE J. Sel. Top. Quant. Electron.* **2012**, *18*, 1244–1259.

147. Yuan, S.; Roney, C.A.; Wierwille, J.; Chen, C.W.; Xu, B.; Griffiths, G.; Jiang, J.; Ma, H.; Cable, A.; Summers, R.M.; Chen, Y. Co-registered optical coherence tomography and fluorescence molecular imaging for simultaneous morphological and molecular imaging. *Phys. Med. Biol.* **2010**, *55*, 191–206.

148. Harms, F.; Dalimier, E.; Vermeulen, P.; Fragola, A.; Boccara, A. Multimodal Full-Field Optical Coherence Tomography on biological tissue: Toward all optical digital pathology. In *Prcoeedings of the SPIE BiOS*, San Francisco, CA, USA, 21–26 January 2012; p. 821609.

149. Coron, E.; Auksorius, E.; Pieretti, A.; Mahé, M.; Liu, L.; Steiger, C.; Bromberg, Y.; Bouma, B.; Tearney, G.; Neunlist, M.; et al. Full-field optical coherence microscopy is a novel technique for imaging enteric ganglia in the gastrointestinal tract. *Neurogastroenterol. Motil.* **2012**, *24*, e611–e621.

150. Chen, T.W.; Wardill, T.J.; Sun, Y.; Pulver, S.R.; Renninger, S.L.; Baohan, A.; Schreiter, E.R.; Kerr, R.A.; Orger, M.B.; Jayaraman, V.; et al. Ultrasensitive fluorescent proteins for imaging neuronal activity. *Nature* **2013**, *499*, 295–300.

151. Gee, K.; Brown, K.; Chen, W.N.; Bishop-Stewart, J.; Gray, D.; Johnson, I. Chemical and physiological characterization of fluo-4 Ca^{2+}-indicator dyes. *Cell Calcium* **2000**, *27*, 97–106.

152. Siegel, M.S.; Isacoff, E.Y. A genetically encoded optical probe of membrane voltage. *Neuron* **1997**, *19*, 735–741.

153. Imamura, H.; Nhat, K.P.H.; Togawa, H.; Saito, K.; Iino, R.; Kato-Yamada, Y.; Nagai, T.; Noji, H. Visualization of ATP levels inside single living cells with fluorescence resonance energy transfer-based genetically encoded indicators. *Proc. Natl. Acad. Sci. USA* **2009**, *106*, 15651–15656.

154. Kit for Apoptosis Detection from ThermoFisher Scientific. Available online: https://www.thermofisher. com/order/catalog/product/C10617 (accessed on 20 January 2017).

155. Farhat, G.; Mariampillai, A.; Yang, V.X.; Czarnota, G.J.; Kolios, M.C. Detecting apoptosis using dynamic light scattering with optical coherence tomography. *J. Biomed. Opt.* **2011**, *16*, 0705055.

156. Rappaz, B.; Marquet, P.; Cuche, E.; Emery, Y.; Depeursinge, C.; Magistretti, P. Measurement of the integral refractive index and dynamic cell morphometry of living cells with digital holographic microscopy. *Opt. Express* **2005**, *13*, 9361–9373.

157. Brown, E.; McKee, T.; diTomaso, E.; Pluen, A.; Seed, B.; Boucher, Y.; Jain, R.K. Dynamic imaging of collagen and its modulation in tumors in vivo using second-harmonic generation. *Nat. Med.* **2003**, *9*, 796–800.

158. Débarre, D.; Supatto, W.; Pena, A.M.; Fabre, A.; Tordjmann, T.; Combettes, L.; Schanne-Klein, M.C.; Beaurepaire, E. Imaging lipid bodies in cells and tissues using third-harmonic generation microscopy. *Nat. Methods* **2006**, *3*, 47–53.

159. Aptel, F.; Olivier, N.; Deniset-Besseau, A.; Legeais, J.M.; Plamann, K.; Schanne-Klein, M.C.; Beaurepaire, E. Multimodal nonlinear imaging of the human cornea. *Investig. Ophthalmol. Vis. Sci.* **2010**, *51*, 2459–2465.

160. Olivier, N.; Luengo-Oroz, M.A.; Duloquin, L.; Faure, E.; Savy, T.; Veilleux, I.; Solinas, X.; Débarre, D.; Bourgine, P.; Santos, A.; et al. Cell lineage reconstruction of early zebrafish embryos using label-free nonlinear microscopy. *Science* **2010**, *329*, 967–971.

161. Wang, B.G.; König, K.; Halbhuber, K.J. Two-photon microscopy of deep intravital tissues and its merits in clinical research. *J. Microsc.* **2010**, *238*, 1–20.

162. Sun, W.; Chang, S.; Tai, D.C.; Tan, N.; Xiao, G.; Tang, H.; Yu, H. Nonlinear optical microscopy: Use of second harmonic generation and two-photon microscopy for automated quantitative liver fibrosis studies. *J. Biomed. Opt.* **2008**, *13*, 064010.

163. Commercial Multiphoton Platform for Histology. Available online: http://www.histoindex.com/Genesis-200 (accessed on 20 January 2017).

164. Evans, C.L.; Xie, X.S. Coherent anti-Stokes Raman scattering microscopy: Chemical imaging for biology and medicine. *Annu. Rev. Anal. Chem.* **2008**, *1*, 883–909.

165. Bégin, S.; Burgoyne, B.; Mercier, V.; Villeneuve, A.; Vallée, R.; Côté, D. Coherent anti-Stokes Raman scattering hyperspectral tissue imaging with a wavelength-swept system. *Biomed. Opt. Express* **2011**, *2*, 1296–1306.

166. Lin, C.Y.; Suhalim, J.L.; Nien, C.L.; MiljkoviÄ, M.D.; Diem, M.; Jester, J.V.; Potma, E.O. Picosecond spectral coherent anti-Stokes Raman scattering imaging with principal component analysis of meibomian glands. *J. Biomed. Opt.* **2011**, *16*, 021104.

167. Mahou, P.; Olivier, N.; Labroille, G.; Duloquin, L.; Sintes, J.M.; Peyriéras, N.; Legouis, R.; Débarre, D.; Beaurepaire, E. Combined third-harmonic generation and four-wave mixing microscopy of tissues and embryos. *Biomed. Opt. Express* **2011**, *2*, 2837–2849.

168. Michels, S.; Pircher, M.; Geitzenauer, W.; Simader, C.; Götzinger, E.; Findl, O.; Schmidt-Erfurth, U.; Hitzenberger, C. Value of polarisation-sensitive optical coherence tomography in diseases affecting the retinal pigment epithelium. *Br. J. Ophthalmol.* **2008**, *92*, 204–209.

169. Wang, J.; Léger, J.F.; Binding, J.; Boccara, A.C.; Gigan, S.; Bourdieu, L. Measuring aberrations in the rat brain by coherence-gated wavefront sensing using a Linnik interferometer. *Biomed. Opt. Express* **2012**, *3*, 2510–2525.

170. Xiao, P.; Fink, M.; Boccara, A.C. Adaptive optics full-field optical coherence tomography. *J. Biomed. Opt.* **2016**, *21*, 121505.

171. Auksorius, E.; Boccara, A.C. Dark-field full-field optical coherence tomography. *Opt. Lett.* **2015**, *40*, 3272–3275.

172. Badon, A.; Li, D.; Lerosey, G.; Boccara, A.C.; Fink, M.; Aubry, A. Smart optical coherence tomography for ultra-deep imaging through highly scattering media. *arXiv* **2015**, arXiv:1510.08613.

173. Srinivasan, V.J.; Sakadžić, S.; Gorczynska, I.; Ruvinskaya, S.; Wu, W.; Fujimoto, J.G.; Boas, D.A. Quantitative cerebral blood flow with optical coherence tomography. *Opt. Express* **2010**, *18*, 2477–2494.

Article

Optical Coherence Tomography (OCT) for Time-Resolved Imaging of Alveolar Dynamics in Mechanically Ventilated Rats

Christian Schnabel [1,*], Maria Gaertner [1,2] and Edmund Koch [1]

1   Department of Anesthesiology and Intensive Care Medicine, Clinical Sensoring and Monitoring, Faculty of Medicine CGC, Technische Universität Dresden, Dresden 01307, Germany; maria.gaertner@me.com (M.G.); edmund.koch@tu-dresden.de (E.K.)
2   InfraTec GmbH, Gostritzer Straße 61-63, Dresden 01217, Germany
*   Correspondence: christian.schnabel@tu-dresden.de; Tel.: +49-351-458-6133

Academic Editor: Michael Pircher
Received: 19 January 2017; Accepted: 10 March 2017; Published: 15 March 2017

Abstract: Though artificial ventilation is an essential life-saving treatment, the mechanical behavior of lung tissue at the alveolar level is still unknown. Therefore, we need to understand the tissue response during artificial ventilation at this microscale in order to develop new and more protective ventilation methods. Optical coherence tomography (OCT) combined with intravital microscopy (IVM) is a promising tool for visualizing lung tissue dynamics with a high spatial and temporal resolution in uninterruptedly ventilated rats. We present a measurement setup using a custom-made animal ventilator and a gating technique for data acquisition of time-resolved sequences.

Keywords: optical coherence tomography; OCT; mechanical ventilation; lung imaging

1. Introduction

Lung diseases and related affections of the body are one of the most common diseases in the industrial world [1]. Therefore, suitable treatment strategies are necessary using protective mechanical ventilation that shows a significant reduction in morbidity and mortality of affected patients [2,3]. Beside this effort, a detailed understanding of the mechanical transactions in alveolar lung tissue is still lacking. Hence, the choice of appropriated ventilation strategies and parameters during mechanical ventilation is essential to avoiding vast damage in sensitive alveolar tissue and to assure a patient's treatment success. A proper decision should be based on the detailed knowledge of tissue behavior during ventilation with respect to the particular disease. Common imaging techniques such as CT or MRT used in human diagnostics are not appropriate for investigating the ventilation effects on the microscale of alveoli. Therefore, it is of great medical and scientific interest how mechanical ventilation affects lung tissue on the alveolar level of gas exchange. Detailed insights into lung tissue behavior can be achieved in animal experiments using high-resolution imaging methods such as optical coherence tomography (OCT). In combination with intravital microscopy (IVM), both imaging techniques allow for the visualization of alveolar structures with a high spatial resolution [4,5]. Further information can be gathered investigating the tissue dynamics. This challenging task is commonly solved by high-speed OCT systems using Fourier-domain mode locked lasers [6] with the disadvantage of either a small scan area or a bad spatial resolution. Therefore, in the presented study, we used a custom-made ventilator and a new gating algorithm for image data acquisition to get new insights into lung tissue behavior on the alveolar level during uninterrupted mechanical ventilation in rats. These measurements can promote the understanding of alveolar mechanics and the effort for the development of more protective ventilation strategies.

2. Materials and Methods

2.1. Imaging Setup

The imaging system used for the investigation of lung tissue dynamics during ventilation is a combination of OCT and IVM using the same beam path in one scanner head. The custom-made OCT setup is described in detail by Meissner et al. [7]. Shortly, the system consists of a superluminescence diode (Superlum, Moscow, Russia) centered at 845 nm with a FWHM of 50 nm. As OCT is based on the use of a Michelson interferometer, the emitted near-infrared light is separated into a reference beam and a sample beam. The sample beam is deflected using two galvanometer scanners (Cambridge Technology Inc., Cambridge, MA, USA) to perform a 2D-scan pattern (x and y) over the sample. The scattered and reflected light from different sample depths (z) is collected and superimposed with the reference beam that was reflected on a mirror. This interference signal is spectrally resolved by a diffraction grating and measured by a line-detector (Teledyne DALSA Inc., Bromont, QC, Canada). The depth-dependent information is calculated by a Fast Fourier Transformation of the spectrum and results in a depth profile of backscattered light called an A-scan. Displaying the logarithm of the intensity from several A-scans at different transversal positions on an 8 bit grayscale results in a cross-sectional view called a B-scan. The axial and lateral resolution are 11 μm and 6 μm in air, respectively. The line scan rate is 12 kHz, which allows data acquisition of typical cross sections within 29 ms (320 × 512 pixels corresponding in this system to 1.28 × 2.8 mm (width × depth)).

For IVM, a high-speed video camera (acA2000-340 kc, Basler, Ahrensburg, Germany) in combination with an adjustable liquid lens (EL-10-30, Optotune AG, Dietikon, Swizerland) is used to acquire 2D surface images with a maximum framerate of 340 fps (frames per second) at a full resolution of 2040 × 1088 pixels, which corresponds to a field of view of 3 × 1.5 mm using a four-fold magnification (refer to Figure 1). Therefore, the IVM beam path is coupled into the same beam path used for OCT by a dichroic mirror. Illumination for IVM is achieved by using a custom-made white light LED ring at the bottom of the scanner head. The electrical lens is driven by a 20 kHz pulse width modulated current signal allowing a focal shift of 5 mm within a 15 ms response time. This effectively reduces the video image acquisition down to a maximum of 66 fps.

2.2. Image Acquisition and Data Processing

The focal plane of IVM was adjusted during artificial ventilation by tracking the pleural surface using the depth information from OCT cross sections. Previous experiments with our animal model have shown that lung tissue movement in the axial (z-) direction is limited to a maximum of 1 mm. Using conventional ventilation settings (especially avoiding high inspiration pressure) in addition to the physiological conditions of the thorax after surgery (refer to Section 2.3), the axial shift of the lung tissue is typically 0.6 mm. In OCT cross sections, the signal-to-noise ratio (SNR) is reduced with increasing depth (so-called "roll-off"). In former studies, the roll-off of our system was measured with 8 dB over 1.7 mm depth with an overall SNR of 87 dB. It means that an axial lung tissue movement smaller than 1 mm has just minor effect on OCT image quality and vice versa that the whole tissue movement can be visualized in the OCT cross sections (overall imaging depth 2.8 mm) without the need of refocusing. Therefore, the size of the OCT cross sections is reduced to 64 × 512 pixels (width × depth) leading to 8 ms per B-scan (125 B-scans per second). With this high temporal resolution for visualizing tissue movement, the OCT signal can be used to keep the lung in focus for IVM. Hence, the maximum intensity index, which always corresponds to the lung surface (brightest reflection at air-tissue interface), is extracted from two successive OCT cross sections and the difference of the index position in depth is calculated. Using a lens calibration curve obtained from preliminary tests, the new lens current is set to match the new focus position to grab the IVM image. Because one breath cycle is sufficient for an IVM recording, these video sequences were taken prior to the OCT data acquisition and saved as a video file.

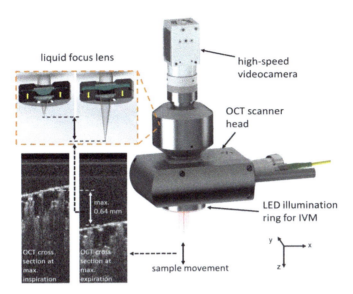

Figure 1. Scanner head for combined image acquisition with optical coherence tomography (OCT) and intravital microscopy (IVM). The adjustment of the focal plane for IVM is achieved by using a current driven liquid lens. The control signal to track the lung surface is generated by calculating the axial shift (in the z-direction) from successive OCT cross sections with a time gap in between of 8 ms (OCT image size of 64 × 512 pixels). The two exemplary OCT cross sections show a lung tissue at max. inspiration and max. expiration. The difference of lung surface in the z-direction between these two B-scans allows for the estimation of maximal lung movement, which was 0.64 mm in the data shown in this paper and up to 1 mm in former studies (not shown here). Both imaging techniques use the same beam path and show the same region of interest. Illumination for IVM is performed by using a LED ring.

Three-dimensional OCT data were acquired using two galvanometer scanners mounted perpendicularly to each other. For one single cross section, the first scanner deflects the light in one direction to measure several A-scans forming the OCT B-scan. Adjacent B-scans were recorded by tilting the second scanner, resulting in a three-dimensional image stack. As mentioned before, the line scan rate of this system is fixed to 12 kHz, resulting in 29 ms for a single cross section (320 × 512 pixels) and an overall measurement time of 9.2 s for an 3D image stack (320 × 320 × 512 pixels). As this takes much longer than a ventilation cycle in rats, other ways to acquire 3D image stacks are needed. Because the ventilation is periodic and changes in the viscoelastic properties of the lung are rather slow, we collected image data over a longer time period and rearranged the image data afterwards. Therefore, during one ventilation cycle, B-scans were acquired at the same line showing the axial tissue movement in one plane with a temporal resolution of 29 ms (using an image size of 320 × 512 pixels). That means that 34 cross sections can be acquired within a ventilation cycle time of 1 s (at an exemplary ventilation rate of 60 breaths per minute). After each ventilation cycle, a trigger signal from the ventilator sets the second scanner to the next position, and cross sections are grabbed continuously at this new position. Each acquired B-scan within a ventilation cycle is consecutively numbered and additionally marked by the corresponding cycle number. In a post-processing step, shown as scheme in Figure 2, all B-scans are re-sorted by using the index number and the trigger number to form the different 3D OCT stacks, each showing a different phase of the ventilation cycle. The general reconstruction is shown in Figure 2 where the rearrangement of OCT cross sections from successive ventilation cycles for the first 3D stack at the beginning of inspiration and the 14th OCT stack at maximum pressure are visualized (inspiration to expiration ratio I:E = 1:1 see below). The overall amount of 3D stacks is 34 with a time gap in between of 29 ms corresponding to the acquisition time for one single B-scan. All 3D stacks are combined to the resulting 4D OCT stack, which shows the complete tissue movement during one breath although measured over several consecutive ventilation cycles. As a consequence of this procedure, the number of volume data per cycle can be increased either by decreasing the ventilation rate or by decreasing the size of the B-scans.

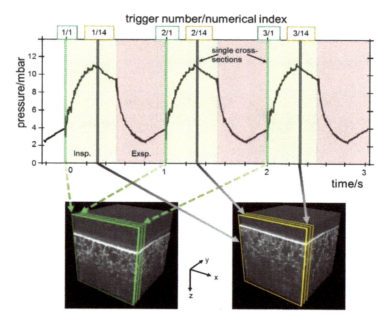

Figure 2. Example of 3D image reconstruction after measurement during uninterrupted artificial ventilation using the numerical index and the corresponding trigger number. Inspiration (light green) to expiration (light red) time ratio was 1:1. Each reconstructed 3D OCT stack shows a different phase of the tissue movement within the ventilation cycle. This procedure is shown in this figure for the first and the 14th 3D stack. Each first B-scan (marked by the numerical index 1) of the consecutive ventilation cycle (marked by the increasing trigger number) is rearranged to form the first 3D stack showing the tissue state at the beginning of the ventilation cycle. The 14th B-scan of each cycle is rearranged to form the 14th 3D stack showing the tissue at maximum pressure.

2.3. Animal Preparation and Artificial Ventilation

All experiments were approved by the animal care and use committee of the local government authorities (Landesdirektion Sachsen; AZ 24-9168.11-1/2012-17) and were performed in accordance with the Guide for Care and Use of Laboratory Animals (Institute of Laboratory Animal Resources, 7th edition, 1996). An in vivo rat model was used for imaging subpleural alveoli. Therefore, female rats (Sprague-Daley, 300 g) were anesthetized to a level at which spontaneous breathing effort is suppressed, intubated with a tracheal tube, and ventilated with 60 breaths per minute in a pressure-controlled manner from 2 to 11 mbar end-expiratory and end-inspiratory pressure, respectively. This deep anesthesia also helps to prevent disturbances in image acquisition. Nevertheless, plateau pressure controlled by servo loop and lung tissue response on overpressure ventilation can vary from cycle to cycle, resulting in small variations of tissue position. Optical access to the lung tissue is necessary for visualizing alveolar structures with OCT and IVM. Therefore, skin and muscle tissue between two ribs was removed by keeping the pleura visceralis intact to maintain the physiological thoracic pressure conditions. This procedure enabled a field of view of about 10 × 5 mm.

Artificial ventilation was performed by a custom-made ventilator described in detail by Schnabel et al. [5]. It consists of two independent custom-made syringe pumps and four magnetic valves to control flow direction. The user interface allows online interactions and a change of ventilator settings. It also provides measured parameters such as ventilated pressure and volume. A 5 V TTL trigger signal is generated after each ventilation cycle and used for gated OCT data acquisition, as described in Section 2.2. The measurement setup is illustrated in Figure 3. IVM and OCT are controlled by one PC running a LabVIEW user interface. A second PC controls the animal ventilator also using a custom-made LabVIEW program. The trigger signal is generated after each ventilation cycle and provided as 5 V TTL signal to a data acquisition device (USB 6009, National Instruments Germany GmbH, Munich, Germany) connected to the imaging PC, where it is used for continuous OCT data acquisition.

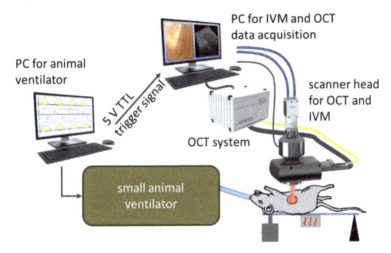

Figure 3. Measurement setup for investigating lung tissue dynamics. The animal is anesthetized and connected to the ventilator. Optical access to alveolar structures for OCT and IVM is achieved by resecting a thoracic window without harming the pleura visceralis in order to keep the physiological thoracic pressure conditions intact. The 5 V trigger signal generated from the ventilator is acquired by the imaging PC using an USB data acquisition board. The IVM image sequence is measured immediately before starting OCT data acquisition.

3. Results

The data acquisition during uninterrupted artificial ventilation with OCT and IVM was tested prior to the animal experiments. A periodic pattern from a function generator mimicking the movement of the pleura up to 1.5 Hz was applied to a loudspeaker with a structured target (data not shown). This enabled us to verify the functionality of the surface-tracking algorithm for IVM and the continuous OCT scanning mode with post-processing rearrangement of the acquired images.

Figure 4 shows exemplary measurement data during in vivo experiments. A collection of the measurement data for IVM and OCT is provided as a supplemental video file online (Video S1: 4D imaging of alveolar dynamics in ventilated rats). OCT and IVM images at four representative time points (during inspiration, at max. inspiration, during expiration and at max. expiration) were chosen to better illustrate the different positions and structures of the alveolar tissue. The pressure-time curve at the top of Figure 4 is a schematic of the used pressure-controlled ventilation mode during experiments performed with a frequency of 60 breaths per minute (f = 1 Hz) from 2 mbar (positive end-expiratory pressure) to 11 mbar (end-inspiratory pressure) with an inspiration-to-expiration ratio (I:E) of 1:1.

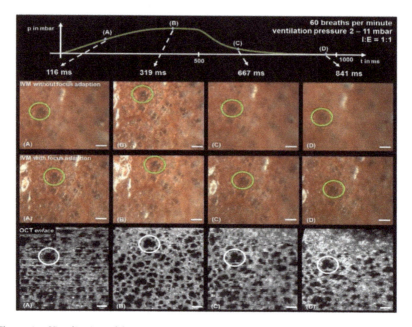

Figure 4. Visualization of lung tissue dynamics during artificial ventilation with OCT and IVM. Pressure-controlled ventilation was performed from 2–11 mbar at a frequency of 1 Hz and an inspiration-to-expiration ratio (I:E) of 1:1 showing a pressure-time course illustrated in the upper diagram. The advantage of the tunable liquid focus lens for IVM can be seen by comparing first and second rows. Most of the tissue movement cannot be imaged properly without focus adaption, so structural information was lost. By using the focus lens, the lung surface can be kept in focal plane over the whole ventilation cycle. The last row shows OCT enface images beneath the pleura calculated from the 3D image stacks. Image reconstruction works well after continuous OCT data acquisition especially in plateau phases (B and D). Small misalignments occur during phases of fast tissue movement (high slope of the pressure curve). One example of corresponding alveolar structures in both imaging techniques (OCT and IVM) is marked by the green and white circles. Scale bar is 150 μm.

It becomes clearly visible that the use of the tunable focus lens is essential to keeping the lung surface in focal plane for IVM over the whole ventilation cycle (compare to first row in Figure 4 "IVM without focus adaption"). Without tracking the lung surface, just a part of the video sequence (in this example at end-inspiratory plateau) is sharp enough to distinguish tissue structures and alveolar boarders. Compared to the second row of Figure 4, one can see the advantages of the electrically tunable focus lens for visualization of the tissue movement in axial direction over the whole ventilation cycle.

The last row of Figure 4 shows the corresponding OCT data stacks after rearrangement of the measured cross sections during uninterrupted artificial ventilation. OCT images show the average tissue movement during one ventilation cycle as enface view calculated from the 3D OCT stacks 25 μm beneath the pleura visceralis. Thereby, the alveoli become visible and the tissue structure can be compared to the IVM images (indicated by the green circle). During phases of high tissue movement in inspiration and expiration, image reconstruction is affected by disturbances as a result of jitter, as the B-scan was not synchronized to the ventilation or of the long overall measurement duration of 5.3 min (34 3D stacks with 320 × 320 × 512 pixels). Therefore, tissue movement is not exactly identical in each ventilation cycle, resulting in slight differences of the tissue structures within the OCT cross sections. This problem was mostly fixed by alignment of the pleura position during post-processing using open source software FIJI [8]. With standard measurement parameters, the imaged area is 1.3×1.3 mm^2 and the pressure step resolution of adjacent 3D stacks is 0.53 mbar (9 mbar/17 stacks half and half for inspiration and expiration).

The advantage of OCT consists in the acquisition of real 3D image data, which provides more accurate information about structure and structural changes during ventilation than 2D images can. As an example, it is not completely clear from 2D IVM images whether all alveoli shrink during expiration or whether some of them move deeper into the tissue and therefore disappear from the image plane. This can be verified with 3D OCT data. For future experiments, the 3D volume of alveoli in the different phases of the ventilation cycle will be measured. In this experiment, the analysis of the area changes of the alveolar structures during ventilation was performed manually at five exemplary alveoli from OCT enface images, due to the large amount of image data and to the fact that no automatic algorithm for structure segmentation was yet available. All enface images were taken at the same position, 25 µm beneath the pleura, to assure comparability of the measured area. The segmentation was done manually in FIJI for five alveoli (marked by the numbers in the first image of Figure 5). Cross sections of Alveolus 1 (marked with the green dashed line) were taken to verify the findings from the enface segmentation regarding the area changes and the pressure-area loops (diagram of Figure 5). To illustrate these changes during ventilation, all measured alveolar areas were normalized to the area at end-expiratory pressure of 2 mbar. Therefore, all calculated pressure-area loops start at 1 (axis of ordinate) and 2 mbar (axis of abscissae). The measured changes from expiration to inspiration are within the range of 15%–25%, which is in good agreement to the results of other studies [9,10].

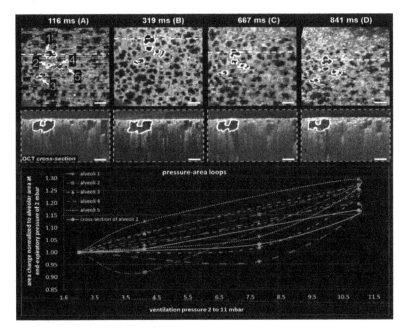

Figure 5. Segmentation of alveolar structures from OCT images and visualization of the pressure-area curves. First row shows OCT enface images 25 µm beneath the pleura and the five alveolar structures chosen for area measurement over the ventilation cycle. To verify these measurements, the cross-sectional view (cross sections taken along the dashed green line) was also taken and the area was measured, too. The pressure-area loops were plotted by normalizing the areas to the initial area at end-expiratory pressure of 2 mbar. The diagram shows the same course for all measurements with a small gap between inspiratory and expiratory area change. Scale bar is 150 µm.

4. Discussion

Investigating dynamic processes with high spatial and temporal resolution is a challenging task. Therefore, suitable imaging techniques and data acquisition protocols are necessary. We showed that OCT and intravital microscopy are promising tools for matching the effort of appropriated imaging techniques for the visualization of lung tissue movement during uninterrupted artificial ventilation in rats. A high-speed video camera and an electrically tunable focus lens controlled by the depth information obtained from fast scanning OCT cross sections were used to keep the lung surface in the focal plane during axial movement. This, allowed acquisition of video sequences over the whole ventilation cycle. The use of an external trigger signal from the animal ventilator in combination with a continuous OCT data acquisition mode enabled to employ a conventional OCT system (line scan rate 12 kHz) for the investigation of fast dynamic processes in vivo without the need of specialized high-speed OCT systems. Some limitations still remain beside these advantages. The continuous scanning mode enables the visualization of tissue movement over one ventilation cycle by combining images that were measured over several cycles. While this scheme works well during plateau phases (end of inspiration, end of expiration), jitter and other disturbances (from the tissue response on artificial ventilation) arising during phases of fast tissue movement result in small misalignments during the reconstruction of each individual 3D stack. This can be seen during the fast tissue movements in inspiration and expiration. Further work will deal with these disturbances on the one hand by improving the handshaking between the animal ventilator and the OCT system and on the other hand by developing a post-processing algorithm for the automatic alignment of adjacent cross sections. Especially, the first point has a major impact on the jitter of the OCT data shown in this paper. As OCT data capture and ventilation are not synchronized, images with a shift of up to one frame period (29 ms) are combined to one volume. This problem will be avoided in future measurements by implementing bidirectional handshaking. For this purpose, the OCT system will also give a trigger signal to the ventilator when an image is still acquired, and the ventilator will wait up to 29 ms before triggering the OCT system. This additional time of 29 ms will influence the breathing rate only marginally, but will help prevent the jitter in the rearranged 3D OCT stacks. Furthermore, we will combine other imaging techniques such as confocal fluorescence microscopy to gather functional information of elastic fibers surrounding the alveoli during artificial uninterrupted ventilation. Finally, new segmentation algorithms will be implemented to handle the mass of image data, and automatically extract alveolar volume changes and tissue characteristics from the 3D OCT stacks.

5. Conclusions

This feasibility study shows that emerging imaging techniques such as OCT in combination with IVM are promising tools for gaining new insights into alveolar dynamics during artificial ventilation in animal experiments. Such measurements can provide necessary information about tissue behavior and promote an understanding of micromechanics that can lead to the development of more protective ventilation strategies.

Supplementary Materials: The following are available online at http://www.mdpi.com/2076-3417/7/3/287/s1, Video S1: 4D imaging of alveolar dynamics in ventilated rats.

Acknowledgments: This project was supported by the German Research Foundation (DFG) "Protective artificial Respiration" (PAR)—KO 1814/6-1 and KO 1814/6-2.

Author Contributions: All authors contributed equally to the paper. Christian Schnabel and Edmund Koch developed the OCT data acquisition and the small animal ventilator. Christian Schnabel performed all the measurements and segmentations and wrote the paper. Maria Gaertner prepared the animal experiments and developed a method of visualizing the 4D OCT data.

Conflicts of Interest: The authors declare no conflict of interest. The funding sponsors had no role in the design of the study; in the collection, analyses, or interpretation of data; in the writing of the manuscript; or in the decision to publish the results.

References

1. WHO. *World Health Statistics*; WHO: Geneva, Switzerland, 2016.
2. Petrucci, N.; Feo, C.D. Lung protective ventilation strategy for the acute respiratory distress syndrome. *Cochrane Database Syst. Rev.* **2013**, *2*.
3. Serpa Neto, A.S.; Cardoso, S.O.; Manetta, J.A.; Pereira, V.G.M.; Espósito, D.C.; Pasqualucci, M.O.; Damasceno, M.C.T.; Schultz, M.J. Association between Use of Lung-Protective Ventilation with Lower Tidal Volumes and Clinical Outcomes among Patients Without Acute Respiratory Distress Syndrome a Meta-analysis. *JAMA* **2012**, *308*, 1651–1659. [CrossRef] [PubMed]
4. Unglert, I.C.; Namati, E.; Warger, C.W., II; Liu, L.; Yoo, H.; Kang, D.; Bouma, E.B.; Tearney, J.B. Evaluation of optical reflectance techniques for imaging of alveolar structure. *JBO* **2012**, *17*, 071303. [CrossRef] [PubMed]
5. Schnabel, C.; Gaertner, M.; Kirsten, L.; Meissner, S.; Koch, E. Total liquid ventilation: A new approach to improve 3D OCT image quality of alveolar structures in lung tissue. *Opt. Express* **2013**, *21*, 31782–31788. [CrossRef] [PubMed]
6. Kirsten, L.; Gaertner, M.; Schnabel, C.; Meissner, S.; Koch, E. Four-dimensional imaging of murine subpleural alveoli using high-speed optical coherence tomography. *J. Biophotonics* **2013**, *6*, 148–152. [CrossRef] [PubMed]
7. Meissner, S.; Knels, L.; Krueger, A.; Koch, T.; Koch, E. Simultaneous three-dimensional optical coherence tomography and intravital microscopy for imaging subpleural pulmonary alveoli in isolated rabbit lungs. *JBO* **2009**, *14*, 051020. [CrossRef] [PubMed]
8. Schindelin, J.; Arganda-Carreras, I.; Frise, E.; Kaynig, V.; Longair, M.; Pietzsch, T.; Preibisch, S.; Rueden, C.; Saalfeld, S.; Schmid, B.; et al. Fiji: An open-source platform for biological-image analysis. *Nat. Methods* **2012**, *9*, 676–682. [CrossRef] [PubMed]
9. Mertens, M.; Tabuchi, A.; Meissner, S.; Krueger, A.; Schirrmann, K. Alveolar dynamics in acute lung injury: Heterogeneous distension rather than cyclic opening and collapse. *Crit. Care Med.* **2009**, *37*, 2604–2611. [CrossRef] [PubMed]
10. Gaertner, M.; Cimalla, P.; Meissner, S.; Kuebler, W.M.; Koch, E. 3D simultaneous optical coherence tomography and confocal fluorescence microscopy for investigation of lung tissue. *J. Biomed. Opt.* **2012**, *17*, 071310. [CrossRef] [PubMed]

Article

Spectral Domain Optical Coherence Tomography for Non-Destructive Testing of Protection Coatings on Metal Substrates

Marcel Lenz [1,*], Cristian Mazzon [2], Christopher Dillmann [3], Nils C. Gerhardt [1], Hubert Welp [3], Michael Prange [2] and Martin R. Hofmann [1]

[1] Department of Photonics and Terahertz Technology, Ruhr Universität Bochum, Universitätsstraße 150, 44780 Bochum, Germany; Nils.Gerhardt@rub.de (N.C.G.); Martin.Hofmann@rub.de (M.R.H.)

[2] Deutsches Bergbau-Museum Bochum, Material Sciences, Herner Straße 45, 44787 Bochum, Germany; Christian.Mazzon@bergbaumuseum.de (C.M.); Michael.Prange@bergbaumuseum.de (M.P.)

[3] Faculty of Electrical Engineering, Information Technology and Business Engineering, Technische Hochschule Georg Agricola, Herner Straße 45, 44787 Bochum, Germany; Christopher.Dillmann@thga.de (C.D.); Hubert.Welp@thga.de (H.W.)

* Correspondence: Marcel.Lenz@rub.de; Tel.: +49-234-322-9334

Academic Editor: Michael Pircher
Received: 1 March 2017; Accepted: 31 March 2017; Published: 6 April 2017

Abstract: In this paper we demonstrate that optical coherence tomography (OCT) is a powerful tool for the non-destructive investigation of transparent coatings on metal substrates. We show that OCT provides additional information which the common practice electrical impedance spectroscopy (EIS) cannot supply. First, coating layer thicknesses were measured and compared with reference measurements using a magnetic inductive (MI) measurement technique. After this validation of the OCT measurements, a customized sectioned sample was created to test the possibility to measure coating thicknesses with underlying corrosion, which cannot be analyzed accurately by MI or EIS measurements. Finally, we demonstrate the benefit of OCT on a standard sample. The OCT measurements provide the correct coating layer thickness with high lateral resolution and even enable metal and corrosion layers to be distinguished from each other.

Keywords: optical coherence tomography; non-destructive testing; layer thickness measurement, protective coatings

---

1. Introduction

In built heritage conservation and for museum objects, long-term conservation (e.g., in old mining facilities) is crucial. Ranging from small instruments to large buildings such as winding towers, protective coatings have to be applied. Therefore, common non-transparent protective coatings are not applicable, because the historical character of these cultural heritages must be maintained. Moreover, further degradation has to be hindered in order to retain the cultural heritage for future generations. To evaluate the long-term capability of different coatings, it is essential to determine the layer thickness of the coatings after they are applied on the sample as well as their durability over time. Additionally, early detection of impurities is necessary to prevent further damage. Moreover, it is desired to differentiate between corroded and non-corroded material. However, so far no perfectly satisfying technique for this analysis is available. Optical coherence tomography (OCT) might be a promising tool to fulfill the mentioned needs. Until now, the acceptance of OCT during the last two decades results from its great benefit in biomedical applications such as ophthalmology and dermatology [1–3].

Nevertheless, many application capabilities of OCT have not yet been exploited. For non-destructive testing, only a few application examples have been published so far; for example, the investigation of art [4–6], polymers [7,8], subsurface defect detection [9], strain field mapping by using polarization-sensitive OCT [10], material observation by en-face OCT [11], pharmaceutical film coatings [12,13], and the investigation of silicon integrated circuits [14–16]. In this case we apply OCT to metal objects with transparent protective coatings in order to obtain information that the actual common practice electrochemical impedance spectroscopy (EIS) cannot deliver [17–29]. In EIS, an AC voltage is swept, and the current is measured. Assuming different layers to be either resistive or capacitive, a Bode or Nyquist plot can be made [30]. From these plots, an educated guess about the functionality of the protective coating can be done. This works very well for layer thicknesses in the atomic scale, but no actual thickness value is provided. However this metrology suffers from the drawback of insufficient lateral resolution due to the fact that the signal is integrated over an area of about 1.5 cm^2, which is a typical size of an EIS probe. Additionally, a long duration for the data acquisition, the questionable accuracy of the equivalent circuit diagram, and the uncertainty about the actual layer thickness are fundamental disadvantages of this method [30]. Furthermore, no imaging capability is typically given. Because we are analyzing the depth profiles of our OCT measurements in order to determine a coating layer thickness, the comparison with EIS measurements is not very reasonable. Therefore, another metrology has to be evaluated.

Currently, a fast and non-destructive method to determine the actual coating layer thickness on steel or iron is to use a magnetic inductive (MI) thickness measurement system [31]. By measuring the retarded magnetic field of a sample covered with a protective coating layer, the thickness can be estimated by comparing it with a reference sample of bare metal. However, this technique allows only a point-wise layer thickness measurement. Moreover, it can only be applied for non-corroded material, because a corrosion layer superimposes with the coating layer, leading to an inaccuracy of the measurement [32]. In a feasibility study, Antoniuk and co-workers proved that OCT is capable of monitoring protective metal coatings [33]. Here we use OCT and apply post-processing algorithms to detect the coating layer and introduce a scattering layer for the underlying material (metal or corrosion). Moreover, we determine their thicknesses, distinguish between the materials, and unveil breaches of the coating. In order to validate our findings, the layer thicknesses were additionally measured with an MI measurement system and compared with expected values.

2. Methodology

2.1. Systems

All OCT measurements were performed with a commercially available OCT system (Thorlabs Ganymede, Dachau/Munich, Germany). Operating with a center wavelength of 930 nm and a bandwidth of 154 nm, the Thorlabs Ganymede Spectral Domain OCT has an axial and lateral resolution of 6 μm and 8 μm, respectively. The maximum penetration depth is 2.73 mm, which was validated by the manufacturer. All these values hold for a refractive index of $n = 1$. Snapshots using the built-in camera with 480 × 640 pixels were taken for orientation. By using an interpolation, these images were resized, in order to have the same dimensionality as the OCT dataset, but not the same resolution (Figure 1) and then compared with a microscopic image taken with a Keyence VHX 2000 microscope (KEYENCE Deutschland GmbH, Neu-Isenburg, Germany).

For the magnetic inductive measurements, the Fischer® Dualscope® (Helmut Fischer GmbH, Sindelfingen-Maichingen, Germany) FMP40 used with the FGAB 1.3 probe was utilized. This probe can be used for non-ferrous and isolating materials. The radius of the measuring tip is 0.75 mm, having an accuracy of ±1 μm and a precision of 0.3 μm for coatings up to 100 μm thickness.

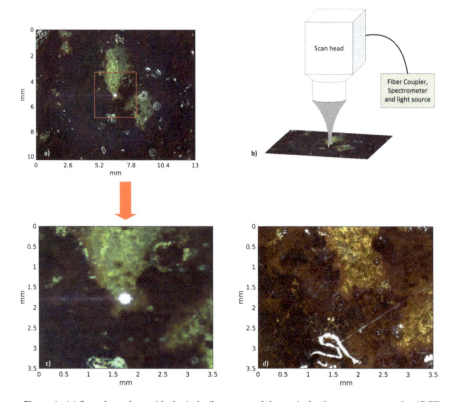

Figure 1. (**a**) Snapshot taken with the in-built camera of the optical coherence tomography (OCT) system of a standard sample; (**b**) measurement schematic; (**c**) interpolated video image; (**d**) microscopic image of the region of interest.

2.2. Samples

In this subsection, we introduce the eight different samples that were investigated during our study. For the validation of our OCT thickness measurement, six samples with bare metal and varying coating layer thicknesses were prepared. These samples were measured with OCT and with the MI system as well. In addition, theoretically expected values were calculated from the weight of the deposited material.

Section 2.2.2 describes the fabrication of a customized sectioned sample. With this sample, we demonstrated that OCT is capable of monitoring the coating layer thickness (even for corroded areas), and that both regions with metal or corrosion can be separated.

At last, a standard sample is shown in Section 2.2.3, where no defined region of corrosion is present. The purpose of this sample was to show that OCT is capable of distinguishing between metal and corrosion, no matter how it is distributed. Moreover, it was demonstrated that breaches of the coating layer can be displayed.

2.2.1. Coated Bare Metal Samples

For the measurements of coating layer thickness, bare metal was covered by the coating material Paraloid B48N (copolymer of methyl methacrylate, Kremer Pigmente GmbH & Co. KG, Aichstetten, Germany), which has a refractive index of 1.89 [34]. At the beginning, six samples with different layer thicknesses were customized in order to see the limits of our OCT system for determining the layer thickness. The coating layer was brushed on five of the samples, and on one sample it was sprayed. The thinnest coating that could be applied was greater than 6 μm by brushing it onto the metal. The theoretically expected value for the layer thickness t can be determined as follows [35]:

$$t = \frac{\Delta w}{\rho * A} \quad , \tag{1}$$

where Δw is the weight difference after the coating was applied, ρ the density of the coating, and A the area of the sample. For orientation, a mask was put on the samples covering the complete sample except for a circular area of 2 cm². The center of the circle was analyzed point-wise by magnetic inductive measurements with the Fischer® Dualscope® FMP40, generating 50 data points. In the center of the circle, an area of 3.5 × 3.5 mm² was imaged with the OCT system. This was the largest area possible with our OCT system in order to fulfill the sampling theorem for the lateral resolution. Video images of the region of interest were taken with the built-in camera of the Ganymede system, as well as with the microscope.

2.2.2. Customized Sectioned Sample

After the different coating layer thicknesses were measured, we created a customized sample which was divided into four regions containing metal and corroded metal with and without a protective coating layer (sample 7, Figure 2).

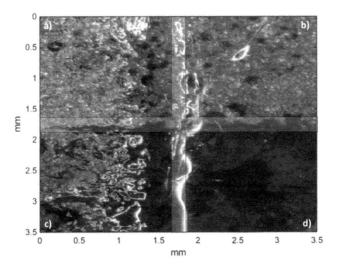

Figure 2. Microscopic image of the customized sectioned sample: (**a**) pure metal; (**b**) pure metal with coating layer on top; (**c**) corroded area; (**d**) corroded material with coating layer on top.

In order to prepare the desired sample, we first corroded the complete sample artificially by embedding the metal into a natrium chloride bath and exposing it in a climatic chamber afterwards. In more detail, the sample was created following the procedure described in [36]:

1. Embedding in a 5% NaCl-solution
2. Exposure to 100% relative humidity for 8 h at 40 °C
3. Drying for 16 h at room temperature
4. Repetition of steps 1 to 3 five times
5. Exposure to outdoor weathering for 2 months inclination 45°, south-side exposure
6. Desalination by immersion with distilled water

7. Soft air abrasive cleaning with walnut shells (pressure 8 bar, distance 50 cm)
8. Additional air abrasive cleaning for half of the sample (upper part) until bare metal was reached.
9. Application of coatings by brush, five to six repetitions (resulting layers), intermediate drying period of 3–4 h for each layer.

To avoid infiltration of the coating, a sellotape stripe was glued before applying the Paraloid B48N onto the sample. However, even by the use of a sellotape stripe, some infiltration zone could be observed.

2.2.3. Standard Sample

Finally, we also prepared a standard sample as is typically used for EIS (sample 8). In contrast to the customized sectioned sample, no defined region for metal or corrosion was present. The preparation was done as described in Section 2.2.2, but without additional air abrasive cleaning (step 8). Furthermore, the complete sample was covered with Paraloid B48N.

2.3. Post-Processing

Post-processing of the OCT data was done using Matlab (version 2015a, MathWorks, Natick, MA, USA). By utilizing the peak find function in Matlab, several peaks in the OCT A-Scans (line scans perpendicular to the surface) were detected. By choosing optimized parameters and adding appropriate constraints (e.g., thresholding), only the three peaks of interest were determined. The minimum peak prominence and the minimum peak height were the parameters that were optimized.

Using a smoothing algorithm, outliers from the peak detection could be suppressed. Phantom layers caused by autocorrelation were not detected by optimizing the algorithm in terms of intensity. The three peaks of interest correspond to the beginning of the coating layer, the end of the coating layer, and the end of the scattering layer (metal or corrosion). Composing OCT A-Scans along a line parallel to the surface provides a so called B-Scan. A typical B-Scan containing a coating and a scattering layer is shown in Figure 3. Assuming a refractive index of 1.89 for Paraloid B48N, the thickness of the coating layer was calculated by computing the path difference between the second and the first peak and dividing by the refractive index of 1.89. The second layer that was calculated was considered as a scattering layer. Thus, the distance between upper and lower boundary of the material was identified as the scattering layer thickness.

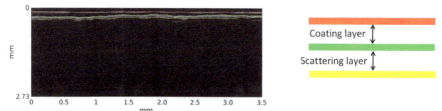

Figure 3. OCT B-Scan of corroded metal (sample 8): **red** line is the top of the coating layer, **green** line determines the end of the coating and the beginning of the corrosion, **yellow** line determines the end of the scattering-induced layer. Axes are given as optical path length $n \times d$.

3. Results

3.1. Coated Bare Metal Samples

In Figure 4, the OCT thickness calculations for samples 1–6 are compared to the thickness measurements using the MI system. The mean value for a three-dimensional OCT data set and an error bar using the standard deviation were plotted. The value of the MI system is an average over 50 sampling points that were chosen manually, trying to be as close as possible to the center of the circle. These values and their standard deviation can also be seen for comparison in Table 1.

Table 1. Comparison between optical coherence tomography measurement, magnetic inductive (MI) measurement, and the expected theoretical values. The sprayed sample is marked with an *. All thickness values are given in µm.

Concept	Sample 1	Sample 2	Sample 3	Sample 4	Sample 5	Sample 6 *
OCT	6.86	22.97	42.27	48.91	97.00	80.24
std. OCT	0.65	1.61	2.01	2.52	2.74	1.85
MI	7.16	21.89	43.85	56.57	83.34	88.70
std. MI	1.14	1.5	3.8	9.81	9.47	2.5
expected th. value	11.27	27.44	53.70	64.10	94.30	86.26

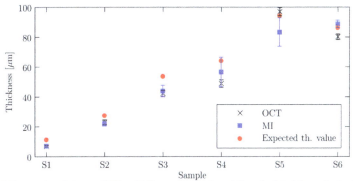

Figure 4. Comparison between OCT and MI measurements and the calculated theoretical value.

For thin coating layers, the results of both techniques are in very good agreement with each other. For thicker layers, the values differ. This is due to the fact that the layer had to be brushed or sprayed multiple times for thicker samples. This leads to much more pronounced height variations, as can be seen in Figure 5. As the MI measurements are performed with a hand-held probe, a larger area might be investigated. This might be the reason for a higher standard deviation. Another possibility for the measurement uncertainty could be a density change, leading to a refractive index change as it was previously observed for thicker coatings [13]. For sample S6, increased scattering could be observed, supporting this theory.

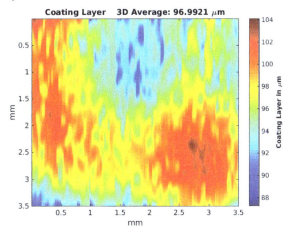

Figure 5. 3D heat map of the calculated layer thickness for the OCT measurement of sample 5.

3.2. Customized Sectioned Sample

Figure 6 shows the coating layer thickness for our customized sectioned sample. As could be guessed from the microscopic image in Figure 2, there is no sharp transition from non-coated to protected material.

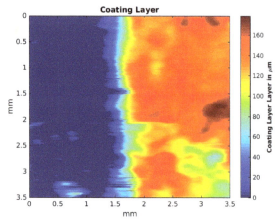

Figure 6. 3D heat map of the calculated coating layer thickness for the OCT measurement of the customized sample (sample 7).

The transition zone can be clearly identified. Moreover, it can be seen that the coating adheres better on the polished metal than on corrosion, as expected. Our measurements of this sample clearly show that OCT not only provides the layer thicknesses but also enables corroded and non-corroded areas to be clearly distinguised (Figure 7).

The scattering layer thicknesses for the customized sample are shown in Figure 7. For the corroded area, a higher scattering layer thickness can be seen. However, these differences are not as pronounced as in Section 3.3. This is because the complete sample was first corroded artificially. Then, abrasive cleaning was performed to remove the corrosion, but it did not provide a surface as smooth as the initial bare metal. Both findings will be further visualized with measurements on the standard sample in the next subsection.

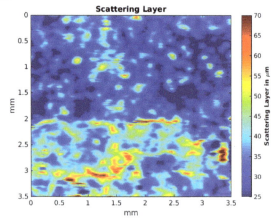

Figure 7. 3D heat map of the calculated scattering layer thickness for the OCT measurement of the customized sample (sample 7).

3.3. Standard Sample

At last, the standard sample described in Section 2.2.3 was investigated with our OCT system. The thickness of the coating and the scattering layer was determined for the complete three-dimensional data set, and a heat map was generated. Figure 8 shows the microscopic image, the coating layer thickness, and the scattering layer thickness. There is a slight shift between the OCT measurement and microscopic image. This is because no additional markers were used in order to prevent any damage to the sample. In Figure 8b, a breach in the coating can be clearly seen (dark blue area). Moreover, Figure 8c shows the difference between pure metal and corroded metal. Pure metal can be recognized as dark blue color. With this sample, we were able to distinguish between metal and corrosion with μm resolution without a priori knowledge of the distribution of the material. Furthermore, small breaches in the transparent protective coating could be displayed.

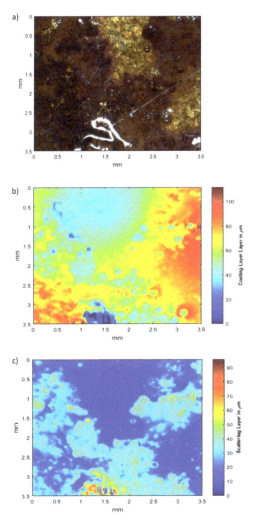

Figure 8. (**a**) Microscopy image of a standard sample (sample 8); (**b**) heat map of the calculated layer thickness for the OCT measurement; (**c**) heat map of the calculated scattering layer thickness for the OCT measurement.

4. Discussion

By employing optical coherence tomography as a new powerful tool for the non-destructive testing of metal coatings, additional valuable information about the sample is available. In contrast to electrochemical impedance spectroscopy, OCT provides an imaging modality. Moreover, lateral information about the sample with µm resolution is available. Further studies with artificial degradation will be performed to evaluate if this lateral resolution is sufficient to estimate the durability of the coatings.

We have proven that the determined thicknesses of the coatings are in good agreement with the theoretically expected values and with the MI measurements. We did not have to change the parameters for different layer thicknesses. However, if a different coating material were to be investigated, these parameters might be adjusted. When analyzing thicker coatings, the standard deviation for the MI measurements tends to rise, but not for the OCT measurements. The differences between the theoretical values and the OCT measurements come from a non-uniform distribution of the material, which cannot be considered at all in the simple theoretical estimation. Another explanation for the uncertainty could be a change in the refractive index, caused by a density change. This deviation appears in particular for thicker samples. Furthermore, a cross-sectional microscopic image would be beneficial to validate our findings. However, the preparation of a cross-sectional sample profile image is difficult to achieve without compromising the sample. During the different preparatory work steps (e.g., cutting, embedding (curing process), and grinding at the end), high temperature can be reached. This is typically destructive for products with low temperature stability, such as Paraloid B48N (glass transition temperature, $Tg = 50\ °C$). Though we have shown that our OCT measurements are already superior to the common MI technique, future applications may induce further demands in terms of axial resolution, field of view, or image acquisition speed:

In our study the thickness of the transparent protective coatings was always more than 6 µm, so the axial resolution of our OCT system was fully sufficient to differentiate between the different layers. Nevertheless, if thinner coatings would have to be analyzed, OCT systems with 1 µm axial resolution have already been demonstrated [37].

In our study, we had a field of view of $3.5 \times 3.5\ mm^2$ in which we were able to see the different regions of interest (i.e., corrosion and metal with and without protective coating). However, in future applications, a considerably larger field of view may be required. For that purpose, an OCT system with an enormous field of view in the meter range has been recently reported by Wang et al. [38].

The A-Scan rate of our system was 29 kHz. The time needed to scan an area of $3.5 \times 3.5\ mm^2$ was approximately one minute, which is much faster than an analysis with common EIS systems. However, much less acquisition time may be desired when large objects are investigated. In that case, high-speed OCT systems with scan rates in the MHz range may be used [39].

We are confident that in the future OCT will enable the investigation of real mining objects that have to be conserved, and that all requirements for that purpose will be covered.

5. Conclusions

In this paper, we have demonstrated the benefit of employing optical coherence tomography as a new tool for the investigation of transparent protection coatings. Our comparison with a magnetic inductive layer thickness methodology (which is a commonly used method) validated that precise thickness measurements with OCT are possible, even with underlying corrosion layers. To our knowledge, no other non-destructive modality has provided comparable information. Moreover, by using OCT, the differentiation between metal and corroded material becomes possible with µm resolution in the lateral dimension. In the future we will perform long-term observations of metal samples with OCT and EIS in order to create additional knowledge in the field of protective layer coatings for preserving cultural heritage in the future. After this first proof of principle, we further want to evaluate the quality of different coatings in terms of performance and long-term stability.

Appl. Sci. **2017**, *7*, 364

Acknowledgments: This work was financially supported by the Stiftung Rheinisch-Westfälischer Technischer Überwachungsverein (RWTÜV, Essen, Germany) (Project Numbers: S189/10024/2015, S189/10025/2015). We acknowledge support by the the DFG Open Access Publication Funds of the Ruhr-Universität Bochum (Bochum, Germany).

Author Contributions: Marcel Lenz performed the OCT measurements, analyzed the data and wrote the paper. Cristian Mazzon prepared the samples and did the MI measurements. Christopher Dillmann worked on the OCT analysis. Nils C. Gerhardt and Hubert Welp supervised the OCT work and the OCT analysis. Michael Prange supervised the sample preparation and MI measurements and revised the manuscript. Martin R. Hofmann supervised the complete work and edited the manuscript.

Conflicts of Interest: The authors declare no conflict of interest.

References

1. Hee, M.R.; Izatt, J.A.; Swanson, E.A.; Huang, D.; Schuman, J.S.; Lin, C.P.; Puliafito, C.A.; Fujimoto, J.G. Optical coherence tomography of the human retina. *Arch. Ophthalmol.* **1995**, *113*, 325–332.
2. Welzel, J.; Lankenau, E.; Birngruber, R.; Engelhardt, R. Optical coherence tomography of the human skin. *J. Am. Acad. Dermatol.* **1997**, *37*, 958–963.
3. Drexler, W.; Fujimoto, J.G. *Optical Coherence Tomography: Technology and Applications*; Springer Science & Business Media: Berlin/Heidelberg, Germany, 2008.
4. Rouba, B.; Karaszkiewicz, P.; Tymioska-Widmer, L.; Iwanicka, M.; Góra, M.; Kwiatkowska, E.; Targowski, P. Optical coherence tomography for non-destructive investigations of structure of objects of art. *J. Nondestruct. Test.* **2008**, *13*, doi:10.1117/12.813477.
5. Targowski, P.; Iwanicka, M.; Tymioska-Widmer, L.; Sylwestrzak, M.; Kwiatkowska, E.A. Structural examination of easel paintings with optical coherence tomography. *Acc. Chem. Res.* **2009**, *43*, 826–836.
6. Targowski, P.; Iwanicka, M. Optical coherence tomography: Its role in the non-invasive structural examination and conservation of cultural heritage objects—A review. *Appl. Phys. A* **2012**, *106*, 265–277.
7. Wiesauer, K.; Pircher, M.; Götzinger, E.; Hitzenberger, C.K.; Oster, R.; Stifter, D. Investigation of glass–fibre reinforced polymers by polarisation-sensitive, ultra-high resolution optical coherence tomography: Internal structures, defects and stress. *Compos. Sci. Technol.* **2007**, *67*, 3051–3058.
8. Stifter, D.; Wiesauer, K.; Wurm, M.; Schlotthauer, E.; Kastner, J.; Pircher, M.; Götzinger, E.; Hitzenberger, C. Investigation of polymer and polymer/fibre composite materials with optical coherence tomography. *Measur. Sci. Technol.* **2008**, *19*, 074011, doi:10.1088/0957-0233/19/7/074011.
9. Duncan, M.D.; Bashkansky, M.; Reintjes, J. Subsurface defect detection in materials using optical coherence tomography. *Opt. Express* **1998**, *2*, 540–545.
10. Stifter, D.; Burgholzer, P.; Höglinger, O.; Götzinger, E.; Hitzenberger, C.K. Polarisation-sensitive optical coherence tomography for material characterisation and strain-field mapping. *Appl. Phys. A* **2003**, *76*, 947–951.
11. Wiesauer, K.; Pircher, M.; Götzinger, E.; Bauer, S.; Engelke, R.; Ahrens, G.; Grützner, G.; Hitzenberger, C.; Stifter, D. En-face scanning optical coherence tomography with ultra-high resolution for material investigation. *Opt. Express* **2005**, *13*, 1015–1024.
12. Zhong, S.; Shen, Y.C.; Ho, L.; May, R.K.; Zeitler, J.A.; Evans, M.; Taday, P.F.; Pepper, M.; Rades, T.; Gordon, K.C.; et al. Non-destructive quantification of pharmaceutical tablet coatings using terahertz pulsed imaging and optical coherence tomography. *Opt. Lasers Eng.* **2011**, *49*, 361–365.
13. Lin, H.; Dong, Y.; Shen, Y.; Zeitler, J.A. Quantifying pharmaceutical film coating with optical coherence tomography and terahertz pulsed imaging: An evaluation. *J. Pharm. Sci.* **2015**, *104*, 3377–3385.
14. Serrels, K.A.; Renner, M.K.; Reid, D.T. Optical coherence tomography for non-destructive investigation of silicon integrated-circuits. *Microelectron. Eng.* **2010**, *87*, 1785–1791.
15. Stifter, D. Beyond biomedicine: A review of alternative applications and developments for optical coherence tomography. *Appl. Phys. B* **2007**, *88*, 337–357.
16. Nemeth, A.; Leiss-Holzinger, E.; Hannesschläger, G.; Wiesauer, K.; Leitner, M. *Optical Coherence Tomography-Applications in Non-Destructive Testing and Evaluation*; INTECH Open Access Publisher: Rijeka, Croatia, 2013.
17. Mazzon, C.; Letardi, P.; Brüggerhoff, S. Development of Monitoring Techniques for Coatings Applied to Industrial Heritage. In Proceedings of the BigStuff 2015–Technical Heritage: Preserving Authenticity-Enabling Identity, Lewarde, France, 3–4 September 2015.

18. Mansfeld, F. Electrochemical impedance spectroscopy (EIS) as a new tool for investigating methods of corrosion protection. *Electrochim. Acta* **1990**, *35*, 1533–1544.

19. Letardi, P.; Spiniello, R. Characterisation of Bronze Corrosion and Protection by Contact-Probe Electrochemical Impedance Measurements. In Proceedings of the Metal 2001: Proceedings of the International Conference on Metals Conservation, Santiago, Chile, 2–6 April 2001; Western Australian Museum: Fremental, Australia, 2004; pp. 316–319.

20. Mottner, P. Investigations of Transparent Coatings for the Conservation of Iron and Steel Outdoor Industrial Monuments. In Proceedings of the Metal 2001: Proceedings of the International Conference on Metals Conservation, Santiago, Chile, 2–6 April 2001; Western Australian Museum: Fremental, Australia, 2004; pp. 279–288.

21. Angelini, E.P.M.V.; Assante, D.; Grassini, S.; Parvis, M. EIS measurements for the assessment of the conservation state of metallic works of art. *Int. J. Circuits Syst. Signal Process.* **2014**, *8*, 240–245.

22. Cano, E.; Crespo, A.; Lafuente, D.; Barat, B.R. A novel gel polymer electrolyte cell for in-situ application of corrosion electrochemical techniques. *Electrochem. Commun.* **2014**, *41*, 16–19.

23. Cano, E.; Lafuente, D.; Bastidas, D.M. Use of EIS for the evaluation of the protective properties of coatings for metallic cultural heritage: A review. *J. Solid State Electrochem.* **2010**, *14*, 381–391.

24. Ellingson, L.A.; Shedlosky, T.J.; Bierwagen, G.P.; de la Rie, E.R.; Brostoff, L. The use of electrochemical impedance spectroscopy in the evaluation of coatings for outdoor bronze. *Stud. Conserv.* **2013**, *49*, 53–62.

25. Hallam, D.; Thurrowgood, D.; Otieno-Alego, V.; Creagh, D. An EIS Method for Assessing Thin Oil Films Used in Museums. In Proceedings of the Metal 2004 National Museum of Australia Canberra ACT, Canberra, Australia, 4–8 October 2004; Volume 4, p. 8.

26. Letardi, P. *Corrosion and Conservation of Cultural Heritage Metallic Artefacts: 7. Electrochemical Measurements in the Conservation of Metallic Heritage Artefacts: An Overview*; Elsevier Inc.: Amsterdam, The Netherlands, 2013.

27. Letardi, P.; Beccaria, A.; Marabelli, M.; D'ERCOLI, G. Application of Electrochemical Impedance Measurements as a Tool for the Characterization of the Conservation and Protection State of Bronze Works of Art. In Proceedings of the Conférence Internationale sur la Conservation des Métaux, Draguignan, France, 27–29 May 1998; pp. 303–308.

28. Mansfeld, F. Use of electrochemical impedance spectroscopy for the study of corrosion protection by polymer coatings. *J. Appl. Electrochem.* **1995**, *25*, 187–202.

29. Price, C.; Hallam, D.; Heath, G.; Creagh, D.; Ashton, J. An Electrochemical Study of Waxes for Bronze Sculpture. In Proceedings of the Conférence Internationale sur la Conservation des Métaux, Semur en Auxois, France, 25–28 September 1995; pp. 233–241.

30. Amirudin, A.; Thieny, D. Application of electrochemical impedance spectroscopy to study the degradation of polymer-coated metals. *Prog. Org. Coat.* **1995**, *26*, 1–28.

31. Fischer, H. Electromagnetic Probe for Measuring the Thickness of Thin Coatings on Magnetic Substrates. US Patent 4,829,251, 9 May 1989.

32. Wolfram, J.; Brüggerhoff, S.; Eggert, G. Better Than Paraloid B-72? Testing Poligen Waxes as Coatings for Metal Objects. In Proceedings of the METAL 2010: Proceedings of the Interim Meeting of the ICOM-CC Metal Working Group, Clemson University, Clemson, SC, USA, 11–15 October 2010; pp. 124–131.

33. Antoniuk, P.; Strąkowski, M.; Pluciński, J.; Kosmowski, B. Non-Destructive Inspection of Anti-corrosion protective coatings using optical coherent tomography. *Metrol. Measur. Syst.* **2012**, *19*, 365–372.

34. Derrick, M. Paraloid B-48N. Available online: http://cameo.mfa.org/wiki/Paraloid_B-48N (accessed on 16 January 2017).

35. Decker, P.; Brüggerhoff, S.; Eggert, G. To coat or not to coat? The maintenance of Cor-Ten® sculptures. *Mater. Corros.* **2008**, *59*, 239–247.

36. Seipelt, B.; Pilz, M.; Kiesenberg, J. Transparent Coatings-Suitable Corrosion Protection for Industrial Heritage Made of Iron? In Proceedings of the Conférence Internationale sur la Conservation des Métaux, Draguignan, France, 27–29 May 1998; pp. 291–296.

37. Dubois, A.; Vabre, L.; Boccara, A.C.; Beaurepaire, E. High-resolution full-field optical coherence tomography with a Linnik microscope. *Appl. Opt.* **2002**, *41*, 805–812.

38. Wang, Z.; Potsaid, B.; Chen, L.; Doerr, C.; Lee, H.C.; Nielson, T.; Jayaraman, V.; Cable, A.E.; Swanson, E.; Fujimoto, J.G. Cubic meter volume optical coherence tomography. *Optica* **2016**, *3*, 1496–1503.
39. Klein, T.; Huber, R. High-speed OCT light sources and systems [Invited]. *Biomed. Opt. Express* **2017**, *8*, 828–859.

Article

Scan-Less Line Field Optical Coherence Tomography, with Automatic Image Segmentation, as a Measurement Tool for Automotive Coatings

Samuel Lawman [1,2], Bryan M. Williams [2], Jinke Zhang [1], Yao-Chun Shen [1,*] and Yalin Zheng [2]

[1] Department of Electrical Engineering and Electronics, University of Liverpool, Liverpool L69 3GJ, UK;
     S.Lawman@liverpool.ac.uk (S.L.); sgjzhan5@student.liverpool.ac.uk (J.Z.)
[2] Department of Eye and Vision Science, University of Liverpool, Liverpool L7 8TX, UK;
     Bryan.Williams@liverpool.ac.uk (B.M.W.); Yalin.Zheng@liverpool.ac.uk (Y.Z.)
*   Correspondence: Y.C.Shen@liverpool.ac.uk; Tel.: +44-(0)151-794-4575

Academic Editor: Michael Pircher
Received: 3 February 2017; Accepted: 29 March 2017; Published: 1 April 2017

Abstract: The measurement of the thicknesses of layers is important for the quality assurance of industrial coating systems. Current measurement techniques only provide a limited amount of information. Here, we show that spectral domain Line Field (LF) Optical Coherence Tomography (OCT) is able to return to the user a cross sectional B-Scan image in a single shot with no mechanical moving parts. To reliably extract layer thicknesses from such images of automotive paint systems, we present an automatic graph search image segmentation algorithm. To show that the algorithm works independently of the OCT device, the measurements are repeated with a separate time domain Full Field (FF) OCT system. This gives matching mean thickness values within the standard deviations of the measured thicknesses across each B-Scan image. The combination of an LF-OCT with graph search segmentation is potentially a powerful technique for the quality assurance of non-opaque industrial coating layers.

Keywords: optical coherence tomography; coatings; spectral domain; time domain; quality assurance; line field; image segmentation; graph search

---

1. Introduction

Non-opaque coatings are used to provide protection and aesthetic enhancement to a wide variety of objects in many contexts. Examples of this variety are the heavy industry clear top coats in modern car paint systems [1], the do-it-yourself varnishing of outdoor wooden objects [2], and the highly specialised varnish coatings for valuable works of art [3]. This paper is focused on the economically important industrial coating context [1,4]. The term Quality Assurance (QA) [5] refers to the broad systems that ensure consistent quality of any (tangible or intangible) produced product. The appropriate processes and measures required are dependent on what that product is. To ensure that the layers of industrial coatings, such as automotive paint systems, give the desired protection and appearance, QA inspection methods that are capable of quantifying, and thus checking, the applied coating thickness and uniformity are required. However, it would be impractical to measure the thickness of the coating layers over the whole car body surface of every car produced on an automobile production line. Instead, to catch errors in the production process, sample measurements of thicknesses and consistency at appropriate points (spatial and temporal) are sufficient. To meet this need, a variety of instruments have been developed and are now available commercially. These commercially available instruments fall into two main categories. Magnetic and eddy current systems give a contact, single point measurement of coatings on metallic substrates [6]. They cannot be used with non-metallic substrates and cannot give the thicknesses of multiple layers nor the uniformity of the coatings. The other category of commercially available systems are Ultrasound probe systems [6], which are able to measure independently of coatings on various substrates and can measure multiple layers. Though used widely for imaging in medicine [7], the ultrasound probe systems currently commercially available for coating measurements remain limited to a single point A-Scan. Manual scanning of probe can be performed by the user during repeat measurements. Remote ultrasound, such as laser ultrasonic methods [8], has been an area of research for coating measurements [9]. However, the added complexity of the device would add significantly to the cost and reduce portability.

Terahertz (THz) imaging has the ability to image the majority of coatings [10] but would be a relatively expensive solution due to the specialised band of electromagnetic (EM) radiation used. Optical confocal microscopy [11] instruments are able to characterise coatings, but remain expensive. Reflectometry [12] and ellipsometry [13] are prevalent in high tech thin film applications for measuring the thicknesses of multiple clear layers from interference effects. However, they are not prevalent for the measurement of industrial coatings.

Amongst Optical Coherence Tomography (OCT) applications outside of bio-medicine [14] is the non-contact measurement of the layers of paintings [15] and wooden artefacts [16], both commonly including optically transparent (varnish) layers. It has also been used to for the QA of transparent layer composited gradient refractive index optical components [17]. Following its use for measuring coating layers in the art conservation field, OCT has recently been considered for its potential QA applications [18] and for forensic use [19] for measuring the coating layers of automotive paints (an industrial field). Like THz imaging, the cost of general laboratory OCT systems is relative high (>£10,000). However, unlike THz imaging, OCT works at visible or near infrared wavelengths with a wide variety of low cost components available off the shelf. For the specific undemanding task of resolving thicknesses of transparent and semi-transparent coatings, much of an OCT device could be down-engineered to reduce cost.

Although a single A-Scan OCT device could be constructed to reduce cost, this would be undesirable for two reasons. Firstly, in this paper we will demonstrate a scenario in which a single A-Scan measurement gives erroneous results and only partial layer information. Multiple measurements of a sample may be required from the user for reliable results. Secondly, a B-Scan image gives more information than single A-Scans, such as a one-dimensional map of coating thickness and surface profile over the length of the measurement [20] and the ability to resolve additional scattering layers that would not be discernible from a single A-scan. However, the majority of OCT systems are optical fibre based measuring a single lateral point at a time. This conventional point by point raster scanning format requires a galvo mirror system(s). This adds to the mechanical complexity, including management of electrical and mechanical response issues [21], and cost of the system. Several alternative formats of OCT devices exist that measure more than one lateral position at a time without the need for mechanical scanning [22–24]. The form considered here is spectral domain Line Field (LF) OCT [25] which measures a B-Scan image in a single shot without moving parts. Though this format has been previously used for imaging in human [26,27] and animal [28,29] in vivo studies, and more recently for the QA of glass [30], the technique has not been previously applied in the measurement of industrial coatings.

Image segmentation [31–34] can break up images into their different regions, thus allowing their automated measurement. In practice, segmentation is often achieved by the thresholding of intensity values [35] at a parameter which may be selected either manually and empirically or automatically [36]. While this can yield results quickly and may be favoured particularly for large datasets, such parameters often require trial-and-error testing and vary between images. It is not well suited to problems involving poorly-defined boundaries, varying contrast, and visibility of noise, which are all characteristic of logarithmic OCT images in general. To address this limitation, many solutions have been proposed such as variational modelling [32,37–39] and machine learning approaches [33,40]. Recently, graph search algorithms have emerged as a popular [41–44] approach for image segmentation due to the low computational energy required, reduced requirements for optimisation, and no need for training data.

Here we report the combined use of LF-OCT to acquire a cross sectional B-Scan image without electro-mechanical scanning, and a graph search sectioning algorithm to measure the clear coat and metallic base coat thickness for four automobile paints. The LF-OCT means that the image is taken in a single shot, with no scanning required. We show that graph search is a robust segmentation method for the measurement of layer thicknesses by OCT devices. The image quality and segmentation algorithm robustness was verified using images taken from the same car paint samples using a time domain Full Field (FF) OCT system. As an independent system based on the same physical principles as the LF-OCT system, the FF-OCT, which has previously been validated and calibrated using micro-X-CT [45], results served to show that no systematic experimental setup or calibration error was present in the system, and that any modality of OCT system will give similar results, with the graph search method applicable to them all. Correspondence with ultrasound and microscopy measurements has previously been shown for one type of OCT [18].

2. Materials and Methods

2.1. Optical Coherence Tomography Systems

The details of the LF-OCT used for this study has been previously described [46]. In brief, in order to measure an OCT B-Scan image in a single shot, a line was illuminated on the sample and reference surface, as shown in Figure 1 (left). This line was then imaged onto the slit of an imaging spectrograph, which resolved the interference spectrum at each position. A single (unaveraged) image was taken with the Back Illuminated Charged Couple Device (BI CCD) camera (iVac, Andor, Belfast, UK) and the Fourier phase differential mask method was used to supress aberration induced image artefacts. The band-pass filtered supercontinuum light source provided an axial resolution of 2.1 $n_G \cdot \mu m$, the plot of the point spread function is given in Figure 2 of [46]. With a 4f arrangement of 75 mm Achromatic pair (AC) objective (Linnik pair) and 100 mm AC collection lenses, the system's lateral resolution was measured to be 18 μm (United States Air Force 1951 resolution test chart (MIL-STD-150A), line pair width resolved).

To cross validate the OCT B-Scan images, a development of a previously presented [45] FF-OCT system was also used and is shown in Figure 1 (right). This time domain OCT system used an inexpensive infrared LED light source with a central wavelength of 850 nm and a bandwidth of 80 nm. The light was collimated for the illumination of the sample and reference mirror. The imaging optics was again a 4f arranged pair (objective $F = 50$ mm and collection $F = 250$ mm) of achromatic lenses, but with both situated behind the beam splitter. The panels were mounted on a high precision piezo motor positioner (LPS-65/N-565, Physik Instrumente, Bedford, UK) to scan in the axial direction. A high-speed USB 3.0 Complementary Metal Oxide Semiconductor (CMOS) camera (GS3-U3-23S6M-C, Point Grey, Richmond, British Columbia, Canada) was then used to collect the three-dimensional time domain interference data. During processing of the time domain FF-OCT data, the raw axial signal was convolved with a near ideal measured signal from a mirror. This increased the Signal to Noise Ratio (SNR), but at the cost of a slight detriment to axial resolution. The achieved lateral resolution of this system was significantly better (4.4 μm) than the LF-OCT system, though the axial resolution was worse (3.6 $n_G \cdot \mu m$).

Table 1 gives a summary of the performance of the two instruments. The time to take a measurement with the LF-OCT system was 60 ms, which is more than sufficient for a QA thickness measurement tool. The speed was limited by the mechanical shutter for the CCD camera, with the total illuminating light power physically attenuated accordingly. If required, the acquisition time could be shortened by orders of magnitude by using electronically shuttered CMOS cameras or using a single super-continuum pulse [46]. The 40 s acquisition time for the FF-OCT system would be too long for a practical QA tool. The FF-OCT provided a 3D volume measurement, however, for thickness measurement we show here that the 2D B-Scan of the LF-OCT was sufficient to provide reliable layer thickness measurement values. The raw SNR and sensitivity values of the Fourier domain LF-OCT were higher than the time domain FF-OCT.

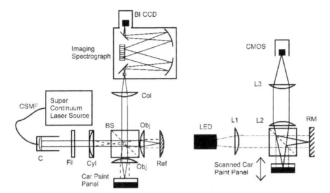

Figure 1. (**Left**) Line Field Optical Coherence Tomography (LF-OCT) setup. CSMF—Continuously Single Mode Fibre. C—Collimator, Fil—Optical Filters (Bandpass and Neutral Density), Cyl—Cylindrical Lens, BS—Cube Beam Splitter. Obj—Objective Lenses, Ref—Reference Interface (Flat glass surface), Col—Collection Lens, BI CCD—Back Illuminated Charged Couple Device camera (iVac, Andor, UK); (**Right**) FF-OCT setup, LED—Light Emitting Diode, L1, L2, and L3—Achromatic lenses, RM—Reference Mirror, CMOS—Complementary Metal Oxide Semiconductor camera (Point Grey, GS3-U3-23S6M-C).

Table 1. Comparison of the specification of the LF and FF (full field) OCT devices, as used in the study.

Performance Parameter	LF-OCT	FF-OCT
Acquisition time (ms)	60	40,000
Acquired Image Pixels	$1 \times 256 \times 1000$	$1920 \times 1200 \times 4000$
Axial Resolution ($n_G \cdot \mu m$)	2.1	3.6
Lateral Resolution (μm)	18	4.4
Signal to Noise Ratio (dB)	79	54
Sensitivity (dB)	93	54

2.2. Samples

Four metallic car paint demo panels, one Indus silver, one Mauritius blue, one Barolo black, and one Santorini black were used in this study. These panels were examples of a common automotive multilayer coating system. The top layer was a clear coat with thickness to be measured. The second layer contains metallic flakes at random depths. For the silver and blue panels, a 2.25 mm $\times$ 1.4 mm area was masked off, and for cross validation purpose both instruments measured a B-Scan across the middle. The difference in positions between the scans was less than 1 mm. For a broader range of samples, the two different black samples were measured with the LF-OCT system only.

2.3. Software and Algorithm

The thicknesses of the clear coat and basecoat layers were calculated by determining the interface contours of the visible layers from the resulting B-scan images using a hierarchical graph-search approach. This assumed that volume scattering rather than interface reflection dominated the imaged layer boundaries, which was correct for all interfaces other than the surface in our data. The surface position was found by taking the average of the top two found profiles. Variational modelling inspired post-processing then improved robustness of the graph search approach by building in region intensity information. We first constructed a graph of nodes and edges using the image such that each pixel corresponded to a node. We thus redefined the segmentation as a minimal cost graph cut problem; that is, the total cost of moving from one side of the graph to the other across the nodes whose values were determined by an energy functional aiming to split the image based on the gradient. To reduce the likelihood of jumping between interfaces, we defined our energy functional as $\mathcal{E}_j(x) = (-1)^j(d_1\partial z(x)/\partial x_1 + d_2\partial z(x)/\partial x_2)$ where $z(x)$ denoted the image to be segmented, $x = (x_1, x_2)^\top$, x_1 and x_2 are the coordinates in the lateral and depth directions, respectively, $d_1, d_2 \in \mathbb{R}_{>0}$ are non-negative parameters controlling the influence of the lateral and depth gradients, and $j \in \{0, 1\}$ determines whether we are seeking an upper or lower boundary.

Dynamic programming was used to find the path of minimal energy cost leading from the left-hand side of the image to the right. Calculating the cost of every possible path across the image would have been time consuming and necessitated storing a large amount of information. To avoid this, the paths were calculated in a structured manner with constraints inbuilt. From each (starting) pixel in the far-left A-Scan, we considered the cost of traversing a sub-path to another pixel in the following column, restricting the set of possible destination pixels to include only those within two rows. For each column, we stored the minimum cost movement only, and this process was repeated until the right-hand side of the image was reached. As well as resulting in a significant reduction in the number of required computations, the restrictions on the potential paths allowed also acted as regularisation on the length of the contour. Once the cumulative costs of the paths from each starting pixel were calculated, the overall minimum cost path could be found. This was done for each interface in a hierarchical way, excluding the nodes assigned to a contour to rule out crossing, defining the segmentation contours $\mathcal{S}_1, \ldots, \mathcal{S}_5 \in \mathbb{R}_>^n$.

Following this, we aimed to remove one surplus contour, leaving us with the best-fit segmentation of the image. We adopted the logic used for contour-fitting in variational methods such as [32,47]. In particular, we referred to the multi-phase work of Vese and Chan [47], which defined contours as level-sets [48] of a single function $\phi(x)$, and considered the correct segmentation as that which minimised the fitting energy only, since we did not need regularisation of the contour length. We identified the contour $\mathcal{S}_\kappa$ to be removed by determining the index κ which solved the minimisation problem

$$\min_\kappa \left\{ \sum_{\substack{i=0 \\ i \neq \kappa+1}}^{5} \int_\Omega (z - c_i^\kappa)^2 \mathcal{H}_i(\phi, l)dx \right\}, \quad c_i^\kappa = \frac{\int_\Omega z\mathcal{H}_i(\phi,l)dx + \delta_i(\kappa)\int_\Omega z\mathcal{H}_{i+1}(\phi,l)dx}{\int_\Omega \mathcal{H}_i(\phi,l)dx + \delta_i(\kappa)\int_\Omega \mathcal{H}_{i+1}(\phi,l)dx}, \tag{1}$$

where $\mathcal{H}_i(\phi, l) = H^{\chi_1}(\phi - l_i)H^{\chi_2}(l_{i+1} - \phi)$ where $\chi_1 = 0$ for $i = 0$ and $\chi_1 = 1$ otherwise, $\chi_2 = 0$ for $i = 5$ and $\chi_2 = 1$ otherwise. Excluding $\mathcal{S}\_\kappa$ and ordering the sets in decreasing total sum, we obtained the list of contours $\mathcal{T}_1, \ldots, \mathcal{T}_4$. We then took the average of the top two contours which gave the upper and lower boundaries of the clear coat-air interface as its location. We could then give the thickness measurements of the clear coat ($\mathcal{T}_c$) and base coat ($\mathcal{T}_b$) layers as the pointwise difference of the sets $\mathcal{T}_c = \wp \cdot \left(\frac{\mathcal{T}_1 + \mathcal{T}_2}{2} - \mathcal{T}_3\right)$, $\mathcal{T}_b = \wp \cdot (\mathcal{T}_3 - \mathcal{T}_4)$, where $\wp$ was the pixel-size in the depth direction. The overall algorithm is shown in Figure 2 below.

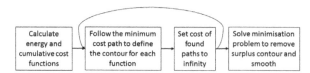

Figure 2. Flow chart of the segmentation process. The energy functional and cumulative cost functions are calculated to determine the minimal cost paths. Once all paths are found, the minimisation problem given in Equation (1) is solved to remove one contour, resulting in the set of segmentations.

3. Results

3.1. OCT Images

Figure 3 shows the OCT B-Scan images obtained for each panel. For Indus silver and Mauritius blue, the images from both LF-OCT and FF-OCT systems are compared. Both instruments give similar images, showing a larger optical thickness for the clear coating on the blue panel. Both instruments also show similar trends in the differences in appearance of the second layer of the two samples. Within the second layers, the signal contrast between the bright flakes and points in between is lower for the blue panel than the silver panel. To give the blue colour, the blue panel second layer may contain blue pigment particles, which would add to the scattering signal between flakes. The main difference between the images from the two instruments is the lateral resolution. The resolved flakes from the high lateral resolution FF-OCT appear narrower than for the LF-OCT, where the lower lateral resolution has broadened their apparent width. Both (i.e., Barolo and Santorini) black LF-OCT images are significantly different, compared with silver and blue, with the visible base coat thickness being much thinner. The clear coat of Santorini black is the thinnest of all the samples.

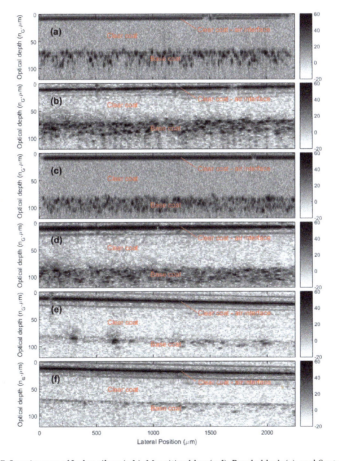

Figure 3. B-Scan images of Indus silver (**a**,**b**), Mauritius blue (**c**,**d**), Barolo black (**e**), and Santorini black (**f**) automotive metallic paint samplers. Taken with FF-TD (**a**,**c**) and LF-FD (**b**,**d**–**f**) parallel OCT systems. The FF-OCT B-Scans are taken from within 1 mm of the location of the corresponding LF-OCT images. Scaling is the same for all images. Colour bar units are dB.

Though the relatively lower lateral resolution of the LF-OCT has partially blurred the second layer, it is still sufficient to resolve the image texture. Figure 4 (top) shows the LF-OCT B-Scan of the silver panel (same as Figure 3b) with two A-Scan locations (red and blue) marked on. Figure 4 (bottom) is the plot of the intensity of these two A-Scans, and the apparent thickness estimated from obvious peaks. Considering these (red and blue) A-Scans separately and without prior knowledge of the sample, as would be the case in a single point measurement, their interpretations give a large difference in the thickness measurement of the clear coat. In this case, for the red A-Scan the apparent signal for the second interface comes from far too deep. No reasonable estimate of the base coat layer could be made without prior knowledge of the A-Scan properties for the sample. In contrast to this lack of information in the 1D single point measurement, in the 2D B-Scan image the layers and their boundaries are immediately visibly apparent without any prior knowledge. In this case the source of the error for the clear coat thickness in the single point measurement is the dominating signal from the metal flakes, which at any given lateral position could be present at any depth within the second layer. For a single point system, to reliably overcome this, the user could take multiple measurements at different sample positions and use the averaged A-scan signal for thickness calculation. However, we will show that reliable coating thickness and uniformity information could be obtained by using the single shot LF-OCT system and automated image segmentation algorithms, without the need for multiple measurements.

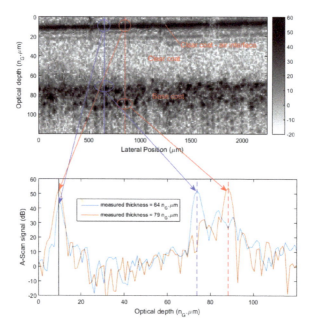

Figure 4. (**Top**) LF-FD-OCT image of silver metallic paint sampler (Figure 3b) with two (blue and red) A-Scan locations marked. (**Bottom**) Plot of the red and blue A-Scans, with the apparent thickness measured using only the single A-Scan data. The arrows show the correspondence of the A-Scan peaks with location in the B-Scan image.

3.2. Automated Segmentation of Layer Thickness

We now calculate the thickness of the clear coat and base coat layers of the paint samples across the whole B-scan image. We first partition the image to identify the layers automatically using the graph search approach described in the previous section. Using forward first order finite differences as a discrete approximation to the derivative, we calculate the energy functional as an important step in identifying the contours. An example of this using the silver sample is shown for $j = 0$ and $j = 1$ in Figure 5a,b respectively. The lower values corresponding to the minimiser appear as the darker colour in the figure and are to be traced automatically. While the lowest intensity path may appear clear in each figure, the remaining paths are difficult to trace by sight. Proceeding hierarchically with the graph search approach, we obtain a set of potential interface contours (Figure 5c) which closely match the visible edges. An additional phantom contour is also found which does not represent a layer boundary but does correspond to a solution of the graph-search problem. We now employ our data-fitting step to remove this contour (Figure 5d). At this point, we have a close segmentation of each layer given the presentation of the paint samples in the image.

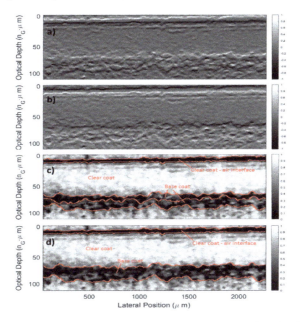

Figure 5. Segmentation process of LF-OCT B-scan of silver car paint sample. (**a,b**) The joint energy functional used to find the initial contours; (**c**) The initial result using graph-search segmentation; (**d**) The result after automated removal of the phantom contour.

As a final step, we compensate for the visible blurring of the air-clear coat layer interface by replacing the two boundary contours with their mean value to obtain the final segmentation of the silver sample shown in Figure 6a. The same procedure is applied to the blue paint sample and the contours are correctly identified in Figure 6c. It is now a simple procedure to obtain the thickness measurements of each layer across the whole B-Scan (Figure 6b,d). Table 2 gives the mean and standard deviation of these measured thicknesses of the clear coating and base coat, within a B-Scan measurement, for the two samples with both OCT instruments. However, it should be noted that this standard deviation does not have a clear relationship with accuracy or precision of the mean value. The variance is caused by the apparent interface texture in the image, which may be real or, as quite likely in this case, a result of the volume scattering properties of one of the layers. Here reflections from the semi-sparse metallic flakes within the base coat layer dominate giving a lumpy image signal, which the graph search algorithm traces over. The repeatability (precision), for the LF-OCT, of the mean clear and base coat thickness values was quantified by measuring at eight locations on the Indus silver sample, the standard deviation was 2 and 5 $n_G \cdot \mu m$, respectively. The accuracy is determined by the axial resolution (systematic biasing of graph search algorithm to one side of the interface will be of this order of magnitude) and/or scattering homogeneity of the layers (non-homogeneous properties causing the apparent profile undulate into the layer, as is likely in these results). If we treat these as standard error, the accuracy for the given Indus Silver sample LF-OCT measurement is then calculated to be 3 and 4 $n_G \cdot \mu m$ for the clear and base coat, respectively. In Table 2, the largest difference in mean thickness of both techniques is 3 $n_G \cdot \mu m$ for the base coat, which is within the measured undulation, precision, and calculated accuracy values of the LF-OCT. The differences in the mean values for the clear coat are negligible. Therefore, we conclude that the segmentation algorithm gives repeatable (within the technique's measured errors) measurements independent of the OCT instrument. It should be noted that for optical techniques, the precision and accuracy of the measurement of an interface is highly dependent on the sample. For an interface that gives a significant signal, profilometry techniques are able to measure it with high accuracy. However, in this case the signal from the interface is negligible, with the image contrast being dominated by volume scattering; as such, profilometry methods would not work, and image segmentation is the preferred technique. The alternative would be to correct the B-Scan image by rotation for any tilt, and then average the A-Scans. However, this would eliminate lateral spatial information that may be present.

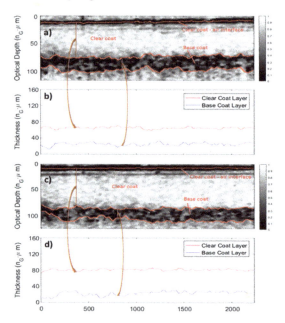

Figure 6. Final segmentation results of the silver (**a**) and blue (**c**) paint samples. The respective thickness profiles of each sample (**b,d**) as demonstrated by the brown arrows. Although they do look non-smooth, it should be noted that the thickness profile figures are presented at an aspect ratio of approximately 1:3.

Table 2. Thickness measurements (mean ± standard deviation) of the base coat and clear coat layers of the paint samples scanned using full field time domain (FF) and line field spectral domain (LF) OCT.

Sample	Method	Base Coat Layer Thickness ($n_G \cdot \mu m$)	Clear Coat Layer Thickness ($n_G \cdot \mu m$)
Indus Silver	FF-OCT	27.88 ± 6.11	65.64 ± 1.34
Indus Silver	LF-OCT	24.95 ± 3.85	64.98 ± 2.44
Mauritius Blue	FF-OCT	25.83 ± 4.62	80.75 ± 1.44
Mauritius Blue	LF-OCT	22.69 ± 5.47	80.68 ± 2.20
Barlo Black	LF-OCT	16.61 ± 4.48	68.96 ± 2.01
Santorini Black	LF-OCT	8.13 ± 3.24	54.50 ± 2.30

4. Discussion

Industrial coatings are economically important. Their QA measurement is important to ensure protection and appearance consistency. To measure the thickness of coating layers, contact single point techniques currently dominate the market. These can only give limited spatial information and may require multiple measurements to ensure accuracy. As shown here, LF-OCT returns to the user a 2D cross-sectional image of non-opaque coatings where both coating thickness and coating uniformity information can be extracted. To collect 2D images, conventional point by point raster scanning OCT requires mechanical scanning of the probe beam across the sample. This adds complexity, and potential fragility, for a challenging industrial environment. In contrast, the proposed LF-OCT imaging requires no mechanical scanning, thus it is an attractive solution for the application. It requires no mechanical moving parts, thus it should increase the robustness of the technique in dirty and rough industrial environments.

Segmentation can be used to extract dimensions automatically from images. To measure the thickness of coating layers from OCT B-Scan images, graph search has been shown to be a robust and fast method. The image segmentation technique can be applied to any B-Scan image of layered structures, independent of the OCT instrument. Here, we showed this robustness by applying an algorithm to data taken using two OCT systems with different imaging specifications and modalities, and recorded matching results within the standard deviation of the thickness profiles. It should also be noted that a 2D B-Scan image, not a full 3D measurement, is required for this reliable thickness measure, which is provided by LF-OCT without any electro-mechanical scanning. Also, the presence and magnitude of any significant non-uniformity of coating thickness, over the lateral dimension of measurement, would be as apparent in this 1D measurement from the 2D B-Scan, as it would in a 2D map from a 3D volume scan. A 2D map of coating thickness in the local area of measurement would not provide any significant further information for the QA measure, which ultimately is required to give a binary pass or fail output.

Currently as presented, the instrument is still of significant cost and size. To be viable against competing techniques, future work must reduce the cost and size of such a device. However, there is potential for reducing the cost of a device by reducing the specification for all the components. Thermal light sources have previously [28] been demonstrated to be suitable for LF-OCT, and during the development of our LF-OCT instrument we successfully used an incandescent bulb of negligible cost to take images. The drawback of thermal sources is that adaption to compensate for low spatial coherence will lead to a slower measurement time. However, this will still be under 1 s so would not be an issue for this application. Also, thermal sources will not be a factor in restricting the bandwidth used, so will allow high axial resolutions to be achieved. The imaging spectrograph part of the device can be constructed, to a reduced size, with off the shelf components of minimum cost. With operation between optical and Near InfraRed (NIR) wavelengths of 400 to 1000 nm, a low cost CMOS camera array can be used. The drawback of this is likely to be on the spectral resolution, which will limit image depth. However, reduction of size favours spectral bandwidth and thus high image axial resolution. The use of a LF-OCT, rather than the more common point by point raster scanning OCT, arrangement would mean no mechanical moving parts would be required. Future work will involve construction and testing of this proposed low cost device. As well as application to automotive coatings, the combination of the low cost LF-OCT device and graph search segmentation could be used for industrial coatings on wood [2] and in aerospace applications, such as thermal barrier coatings for aerospace turbine blades [49]. It may also be used for the automatic measurement of pharmaceutical tablet coating thicknesses [50].

In conclusion, we have demonstrated the use of LF-OCT to take cross sectional (B-Scan) images of optically transparent coatings without moving parts. Combining this with fast automated graph search image sectioning allows reliable thickness profiles to be calculated with a single, scan-less, LF-OCT measurement. We have illustrated that a single point measurement of layer thickness is inherently less reliable than a measurement from a B-Scan image. The combining of these methods

Appl. Sci. **2017**, *7*, 351

gives a clear development path for a new reliable and cost competitive instrument for the QA of certain industrial coatings.

Acknowledgments: The Ultrasensitive Optical Coherence Tomography Imaging for Eye Disease is funded by the National Institute for Health Research's i4i Programme. This paper summarises independent research funded by the National Institute for Health Research (NIHR) under its i4i Programme (Grant Reference Number II-LA-0813-20005). The views expressed are those of the authors and not necessarily those of the NHS, the NIHR, or the Department of Health. This work is partly supported by the UK EPSRC Research Grant No. EP/L019787/1. The authors thank Roy Donga of Jaguar Land Rover Limited for providing the test samples.

Author Contributions: Samuel Lawman wrote the majority of the paper and conducted the LF-OCT measurements; Bryan M. Williams wrote the minority of the paper and conducted the image segmentation; Jinke Zhang conducted the FF-OCT measurements; Yao-Chun Shen provided the original paper idea and oversaw the LF-OCT and FF-OCT hardware; Yalin Zheng was the LF-OCT project instigator and oversaw the image segmentation work.

Conflicts of Interest: The authors declare no conflict of interest. The founding sponsors had no role in the design of the study; in the collection, analyses, or interpretation of data; in the writing of the manuscript, and in the decision to publish the results.

References

1. Akafuah, N.K.; Poozesh, S.; Salaimeh, A.; Patrick, G.; Lawler, K.; Saito, K. Evolution of the automotive body coating process—A review. *Coatings* **2016**, *6*, 24. [CrossRef]
2. Evans, P.D.; Haase, J.G.; Shakri, B.; Kiguchi, M. The search for durable exterior clear coatings for wood. *Coatings* **2015**, *5*, 830–864. [CrossRef]
3. Lawman, S.J. *Optical and Material Properties of Varnishes for Paintings*; Nottingham Trent University: Nottingham, UK, 2011.
4. Matheson, R.R. Paint and coatings technology: Current industrial trends. *Polym. Rev.* **2006**, *46*, 341–346. [CrossRef]
5. Hayes, G.E. *Quality Assurance: Management & Technology*, 2nd ed.; Charger Productions: Capistrano Beach, CA, USA, 1976.
6. Brasunas, J.C.; Cusham, G.M.; Lakew, B. Thickness measurement. In *The measurement, Instrumentation and Sensors: Handbook*; Webster, J.G., Ed.; CRC Press: Boca Raton, FL, USA, 1999.
7. Szabo, T.L. *Diagnostic Ultrasound Imaging: Inside Out*, 2nd ed.; Academic Press: Oxford, UK, 2014.
8. Tam, A.C. Pulsed-laser generation of ultrashort acoustic pulses—Application for thin-film ultrasonic measurements. *Appl. Phys. Lett.* **1984**, *45*, 510–512. [CrossRef]
9. White, J.S.; LaPlant, F.P.; Dixon, J.W.; Emch, D.J.; Datillo, V.P. Non-Contact Real-Time Film Thickness Gage for Automotive Body Painting Applications. *SAE Tech. Paper* **1998**. [CrossRef]
10. Su, K.; Shen, Y.C.; Zeitler, J.A. Terahertz sensor for non-contact thickness and quality measurement of automobile paints of varying complexity. *IEEE Trans. Terahertz Sci. Technol.* **2014**, *4*, 432–439. [CrossRef]
11. Ling, X.; Pritzker, M.D.; Byerley, J.; Burns, C.M. Confocal scanning laser microscopy of polymer coatings. *J. Appl. Polym. Sci.* **1998**, *67*, 149–158. [CrossRef]
12. Miller, D.A. *Optical Properties of Solid Thin Films by Spectroscopic Reflectometry and Spectroscopic Ellipsometry*; City University of New York: New York, NY, USA, 2008.
13. Garcia-Caurel, E.; De Martino, A.; Gaston, J.P.; Yan, L. Application of spectroscopic ellipsometry and mueller ellipsometry to optical characterization. *Appl. Spectrosc.* **2013**, *67*, 1–21. [CrossRef] [PubMed]
14. Stifter, D. Beyond biomedicine: A review of alternative applications and developments for optical coherence tomography. *Appl. Phys. B* **2007**, *88*, 337–357. [CrossRef]
15. Liang, H.; Cid, M.G.; Cucu, R.G.; Dobre, G.M.; Podoleanu, A.G.; Pedro, J.; Saunders, D. En-face optical coherence tomography—A novel application of non-invasive imaging to art conservation. *Opt. Express* **2005**, *13*, 6133–6144. [CrossRef] [PubMed]
16. Latour, G.; Echard, J.P.; Soulier, B.; Emond, I.; Vaiedelich, S.; Elias, M. Structural and optical properties of wood and wood finishes studied using optical coherence tomography: Application to an 18th century italian violin. *Appl. Opt.* **2009**, *48*, 6485–6491. [CrossRef] [PubMed]

17. Meemon, P.; Yao, J.N.; Lee, K.S.; Thompson, K.P.; Ponting, M.; Baer, E.; Rolland, J.P. Optical coherence tomography enabling non destructive metrology of layered polymeric grin material. *Sci. Rep.* **2013**, *3*, 1709. [CrossRef]

18. Dong, Y.; Lawman, S.; Zheng, Y.L.; Williams, D.; Zhang, J.K.; Shen, Y.C. Nondestructive analysis of automotive paints with spectral domain optical coherence tomography. *Appl. Opt.* **2016**, *55*, 3695–3700. [CrossRef] [PubMed]

19. Zhang, N.; Wang, C.M.; Sun, Z.W.; Mei, H.C.; Huang, W.; Xu, L.; Xie, L.C.; Guo, J.J.; Yan, Y.W.; Li, Z.H.; et al. Characterization of automotive paint by optical coherence tomography. *Forensic Sci. Int.* **2016**, *266*, 239–244. [CrossRef] [PubMed]

20. Lawman, S.; Liang, H.D. High precision dynamic multi-interface profilometry with optical coherence tomography. *Appl. Opt.* **2011**, *50*, 6039–6048. [CrossRef] [PubMed]

21. Duma, V.F.; Tankam, P.; Huang, J.X.; Won, J.; Rolland, J.P. Optimization of galvanometer scanning for optical coherence tomography. *Appl. Opt.* **2015**, *54*, 5495–5507. [CrossRef] [PubMed]

22. Watanabe, Y.; Yamada, K.; Sato, M. Three-dimensional imaging by ultrahigh-speed axial-lateral parallel time domain optical coherence tomography. *Opt. Express* **2006**, *14*, 5201–5209. [CrossRef] [PubMed]

23. Povazay, B.; Unterhuber, A.; Hermann, B.; Sattmann, H.; Arthaber, H.; Drexler, W. Full-field time-encoded frequency-domain optical coherence tomography. *Opt. Express* **2006**, *14*, 7661–7669. [CrossRef] [PubMed]

24. Fechtig, D.J.; Schmoll, T.; Grajciar, B.; Drexler, W.; Leitgeb, R.A. Line-field parallel swept source interferometric imaging at up to 1 MHz. *Opt. Lett.* **2014**, *39*, 5333–5336. [CrossRef] [PubMed]

25. Zuluaga, A.F.; Richards-Kortum, R. Spatially resolved spectral interferometry for determination of subsurface structure. *Opt. Lett.* **1999**, *24*, 519–521. [CrossRef] [PubMed]

26. Grajciar, B.; Pircher, M.; Fercher, A.F.; Leitgeb, R.A. Parallel fourier domain optical coherence tomography for in vivo measurement of the human eye. *Opt. Express* **2005**, *13*, 1131–1137. [CrossRef] [PubMed]

27. Zhang, Y.; Rha, J.T.; Jonnal, R.S.; Miller, D.T. Adaptive optics parallel spectral domain optical coherence tomography for imaging the living retina. *Opt. Express* **2005**, *13*, 4792–4811. [CrossRef] [PubMed]

28. Graf, R.N.; Brown, W.J.; Wax, A. Parallel frequency-domain optical coherence tomography scatter-mode imaging of the hamster cheek pouch using a thermal light source. *Opt. Lett.* **2008**, *33*, 1285–1287. [CrossRef] [PubMed]

29. Robles, F.E.; Wilson, C.; Grant, G.; Wax, A. Molecular imaging true-colour spectroscopic optical coherence tomography. *Nat. Photon.* **2011**, *5*, 744–747. [CrossRef] [PubMed]

30. Chen, Z.Y.; Zhao, C.; Shen, Y.; Li, P.; Wang, X.P.; Ding, Z.H. Ultrawide-field parallel spectral domain optical coherence tomography for nondestructive inspection of glass. *Opt. Commun.* **2015**, *341*, 122–130. [CrossRef]

31. Mumford, D.; Shah, J. Optimal approximation by piecewise smooth functions and associated variational problems. *Commun. Pur. Appl. Math.* **1989**, *42*, 577–685. [CrossRef]

32. Chan, T.F.; Vese, L.A. Active contours without edges. *IEEE Trans. Image Process.* **2001**, *10*, 266–277. [CrossRef] [PubMed]

33. Vermeer, K.; Van der Schoot, J.; Lemij, H.; De Boer, J. Automated segmentation by pixel classification of retinal layers in ophthalmic oct images. *Biomed. Opt. Express* **2011**, *2*, 1743–1756. [CrossRef] [PubMed]

34. Garvin, M.K.; Abràmoff, M.D.; Kardon, R.; Russell, S.R.; Wu, X.; Sonka, M. Intraretinal layer segmentation of macular optical coherence tomography images using optimal 3-D graph search. *IEEE T. Med. Imaging* **2008**, *27*, 1495–1505. [CrossRef] [PubMed]

35. Shen, M.; Cui, L.; Li, M.; Zhu, D.; Wang, M.R.; Wang, J. Extended scan depth optical coherence tomography for evaluating ocular surface shape. *J. Biomed. Opt.* **2011**, *16*, 056007. [CrossRef] [PubMed]

36. Otsu, N. A threshold selection method from gray-level histograms. *Automatica* **1975**, *11*, 23–27. [CrossRef]

37. Ambrosio, L.; Tortorelli, V.M. On the approximation of free discontinuity problems. *Bollettino dell'Unione Mathematica Italiana B* **1992**, *7*, 105–123.

38. Rada, L.; Chen, K. A new variational model with dual level set functions for selective segmentation. *Commun. Comput. Phys.* **2012**, *12*, 261–283. [CrossRef]

39. Williams, B.M.; Spencer, J.A.; Chen, K.; Zheng, Y.; Harding, S.P. An effective variational model for simultaneous reconstruction and segmentation of blurred images. *J. Algorithms Comput. Technol.* **2016**, *10*, 244–264. [CrossRef]

40. Fuller, A.; Zawadzki, R.; Choi, S.; Wiley, D.; Werner, J.; Hamann, B. Segmentation of three-dimensional retinal image data. *IEEE Trans. Vis. Comput. Graph.* **2007**, *13*, 1719–1726. [CrossRef] [PubMed]

41. Bae, E.; Tai, X.-C. Graph cut optimization for the piecewise constant level set method applied to multiphase image segmentation. In *Scale Space and Variational Methods in Computer Vision*; Springer: Berlin, Germany, 2009; pp. 1–13.

42. LaRocca, F.; Chiu, S.J.; McNabb, R.P.; Kuo, A.N.; Izatt, J.A.; Farsiu, S. Robust automatic segmentation of corneal layer boundaries in sdoct images using graph theory and dynamic programming. *Biomed. Opt. Express* **2011**, *2*, 1524–1538. [CrossRef] [PubMed]

43. Williams, D.; Zheng, Y.; Bao, F.; Elsheikh, A. Fast segmentation of anterior segment optical coherence tomography images using graph cut. *Eye Vis.* **2015**, *2*, 1. [CrossRef] [PubMed]

44. Boykov, Y.; Kolmogorov, V. An experimental comparison of min-cut/max-flow algorithms for energy minimization in vision. *IEEE Trans. Pattern Anal. Mach. Intell.* **2004**, *26*, 1124–1137. [CrossRef] [PubMed]

45. Li, C.; Zeitler, J.A.; Dong, Y.; Shen, Y.-C. Non-destructive evaluation of polymer coating structures on pharmaceutical pellets using full-field optical coherence tomography. *J. Pharm. Sci.* **2014**, *103*, 161–166. [CrossRef] [PubMed]

46. Lawman, S.; Dong, Y.; Williams, B.M.; Romano, V.; Kaye, S.; Harding, S.P.; Willoughby, C.; Shen, Y.C.; Zheng, Y.L. High resolution corneal and single pulse imaging with line field spectral domain optical coherence tomography. *Opt. Express* **2016**, *24*, 2395–2405. [CrossRef] [PubMed]

47. Vese, L.A.; Chan, T.F. A multiphase level set framework for image segmentation using the mumford and shah model. *Int. J. Comput. Vis.* **2002**, *50*, 271–293. [CrossRef]

48. Osher, S.; Sethian, J.A. Fronts propagating with curvature-dependent speed: Algorithms based on Hamilton–Jacobi formulations. *J. Comput. Phys.* **1988**, *79*, 12–49.

49. Ellingson, W.A.; Visher, R.J.; Lipanovich, R.S.; Deemer, C.M. Optical nde methods for ceramic thermal barrier coatings. *Mater. Eval.* **2006**, *64*, 45–51.

50. Markl, D.; Zettl, M.; Hannesschlaeger, G.; Sacher, S.; Leitner, M.; Buchsbaum, A.; Khinast, J.G. Calibration-free in-line monitoring of pellet coating processes via optical coherence tomography. *Chem. Eng. Sci.* **2015**, *125*, 200–208. [CrossRef]

Article

Extending the Effective Ranging Depth of Spectral Domain Optical Coherence Tomography by Spatial Frequency Domain Multiplexing

Tong Wu [1,*], Qingqing Wang [1], Youwen Liu [1,2,*], Jiming Wang [1], Chongjun He [1] and Xiaorong Gu [1]

1   Department of Applied Physics, College of Science, Nanjing University of Aeronautics and Astronautics, Nanjing 210016, China; qingqing_xiatian@163.com (Q.W.); jimingw@nuaa.edu.cn (J.W.); hechongjun@nuaa.edu.cn (C.H.); xrgu@nuaa.edu.cn (X.G.)
2   Key Laboratory of Radar Imaging and Microwave Photonics, Ministry of Education, Nanjing University of Aeronautics and Astronautics, Nanjing 210016, China
*   Correspondence: wutong@nuaa.edu.cn (T.W.); ywliu@nuaa.edu.cn (Y.L.); Tel.: +86-255-207-5692 (T.W.)

Academic Editor: Michael Pircher
Received: 29 September 2016; Accepted: 10 November 2016; Published: 17 November 2016

Abstract: We present a spatial frequency domain multiplexing method for extending the imaging depth range of a spectral domain optical coherence tomography (SDOCT) system without any expensive device. This method uses two galvo scanners with different pivot-offset distances in two independent reference arms for spatial frequency modulation and multiplexing. The spatial frequency contents corresponding to different depth regions of the sample can be shifted to different frequency bands. The spatial frequency domain multiplexing SDOCT system provides an approximately 1.9-fold increase in the effective ranging depth compared with that of a conventional full-range SDOCT system. The reconstructed images of phantom and biological tissue demonstrate the expected increase in ranging depth. The parameters choice criterion for this method is discussed.

Keywords: fiber optics imaging; medical and biological imaging; optical coherence tomography

---

1. Introduction

Fourier domain optical coherence tomography (FDOCT) is a high-speed, high-resolution and non-invasive biomedical optical imaging technique [1]. It has been applied in the field of biomedical research, clinical diagnostics, and non-destructive material inspection. FDOCT can be categorized into two embodiments. One is based on a wideband laser source and a spectrometer which is called spectral domain OCT (SDOCT), and the other one is based on a frequency sweeping laser source and a balanced photon detector which is called swept source OCT (SSOCT, also called optical frequency domain imaging, OFDI). For both types of the FDOCT, the achievable imaging depth range is an important performance indicator. In some situations, a long ranging depth is highly desirable. One such example is for the endoscopic application where the distance between the distal end of the endoscopic probe and the tissue surface cannot be accurately controlled. Another example is for imaging the whole eye from anterior to retina.

For FDOCT, the ranging depth can be calculated using the formula $\Delta z = \lambda_0^2/4n\delta\lambda$ and is fundamentally limited by the center wavelength λ_0 of the light source, the spectral resolution $\delta\lambda$, and the refractive index n. The spectral resolution is $\delta\lambda = \Delta\lambda/N$ when the spectrometer is pixel-limited. In this equation, $\Delta\lambda$ represents the spectral bandwidth of the light source and N is the data number of the digitized axial spectral interference signal that covers the bandwidth. In fact, there is a sensitivity roll-off issue for FDOCT, which poses a further limitation to the practical imaging depth range. For the case of SDOCT, the sensitivity roll-off effect is due to the finite pixel number of the line scanning camera. Due to this effect, the visibility of the spectral interference fringe is maximized when the optical path difference (OPD) is zero and decays as the path length difference increases. Thus, the effective ranging depth is defined as the imaging depth at which the sensitivity has decayed −6 dB from the peak value [2]. For FDOCT, the complex conjugate ambiguity further limits the imaging depth to one half of the full-range imaging space. Another limitation on the imaging depth of the FDOCT is posed by the finite depth of focus of the objective lens.

For extending the effective imaging depth range, various techniques have been proposed. A novel MEMS-Vertical Cavity Surface Emitting Laser (VCSEL) source-based [3] SSOCT system was developed for in vivo high-speed imaging of the eye with an unprecedented imaging depth of 50 mm in air and little sensitivity loss. The SDOCT system based on the orthogonal dispersion spectrometer [4] or the frequency comb approach [5] was proposed for large ranging depth imaging through enhancing spectral resolution. To solve the complex conjugate ambiguity, many complex Optical Coherence Tomography (OCT) techniques were proposed. Among these complex OCT methods, an approach involving spatial phase modulation induced by the pivot-offset galvo scanner (GS) in the sample arm is regarded as a widely accepted way [6–9]. The technique was initially introduced for en face OCT imaging [10]. In the approach, a slight eccentricity of the laser beam incident upon the galvo scanner causes a path length modulation during lateral scanning, generating the required carrier frequency. Recently, the optimization of galvo scanning for OCT was investigated [11].

Alternatively, based on implementing multiple reference arms in the FDOCT system the extended ranging depth can also be achieved [12–16]. These multiple-reference-arm approaches can be roughly categorized into two sorts: simultaneous acquisition approach and sequential acquisition approach. For the simultaneous acquisition approach, Nezam et al. presented an OFDI system based on two independent interferometer reference arms, each having an acousto-optic frequency shifter operating at a distinct frequency and an independent round-trip delay [12]. Through combination of a dual-channel OCT system focusing on different segments of the eye, Chuanqing Zhou et al. extended the imaging depth using two OCT setups [13,14]. For the sequential acquisition approach, Hui Wang et al. proposed a simple and inexpensive method for increasing the effective imaging depth range of the SDOCT system, which was realized by employing a high-speed fiber optic switch to serially access two reference paths with different offset delays [2]. Recently, Ruggeri al. enabled OCT imaging of multi-frames at four depths without sacrificing axial resolution by implementing a galvo scanner-based mechanical optical switch in the reference arm [15]. They utilized the system to image the whole eye from anterior to retina. More recently, a SDOCT system capable of deep depth imaging using a 3×3 fiber coupler, and a mechanical shutter was demonstrated [16]. To conclude, the simultaneous acquisition approaches can extend the imaging depth of the OCT system without sacrificing the effective imaging rate. However, they use either expensive frequency modulators or two OCT systems, which increase the complexity and cost of the imaging system. Furthermore, due to splitting the light source power into two channels, the simultaneous approaches suffer from a 3dB loss of the imaging sensitivity [9]. On the other hand, the sequential acquisition approaches extend the ranging depth with no sensitivity penalty, but with the penalty of halving the effective image acquisition rate, and it cannot be applied in some applications that need to image the different depth regions of the sample simultaneously.

In this paper, we present a simultaneous acquisition approach for extending the imaging depth range of a SDOCT system. The method, which is termed spatial frequency domain multiplexing method, is based on two pivot-offset galvo scanners in two separate reference arms. Galvo scanners are popular devices that offer a good scanning speed, a good field of scan, high precision, and repeatability [17]. Thus, the presented approach can image the different depth regions of the sample simultaneously without any expensive acousto-optic or electro-optic modulation device. In the following sections, we theoretically derive the parameter choice criteria for the high signal/crosstalk ratio (SCR) reconstruction and demonstrate the extended imaging depth performances of the system experimentally.

2. Methods

2.1. Spatial Frequency Domain Multiplexing Spectral Domain Optical Coherence Tomography (SDOCT) System

Figure 1 depicts the schematic of the spatial frequency domain multiplexing SDOCT system. It comprises a super luminescent diode (SLD), an interferometer with two pivot-offset galvo scanners implemented in the two separate reference arms, a custom-built spectrometer, and a personal computer for signal processing and image display. Reference Arms I and II have different round-trip delays, which correspond to the shallow and deep regions of the sample, shown as Z_A and Z_B in Figure 1. Thus, two sets of the spectral interferogram corresponding to the depths Z_A and Z_B can be acquired. In the sample arm, the light beam just incidents upon the pivot of the GS0 for sample scanning. The light beams in the reference arms incident upon the GS1 and GS2 with the different offset distances s_1 and s_2 away from the pivots, respectively, as shown in Figure 1.

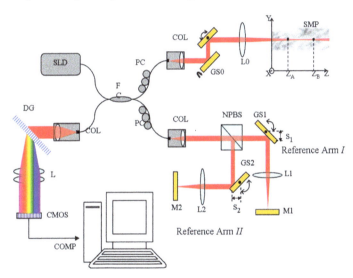

Figure 1. Schematic of the spatial frequency domain multiplexing spectral domain optical coherence tomography (SDOCT) system. SLD: super luminescent diode; FC: fiber coupler; NPBS: non-polarizing beam splitter; M: reference mirror; GS: galvanometer scanner; L: lens; SMP: sample; COL: collimator; PC: polarization controller; DG: diffraction grating; COMP: computer; CMOS: complementary metal oxide semiconductor.

When the galvo scanners in the sample arm and reference arms are driven simultaneously, the linear modulated phase will be added into the two-dimensional (2D) spectral interferograms. By ignoring the signals that do not contribute to the depth localization in the sample for simplicity, the 2D spectral interferogram detected at each wavenumber k and lateral position x can be expressed as

$$
\begin{aligned}
I(x,k) \; = \; & 2S(k)\int \sqrt{R(x,z)}\cos[2nk(z-z_A)+\Phi_1(x)+\phi_A(x,z)]dz \\
& +2S(k)\int \sqrt{R(x,z)}\cos[2nk(z-z_B)+\Phi_2(x)+\phi_B(x,z)]dz
\end{aligned}
\tag{1}
$$

where k represents wavenumber and $k = 2\pi/\lambda$, z is the depth position coordinate of the scatters within the sample, n is the average refractive index of the sample, $R(x,z)$ is the normalized intensity backscattered from the scatter at z, $S(k)$ is the spectral density of the light source, $\Phi_1(x)$ and $\Phi_2(x)$ represent the linear modulated phase induced by GS1 and GS2, and $\phi_A(x,z)$ and $\phi_B(x,z)$ are the phases that relate to the optical heterogeneity of the sample in the focus coherence volume. If the linear modulation phase $\Phi_1(x)$ and $\Phi_2(x)$ have different slopes about the lateral position, the spatial frequency contents of the 2D OCT spectral interferogram corresponding to the depths Z_A and Z_B will be shifted to the different center spatial frequencies away from the zero spatial frequency.

2.2. Demodulation Algorithm for Spatial Frequency Domain Multiplexing SDOCT

As shown in Figure 2, after taking the Fourier transform of the 2D spectral interferogram $I(x,k)$ along the x-direction (step (i) in Figure 2), we obtain the spatial spectrum of the 2D spectral interferogram, which can be expressed as

$$\tilde{I}(v,k) = FT_{x \to v}\{I(x,k)\} \quad = C_A(v+v_{c1},k) + \hat{C}_A(v-v_{c1},k) \\ + C_B(v+v_{c2},k) + \hat{C}_B(v-v_{c2},k) \tag{2}$$

where v is the Fourier transform pair of x. $C(v,k)$ and $\hat{C}(v,k)$ are the pair of Hermit conjugate term due to the Fourier transform of the real valued interference signal and contain the same information on the OCT signal in the spatial frequency domain. The subscripts A and B in Equation (2) indicate the spatial frequency contents corresponding to depths Z_A and Z_B, respectively. v_{c1} and v_{c2} are the center spatial carrier frequencies, which are proportional to the pivot-offset distances. For reconstruction of the complex valued spectral interferogram, two band-pass filters with center frequencies of v_{c1} and v_{c2} are applied to the spatial frequency spectrum (step (ii) in Figure 2). We then obtain the separated spatial frequency content $C_A(v+v_{c1},k)$ and $C_B(v+v_{c2},k)$.

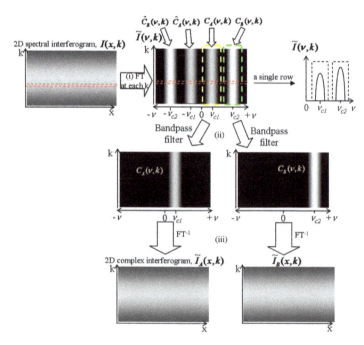

Figure 2. Data processing flowchart of the spatial frequency domain multiplexing SDOCT system. FT: Fourier Transform.

Two complex-valued spectral interferogram $\tilde{I}_A(x,k)$ and $\tilde{I}_B(x,k)$ can now be reconstructed by performing the inverse Fourier transform from the spatial frequency domain back to the spatial domain for the terms $C_A(v+v_{c1},k)$ and $C_B(v+v_{c2},k)$ (step (iii) in Figure 2). For reasons of energy conservation, the data have to be scaled by a factor of 2. Then, after the k space re-sampling the two full-range OCT images corresponding to the depths of Z_A and Z_B can be calculated by inversely Fourier transforming the complex-valued linear-in-k spectral interferogram $\tilde{I}_A(x,k)$ and $\tilde{I}_B(x,k)$ along the wavenumber direction. Finally, an extended range OCT image is formed by concatenating the image regions that correspond to the effective ranging depths of the two full-range OCT images.

2.3. The Center Spatial Frequency and Parameters Design Criterion

Figure 3 is the sketch of the two reference arms with the pivot-offset galvo scanners. The color lines represent the light beams when the galvo scanners are at different scan angles. The pivot-offset distances are indicated as s_1 and s_2. The phase shift between successive axial scans (A-scan) induced by the galvo scanner can be expressed as [9]

$$\delta\phi_i = \frac{8\omega Ts_i\pi}{\lambda_0}, \qquad i=1,2 \tag{3}$$

where ω is the angular velocity of the galvo scanners, T is the exposure time of the camera in the spectrometer, and λ_0 is the center wavelength of the light source.

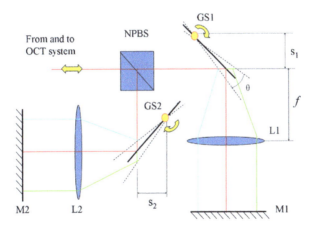

Figure 3. Sketch of the two reference arms in the spatial frequency domain multiplexing SDOCT system. s_1 and s_2: the pivot-offset distances; L1 and L2: the focal lens in the reference arms; M1 and M2: the plane mirrors; f: focal length of the focal lens.

Considering a tomogram of N depth profiles covering a lateral range of Δx, the two center spatial carrier frequencies can be calculated by

$$v_{ci} = \frac{\delta\phi_i}{2\pi} \cdot \frac{N}{\Delta x} \qquad i=1,2. \tag{4}$$

The lateral scanning range Δx is calculated by

$$\Delta x = 2N\omega Tf \tag{5}$$

where f is the focal length of the sample objective.

The positive spatial frequency range can be calculated by

$$\nu_x = \frac{N}{2\Delta x}. \tag{6}$$

For high-quality image reconstruction, another important parameter is the characteristic bandwidth (BW) of the obtained spatial frequency contents, which is determined by the inverse speckle size. The speckle size can be modeled by the transverse resolution, and BW can be expressed as [9]

$$BW = \frac{\pi d}{4\lambda_0 f} \tag{7}$$

where d is the collimated beam diameter entering the sample objective.

In theory, the proposed method could quadruple the effective imaging depth range compared with a conventional SDOCT system. However, practically some factors can affect the improvement of the range depth, which includes the crosstalk between the two spatial frequency contents of the two depths signal, the crosstalk between the spatial frequency content and its complex conjugate content, and the crosstalk between the spatial frequency contents and the interference term between the two reference arms' light fields. Crosstalk between the two spatial frequency contents corresponding to the two depths can occur when the tails of the spatial frequency contents superpose with each other as schematically represented in Figure 4a. In view of the Bedrosian theorem, spatial frequency content should be well separated from the introduced modulation frequency and the other content for successful reconstruction, as shown in Figure 4b. The crosstalk due to the reference arms' interference can be eliminated via background subtraction. According to Equations (3)–(7), the parameters including the angular velocity ω, the complementary metal oxide semiconductor (CMOS) exposure time T, and the offset distance s_i are all responsible to the image reconstruction quality. Thus, to avoid crosstalk and fully utilize the proposed method for extended range imaging, the center spatial carrier frequencies and the system parameters must be chosen carefully.

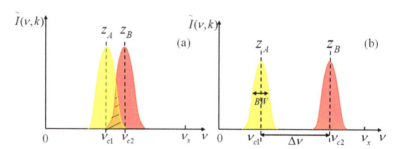

Figure 4. Schematic of the choice of the center spatial carrier frequencies. Crosstalk between the spatial spectra occurs in (**a**), and the two spatial spectra can be well separated through parameters design as shown in (**b**).

To quantify the crosstalk effect, we introduce the SCR defined as a function of the frequency-spacing/bandwidth ratio ($\Delta\nu/BW$) for two reference arms, where $\Delta\nu = \nu_{c2} - \nu_{c1}$ as shown in Figure 4. According to the theoretical value, the SCR can be higher than 100 dB when the frequency-spacing/bandwidth ratio τ satisfies the condition [12]

$$\tau = \frac{\Delta\nu}{BW} \geq 3. \tag{8}$$

Furthermore, to avoid aliasing terms, the bandwidth of the spatial frequency content should be narrow relative to the full spatial frequency range; in this case, the small phase fluctuations will not cause a reduction of the SCR. This condition indicates that [9]

$$\frac{N}{\Delta x} \gg BW. \tag{9}$$

The criteria shown in Equations (8) and (9) will be used to guide the choice of the values of the center spatial carrier frequencies and the other parameters in the experimental SDOCT system.

3. Experiments and Results

For experimental verification of the method, we constructed a fiber-based spatial frequency domain multiplexing SDOCT system as depicted in Figure 1. The imaging system employs a super luminescent diode light source (SLD-371-HP2, Superlum Inc., Carrigtwohill, Ireland) with a center wavelength at 837 nm and a full width at half maximum (FWHM) of 54 nm. The beam diameter d in the sample arm is 1.8 mm, which will produce a 44 μm $1/e^2$ diameter focal spot on the sample with a confocal parameter of 3.7 mm. A 50/50 fiber coupler splits the light beam into the sample arm and the reference arms. In the sample arm, the light beam is deflected by a 2D galvo scanner and focused via a 75 mm achromatic lens L0 to the sample. Within the reference arm, a wide bandwidth non-polarizing beam splitter (NPBS, Daheng Optics Inc., Beijing, China) is used to divide the reference optical power into the two separate reference arms. The fiber collimator on a translation stage in the reference arm is used to adjust the overall delay of the two channels. The single-pass OPD between Reference Arms I and II is set to 3 mm. The GS1 and GS2 are the X and Y mirror of a pair of large beam 2D GS system (GVS012, Thorlabs Inc., Newton, NJ, USA), respectively. The focus lengths of achromatic lens L1and L2 are both 30 mm. The plane mirror M1 and M2 are the wide bandwidth dielectric coated mirrors covering the wavelength range of 800 to 900 nm. The spectral interference signal is detected by a custom-built spectrometer, which consists of a collimator (f = 60 mm, OZ optics Inc., Ottawa, ON, Canada), a transmissive diffraction grating (1800 lines/mm), a camera lens (f =105 mm, Nikon Inc., Tokyo, Japan), and a line-scan CMOS camera (sp2048-70km, Basler AG Inc., Ahrensburg, German). The spectral resolution of the spectrometer is 0.065 nm, allowing for a theoretical effective imaging depth range of 3.5 mm.

In view of avoiding aliasing terms and according to Equation (9), the number of A-scans per B-scan is set to 500, and the lateral scanning range is set to 1.3 mm. The highest positive spatial frequency calculated by Equation (6) is 192 mm^{-1}. According to the values of the sample beam diameter d, the center wavelength, the focus length of the sample objective and Equation (7), the theoretical BW of the spatial frequency content is calculated to be 22.5 mm^{-1}. In order to suppress the crosstalk of the two spatial frequency contents of the two depths signal, the difference of the two center spatial carrier frequencies must be greater than 67.5 mm^{-1} according to Equation (8). Considering that the optimal conjugate suppression ratio occurs when $\delta\varphi$ is $\pi/2$ [9], each of the two center spatial carrier frequencies needs to be close to the central region of the positive spatial frequency range. Thus, the two center spatial carrier frequencies ν_{c1} and ν_{c2} are set to 56.5 and 123.5 mm^{-1}, respectively. According to Equation (4), the phase shift $\delta\varphi_1$ and $\delta\varphi_2$ between successive axial scans are calculated to be 0.3π and 0.7π. The angular velocity of the galvo scanners is set to 0.93 rad/s. The exposure time of the CMOS camera is set to 18 μs. According to Equation (3) and the above parameters, the pivot-offset distances on the GS1 and GS2 mirrors are calculated to be 1.8 and 3.9 mm, respectively. The pivot-offset distance is adjusted via the diagonal translation of the scanners to keep the light beam at the center of the focusing optics.

To experimentally characterize the extension of the imaging depth range of the spatial frequency domain multiplexing SDOCT system, the point spread function is measured by using a plane mirror at various depths in the sample arm. Figure 5 shows the experimentally measured normalized sensitivity as a function of depth for the two reference arms. The measured value of the absolute sensitivity is

59.7 dB. From Figure 5, the separation between the two sensitivity peaks can be used to calibrate the actual relative OPD between the two reference arms, and it can be shown that the actual relative OPD is about 3 mm. The benefit from the use of the spatial frequency domain multiplexing SDOCT system is also demonstrated, and the effective imaging depth range for the proposed system is increased from about 3.4 to 6.4 mm, compared with a conventional full-range SDOCT without depth range enhancement. The depth enhancement factor is 1.9. In this demonstration, the two sensitivity curves are crossed at the −5.5 dB from the sensitivity peaks. Therefore, the effective imaging depth range would be doubled if the relative optical path delay was precisely tuned to make the two sensitivity curves cross at the −6 dB point from the sensitivity peaks.

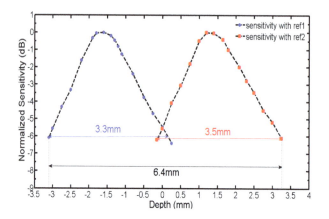

Figure 5. The measured sensitivity curves corresponding to the Reference Arm I and Reference Arm II.

To demonstrate the ranging depth enhancement, we compare image results of the OCT images acquired by the proposed two-reference-arm spatial frequency domain multiplexing approach and the conventional single-reference-arm approach. Firstly, an artificial phantom is imaged by the system. Figure 6a shows the photograph of the phantom taken with a digital camera. The phantom is made by sticking two blocks of sellotape to generate a step 3 mm high. Figure 6b–d are the complex conjugate ambiguity free OCT images of the phantom reconstructed by the data processing algorithm described previously. Figure 6b shows the image acquired with Reference Arm I open while blocking Reference Arm II. As we can see, the upper part of the step is clearly imaged, while the image of the lower part is hard to be discerned. The OPD between the upper part and Reference Arm I resides in the effective imaging range, while that of the lower part and Reference Arm I is larger than 6 mm, out of the effective imaging range. Figure 6c shows the OCT image acquired with Reference Arm II open while blocking Reference Arm I. Likewise, the lower part of the step can be clearly imaged, while the upper part is not clear. Figure 6d shows the reconstructed OCT image of the phantom acquired with both Reference Arms I and II open. The whole phantom including the upper and lower part of the step can be clearly seen. The imaging depth-enhanced OCT image of the artificial phantom demonstrates the feasibility of the proposed method for extending the OCT ranging depth.

The biological tissue of the swine adipose is also in vitro, imaged and compared with the results acquired by the proposed two-reference-arm spatial frequency domain multiplexing approach and the conventional single-reference-arm approach. Figure 7a,b show the images recorded using the two Reference Arms I and II separately. In each case, the area of the tissue around the DC signal component has maximum sensitivity. Figure 7c shows the OCT image of the tissue acquired with both of Reference Arms I and II open. As we can see, the upper and lower part of the biological sample is clearly appeared at the same time. Figure 7d plots an A-scan profile from the same sample position indicated by the red arrows in Figure 7a,b. It can be seen that the signal is about 13 dB stronger than that of the signal 3 mm away due to the sensitivity roll-off. The extended range OCT image of the biological tissue obtained with the system demonstrates the feasibility of the proposed method for extending the OCT ranging depth. Because the galvanometer scanners in the system are used for phase modulation of the lateral interference spectral signal, the phase noise does not affect the quality of the results. In future, the proposed method for extending the OCT ranging depth can be combined with the depth of the focus extension method to improve the quality of the extended range OCT image.

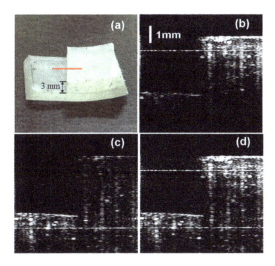

Figure 6. (**a**) The photograph of the artificial phantom of a step comprised of two blocks of sellotape. The red line is the lateral scan position on the phantom; (**b**,**c**) the cross-sectional optical coherence tomography (OCT) image of the phantom acquired with the conventional single-reference-arm approach; (**d**) the extended range OCT image acquired with the proposed two-reference-arm spatial frequency domain multiplexing approach. The blurry horizontal white line is the residual direct current (DC) term.

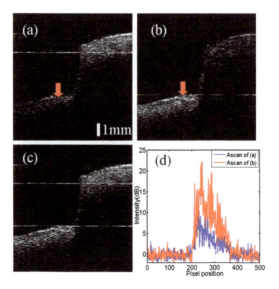

Figure 7. Cross-sectionalOCT image of the swine adipose tissue, (**a,b**) acquired with the conventional single-reference-arm approach. (**c**) The extended range OCT image is obtained with the two-reference–arm spatial frequency domain multiplexing approach. The blurry horizontal white line is the residual DC terms. (**d**) A-scans at the position marked by the red arrows in (**a,b**) demonstrate the advantage of the ranging depth extension.

4. Conclusions

We have proposed and demonstrated a spatial frequency domain multiplexing method to extend the effective imaging depth range for SDOCT system without any expensive acousto-optic or electro-optic modulation device. The spatial frequency domain multiplexing is realized by using two pivot-offset galvo scanners in the two reference arms. The images of shallow and deep regions of a sample can be multiplexed and acquired simultaneously with no image acquisition rate penalty. The proposed two-reference-arm spatial frequency domain multiplexing SDOCT system provides an approximately 1.9-fold increase in the effective imaging range compared with that of a conventional single-reference-arm full-range SDOCT system. We present key system parameter design criteria for ensuring a high signal/crosstalk ratio imaging. The images of an artificial phantom and the swine adipose tissue reconstructed with the proposed method demonstrate the expected increase in the effective imaging depth range. This simple and low-cost method is also applicable to the SSOCT system, and desirable in the in vivo applications requiring a relatively long imaging range.

Acknowledgments: This work was supported by the National Natural Science Foundation of China (61205201, 11174147), the Fundamental Research Funds for the Central Universities, Nanjing University of Aeronautics and Astronautics (NZ2015104), Postdoctoral Research Funds of Jiangsu Province (1201034C), the China Postdoctoral Science Foundation (2013M531345).

Author Contributions: Tong Wu and Youwen Liu conceived the ideas; Tong Wu and Qingqing Wang designed and performed the experiments; Tong Wu and Qingqing Wang processed the data; Jiming Wang, Xiaorong Gu, and Chongjun He contributed with the materials and samples; Tong Wu wrote the paper.

Conflicts of Interest: The authors declare no conflict of interest.

References

1. Huang, D.; Swanson, E.A. Optical Coherence Tomography. *Science* **1991**, *254*, 1178–1181. [CrossRef] [PubMed]

2. Wang, H.; Pan, Y.; Rollins, A.M. Extending the effective imaging range of Fourier-domain optical coherence tomography using a fiber optic switch. *Opt. Lett.* **2008**, *33*, 2632–2634. [CrossRef] [PubMed]
3. Grulkowski, I.; Liu, J.J. Anterior segment and full eye imaging using ultrahigh speed swept source OCT with vertical-cavity surface emitting lasers. *Biomed. Opt. Express* **2012**, *3*, 2733–2751. [CrossRef] [PubMed]
4. Bao, W.; Ding, Z. Orthogonal dispersive spectral-domain optical coherence tomography. *Opt. Express* **2014**, *22*, 10081–10090. [CrossRef] [PubMed]
5. Bajraszewski, T.; Wojtkowski, M. Improved spectral optical coherence tomography using optical frequency comb. *Opt. Express* **2008**, *16*, 4163–4176. [CrossRef] [PubMed]
6. Wang, R. In vivo full range complex Fourier domain optical coherence tomography. *Appl. Phys. Lett.* **2007**, *90*, 054103. [CrossRef]
7. Baumann, B.; Pircher, M. Full range complex spectral domain optical coherence tomography without additional phase shifters. *Opt. Express* **2007**, *15*, 13375–13387. [CrossRef] [PubMed]
8. An, L.; Wang, R. Use of a scanner to modulate spatial interferograms for in vivo full-range Fourier-domain optical coherence tomography. *Opt. Lett.* **2007**, *32*, 3423–3425. [CrossRef] [PubMed]
9. Leitgeb, R.A.; Michaely, R. Complex ambiguity-free Fourier domain optical coherence tomography through transverse scanning. *Opt. Lett.* **2007**, *32*, 3453–3455. [CrossRef] [PubMed]
10. Podoleanu, A.G.; Dobre, G.M.; Jackson, D.A. En-face coherence imaging using galvanometer scanner modulation. *Opt. Lett.* **1998**, *23*, 147–149. [CrossRef] [PubMed]
11. Duma, V.F.; Tankam, P. Optimization of galvanometer scanning for optical coherence tomography. *Appl. Opt.* **2015**, *54*, 5495–5507. [CrossRef] [PubMed]
12. Nezam, S.M.R.M.; Vakoc, B.J. Increased ranging depth in optical frequency domain imaging by frequency encoding. *Opt. Lett.* **2007**, *32*, 2768–2770. [CrossRef]
13. Zhou, C.; Wang, J. Dual channel dual focus optical coherence tomography for imaging accommodation of the eye. *Opt. Express* **2009**, *17*, 8947–8955. [CrossRef] [PubMed]
14. Dai, C.; Zhou, C. Optical coherence tomography for whole eye segment imaging. *Opt. Express* **2012**, *20*, 6109–6115. [CrossRef] [PubMed]
15. Ruggeri, M.; Uhlhorn, S.R. Imaging and full-length biometry of the eye during accommodation using spectral domain OCT with an optical switch. *Biomed. Opt. Express* **2012**, *3*, 1506–1520. [CrossRef] [PubMed]
16. Dai, C.; Fan, S. Dual-channel spectral-domain optical-coherence tomography system based on 3 × 3 fiber coupler for extended imaging range. *Appl. Opt.* **2014**, *53*, 5375–5379. [CrossRef] [PubMed]
17. Montagu, J. Scanners—Galvanometric and resonant. In *Encyclopedia of Optical Engineering*, 1st ed.; Driggers, R.G., Hoffman, C., Eds.; Taylor & Francis: New York, NY, USA, 2003; Volume 3, pp. 2465–2487.

 *applied sciences*

Article

Needle Segmentation in Volumetric Optical Coherence Tomography Images for Ophthalmic Microsurgery

Mingchuan Zhou [1], Hessam Roodaki [1], Abouzar Eslami [2], Guang Chen [1,*], Kai Huang [3,*], Mathias Maier [4], Chris P. Lohmann [4], Alois Knoll [1] and Mohammad Ali Nasseri [4]

[1] Institut für Informatik, Technische Universität München, 85748 München, Germany; mingchuan.zhou@in.tum.de (M.Z.); he.roodaki@tum.de (H.R.); knoll@in.tum.de (A.K.)
[2] Carl Zeiss Meditec AG., 81379 München, Germany; abouzar.eslami@zeiss.com
[3] School of Data and Computer Science, Sun Yat-Sen University, Guangzhou 510006, China
[4] Augenklinik und Poliklinik, Klinikum rechts der Isar der Technische Universit München, 81675 München, Germany; mathias.maier@mri.tum.de (M.M.); chris.lohmann@mri.tum.de (C.P.L.); ali.nasseri@mri.tum.de (M.A.N.)
* Correspondence: guang@in.tum.de (G.C.); huangk36@mail.sysu.edu.cn (K.H.); Tel.: +49-89-289-18112 (G.C.); +49-89-289-18111 (K.H.)

Received: 22 June 2017; Accepted: 18 July 2017; Published: 25 July 2017

Abstract: Needle segmentation is a fundamental step for needle reconstruction and image-guided surgery. Although there has been success stories in needle segmentation for non-microsurgeries, the methods cannot be directly extended to ophthalmic surgery due to the challenges bounded to required spatial resolution. As the ophthalmic surgery is performed by finer and smaller surgical instruments in micro-structural anatomies, specifically in retinal domains, difficulties are raised for delicate operation and sensitive perception. To address these challenges, in this paper we investigate needle segmentation in ophthalmic operation on 60 Optical Coherence Tomography (OCT) cubes captured during needle injection surgeries on ex-vivo pig eyes. Furthermore, we developed two different approaches, a conventional method based on morphological features (MF) and a specifically designed full convolution neural networks (FCN) method, moreover, we evaluate them on the benchmark for needle segmentation in the volumetric OCT images. The experimental results show that FCN method has a better segmentation performance based on four evaluation metrics while MF method has a short inference time, which provides valuable reference for future works.

Keywords: needle segmentation; ophthalmic microsurgery; Optical Coherence Tomography

1. Introduction

Recent research shows that eye pathologies contribute to more than 280 million visual impairments [1]. Therefore, there is an increasing demand for the ophthalmic surgery due to such a large number of claims. Vitreoretinal surgery is a conventional ophthalmic operation consisting of complex manual tasks, as shown in Figure 1. The incisions, created by keratome and trocar at the sclera in a circle and 3.5 mm away from the limbus [2], are made to provide the entrance for three tools: light source, surgical tool, and irrigation cannula [3,4]. The irrigation cannula is used for liquid injection to maintain appropriate intraocular pressure (IOP). The light source is used to illuminate the intended area on the retina, allowing the planar view of the area obtained and analyzed by surgeons through the microscope. To address these challenges, the surgical progress proposes a great challenge of delicate operation and sensitive perception to surgeons. The surgical instrument segmentation is the first step to estimate the needle pose and position, which is extremely important to enhance surgeons' concentration under low illumination intraocular condition.

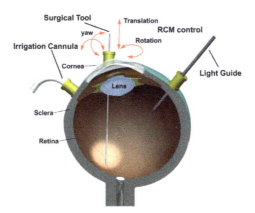

Figure 1. Description of a conventional ophthalmic microsurgery.

Among a variety of surgical tools in ophthalmic operations, the beveled needle is a widely used surgical instrument for delivering the drug into micro-structural anatomies of the eye such as retinal blood vessels and sub-retinal areas. Many studies have been carried out with significant progress in the needle segmentation through microscopic images [5–8]. These works achieved satisfactory results using either color-based or geometry-based features. Nevertheless, due to the limitation of two-dimensional (2D) microscopic images, these detection results in en-face plane view cannot provide enough information to locate the needle pose and position in three-dimensional (3D) space. Many widely used 3D medical imaging technologies, such as computed tomography (CT) scans, fluoroscopy, magnetic resonance (MR) and ultrasound are already applied in brain, thoracic and cardiac surgeries, not only for diagnostic procedures but also as real-time surgical guides [9–13]. However, these imaging modalities cannot achieve a sufficient resolution for ophthalmic interventions. For instance, in MRI-guided interventions with millimeter resolution in breast and prostate biopsies, 18 gauge needles with a diameter of 1.27 mm are used. Yet for ophthalmic surgery, 30 gauge needles with a diameter of 0.31 mm require resolution submillimeter [14].

Optical Coherence Tomography (OCT) is originally used in ophthalmic diagnosis for its micron level resolution [15]. Recently, OCT application has been extented to provide real-time information of intra-operative interactions between the surgical instrument and intraocular tissue [16]. The microscope-mounted intra-operative OCT (iOCT) developed by Carl Zeiss Meditec (RESCAN 700) was firstly described in clinical use in 2014 [17]. The iOCT integrated to on this microscope is capable of sharing the same optical path with the microscopic view and provide real-time cross section information, which is an ideal imaging modality for ophthalmic surgeries. The iOCT device can also obtain a volumetric image cubes with multi B-scans acquirements. The scan area can be adjusted via region-of-interest (ROI) shown in the microscopic images in CALLISTO eye assistance computer system.

When using iOCT to estimate the needle pose and position, the first step is to obtain the needle point cloud and to segment the needle voxels in the volumetric OCT images. Apart from the difficulties caused by image speckle, low contrast, signal loss due to shadowing, and refraction artifacts, the needle in OCT images has much more details of the tip part (needle may have several fragments in the B-scan image) and has illusory needle reflection [18]. All these challenges make needle segmentation in OCT images different from other imaging modalities such as 3D ultrasound. To the best of our knowledge, there is no systematic empirical research addressing the needle segmentation in volumetric OCT images. Previous studies from [19,20] focused on the visualization of volumetric OCT. Rieke et al. [7,21] studied the surgical instrument segmentation in cross-sectional B-scan image. In their work, the position of the cross-section is localized by microscope image, and the pattern of needle in B-scan image is simplified. Therefore, their methods cannot be directly applied to segmenting the needle in volumetric OCT images.

This study focuses on developing the approaches in two popular mechanisms for needle segmentation in volumetric OCT images. We propose two methods, corresponding to mechanisms of manually feature exaction and automatically feature exaction, to tackle difficulties for the needle segmentation in OCT images: (a) with the needle shadow principle [22], a conventional method based on morphological features (MF) comes to the mind; (b) another approach is based on the recently developed fully convolution neural networks (FCN) [23], which has been applied for MRI medical image analysis. The MF method is usually straightforward which needs features with parameters manually input upon the analysis of all situations. The FCN is a complicated network which can learn all features automatically by properly training the model based on dataset. We extend the FCN method to identify the needle in OCT images. The main contributions of this paper are: (a) A specific FCN method is developed and compared with the conventional method for the needle segmentation with different pose and rotation; (b) A benchmark including 60 OCT cubes with 7680 images is set up using ex-vivo pig eyes for evaluation of both methods. The rest of the paper is organized as follows: Section 2 gives the basic configuration of OCT and typical patterns of needle in OCT B-scan images, afterwards, two methods are presented for needle segmentation in detail. We carry out the experiments and describe the results in Section 3. Section 4 gives the discussion and Section 5 concludes the presented work.

2. Method

The schematic diagram of needle segmentation in OCT cube is shown in Figure 2. In order to preserve as much information as possible, the highest resolution of OCT scan on RESCAN 700 is selected which is 128 B-scans each with 512 A-sacns in 3×3 mm. Each A-scan has 1024 pixels for the 2 mm depth information (see Figure 2a). Figure 2b shows the needle in an ex-vivo pig eye experiment, which is an en-face plane view obtained by the ophthalmic microscope. The scan area is decided by the rectangle and can be adjusted by a foot panel connected to the RESCAN 700. Afterwards, the volumetric OCT images are generated (see Figure 2c). Figure 2d is the needle segmentation based on the process of OCT image cube. By taking consideration of different needle rotations and positions in OCT cube, the needle in B-scan can be seen as the following patterns in Figure 3. It shows that most of the needle fragments have the clear shadow except for the needle tip part. The refraction fragment usually has a larger amount of pixels in comparing to normal imaging parts. The qualified method should be able to segment as many needle pixels as possible in various conditions.

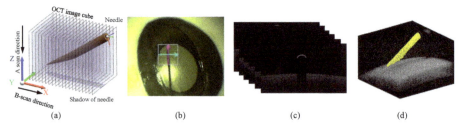

      (a)   (b)   (c)   (d)

Figure 2. The schematic diagram of needle segmentation in an Optical Coherence Tomography (OCT) cube. (**a**) The needle in OCT cube; (**b**) The microscopic image of the needle with ex-vivo pig eyes (the white box indicates the OCT scan area); (**c**) The output of B-scan sequence; (**d**) The needle (yellow) and background (gray) segmentation point cloud.

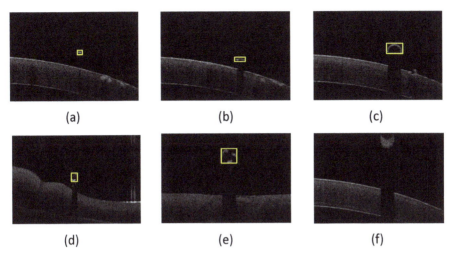

Figure 3. The different patterns of needle seen in OCT B-scan (The needle fragments are labeled with yellow bounding box). Form (**a**) is the needle tip without clear shadow. Since this part of the object is too small, the result of light diffraction leads to unclear shadow on the background. (**b**) shows the needle tip with clear shadow. In this form the needle is downwards-beveled that can be seen as a single fragment piece in the OCT B-scan. In form (**c**) the needle body has clear shadow. In forms (**d**,**e**) several needle tip fragments with clear shadows can be seen here the needle is upwards-beveled; (**f**) shows the needle refraction fragment in this form the needle is out of OCT image range with a clear refraction in B-scan image.

2.1. Morphological Features Based Method

Due to the fact that most of the needle parts in each B-scan are located outside the tissue creating shadows, a morphological feature based method is proposed. The B-scan gray image is transformed into a binary image by thresholding the OCT images. The threshold value is adaptively defined based on statistical measurements of each B-scan. This simple and effective method has been used and evaluated in the automatic segmentation of structures for OCT images [24]. Moreover, a median filter is applied for eliminating the noise. Furthermore, the topmost surface is segmented and considered as the tissue surface. By scanning from left to right, any vertical jump or drop in this surface layer is detected and considered as the beginning and the end of an instrument reflection, respectively. To avoid the misdetection of anatomical features of the eye tissue as a reflection caused by the needle, only reflections with invisible intra-tissue structures are confirmed to be needle reflections. A bounding box is used to cover the region of detected needle part, see Figure 2b–e, and the width of bounding box reflects the width of needle cross-section which can be used to analyze the diameter of the needle. In consideration of the situation for several needle tip parts, the bounding boxes in one B-scan image, whose distance between any two of them is less than a threshold d_b, can be merged into one bounding box. The needle refraction fragment will be then removed from the detected results since it is not the real needle position. Removal operation will be performed when either of the conditions is satisfied: (1) the top edge of the needle fragment bounding box is closed to the upper edge of the image; (2) the number of needle pixels is more than a threshold. Here, in the MF method, features are obtained by observation and summarization of the needle patterns in the B-scan image. Some of the parameters are manually decided which may influence the accuracy of segmentation. In the next section, we will introduce another method that can learn the features by itself with the training dataset.

2.2. Full Convolution Neural Networks Based Method

2.2.1. Network Description

In this section, we present a specifically designed FCN method inspired by the work of Long et al. [23]. Figure 4 shows the schematic representation of our network. We perform convolutions aiming at both extracting features from the data and segmenting the needle out of the image. The left part of the network consists of a compression path, while the right part decompresses the signal until original sizes are reached.

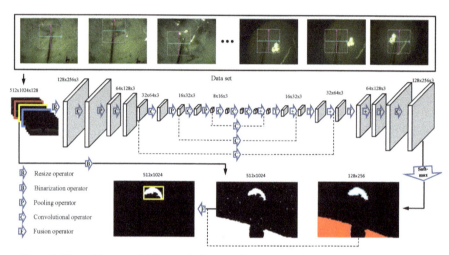

Figure 4. The architecture of full convolution neural networks (FCN) based needle segmentation method.

In order to reduce the size of the network and consider existence of the needle in B-scan image sequence continuously, we take three adjacent B-scan images in OCT cube as input and resize the image with a factor of χ. The left side of the network is divided into different stages operating at different resolutions. Similar to the approach demonstrated in [23], each stage comprises one to two layers. The convolution layer performed in each stage uses volumetric kernels having a size of $7 \times 7 \times 3$ voxels with stride 1. The pooling uses the max-pooling operation in FCN with $7 \times 7 \times 3$ voxels with stride 2, thus the size of the resulting feature maps is halved. PReLu non linearities are applied throughout the network. Downsampling allows us to reduce the size of the input information, and furthermore increase the receptive field of the features being computed in subsequent network layers. Each pair of the convolutional and pooling layers computes two times higher features more than the one in the previous layer.

After three convolutional layers with image size of $4 \times 8 \times 3$, the network increases the low-resolution input by de-convolution combining the pooling results from previous layers [23]. The two feature maps computed from the last convolutional layer, having $1 \times 1 \times 1$ kernel size and producing the same size as input images which will be converted to probabilistic segmentations of the foreground and background regions by using soft-max operation. In order to obtain the original resolution of the needle foreground, the segmentation result from FCN is used to fuse with the original image after binarization and 4-connected components labeling. The labeled area with a certain number of pixels voting from the output foreground of FCN will be considered as the needle fragments. All needle fragments will be covered by one bounding box to indicate the ROI.

2.2.2. Training

The needle fragment always occupies only a small region in B-scan image compared to the background, which leads the network having a strong bias towards the background. In order to avoid the learning progress trapping into the local minima, the dice coefficient based objective function is used to define our subject function. This will increase the weight of the foreground during the training phase. The dice coefficient ℓ of the two binary images can be represented as [25]:

$$\ell = \frac{2 \sum_i^n p_i g_i}{\sum_i^n p^2 + \sum_i^n g^2} \tag{1}$$

where the predicted binary segmentation image $p_i \in P$, the ground truth binary volume $g_i \in G$ and n indicates the amount of pixels. The gradient can be calculated to obtain the gradient with respect to the j-th pixel of the prediction.

$$\frac{\partial \ell}{\partial p_j} = 2\left(\frac{g_j(\sum_i^n p_i^2 + \sum_i^n g_i^2) - 2p_j(\sum_i^n p_i g_i)}{(\sum_i^n p_i^2 + \sum_i^n g_i^2)^2}\right) \tag{2}$$

After the soft-max operator, we obtain a map of probability for the needle and the background. The pixel with a probability of more than 0.5 will be treated as the needle part.

3. Experiments and Results

3.1. Experimental Setup and Evaluation Metrics

The experiments were carried out on ex-vivo pig eyes. A micro-manipulator was used to grip the needle. The CALLISTO eye assistance computer system was established to show the microscopic image and display preview of OCT scans (see Figure 5). A foot pedal connected to the RESCAN700 was settled to relocate the scan area. We captured the needle with different poses and positions in OCT scan area on ex-vivo pig eyes. The OCT cube with resolution of $512 \times 128 \times 1024$ voxels and a corresponding imaging spatial area of $3 \times 3 \times 2$ millimeters, 60 OCT cubes with 7680 B-scan images were manually segmented and marked the needle pixels as the ground truth data set.

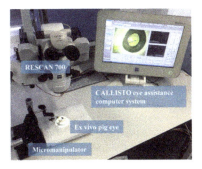

Figure 5. The experimental setup of ophthalmic microsurgery on ex-vivo pig eyes. The micro-manipulator is designed to grasp the syringe and place the needle close to the eye tissue. The CALLISTO eye assistance computer system is set up to display the en-face microscopic image and cross sectional images of OCT cube.

Both of the aforementioned methods were implemented on a Intel(R) Xeon(R) CPU E5-2620 v3 2.40 GHz with a GeForce GTX 980 Ti and memory of 64 GB running Ubuntu 16.04 operating system. The MF based method was implemented with OpenCV 2.4 in C++. The FCN based method used Caffe [26] framework to design network with python and the rest image processing parts were implemented also with OpenCV 2.4 in C++. We used four metrics to evaluate the performance of two methods. Let n_p be the number of needle pixels in prediction, and n_p' be the number of needle pixels in prediction correctly. The details of the four metrics are described as follow: (1) pixel error number can be false positive (FP) needle pixels and false negative (FN) needle pixels in predicted result; (2) pixel accuracy rate equals n_p'/n_p; (3) The average bounding box absence rate indicates the degree of missing needle segment in B-scan image; (4) The width error of bounding box influences the further needle dimension analysis.

3.2. Results

Among 60 OCT cubes, 40 OCT cubes with 5120 images are applied to training the proposed network, while the rest 20 OCT cubes are used to verify the CNN network meanwhile giving a comparison with the MF based method. The needle in each cube has different pose and position, but all B-scan image sequences follow a pattern of no needle appearance to needle appearance since the needle appearance is always continuous. In order to make the segmentation result of each cube comparable, we mapped all indexes of B-scan images in one cube into three piecewise intervals with the increasing index, (1) no needle above the tissue, (2) needle appearance, and (3) needle above the tissue but out of OCT image range thus reflection exists. All indexes of B-scan images were mapped such that 0–25% is the first interval, 25–75% is the second interval and 75–100% is the third interval. Therefore, in case in an OCT cube the index for the first needle appearance image (taken from the ground truth) is at 50 and the index for needle end image is at 100, the currently evaluated image has an index of 60, and its metrics information will be at $25\% + (60 - 50)/(100 - 50) = 45\%$. The metrics for this image are then sorted into buckets of 2% and the values are averaged over other images' metrics value in the given bucket.

The evaluation of two methods under four metrics is shown in Figure 6. Figure 6a shows the comparison of FN and FP for the average needle pixel number using two methods. The average pixel number of FP for MF and FCN is almost equal to 0 which means that few background pixels are classified as needle points. For the performance of average pixel number for FN, both methods have problems with recognition at the beginning of the needle. This artifact is probably caused by unclear shadow of the small needle tip segmentation. However, the FCN performs better than the MF which indicates fewer needle pixels are incorrectly classified to the background. The overall average FN pixel number for two methods are 187.2 and 34.1, respectively. Especially, the FCN can almost segment all of the needle pixels for the needle body part. Figure 6b shows the accuracy rate of needle pixel for two methods, which further indicates that the beginning of needle tip part has a low accuracy rate with 0.19 and 0.33 for MF and FCN, respectively. Both methods give an acceptable accuracy rate during the needle body, while FCN has a better accuracy rate. The comparison of detection accuracy rate for bounding box is shown in Figure 6c. The rate of missing bounding boxes for MF method is quite dramatic in the beginning with around 80%, but rapidly drops to a steady 10% until around the middle. Starting from the middle, almost all bounding boxes are calculated correctly. The FCN method has a similar pattern with better results. It has 66.6% of missing bounding box rate at most for the beginning of needle tip and almost none of missing bounding box detection for the needle body part. Figure 6d gives the average width error of bounding box detection in the same level that the maximum of average width error is under 1 pixel.

We also analyze the performance results of two methods with each processing stage per B-scan image (see Tables 1 and 2). The FCN method has a 121.83 ms average inference time while the MF method has a shorter inference time of 25.6 ms, indicating that MF method performs better for time sensitive purpose usage on the current platform. The variance of FCN method is lower than MF method as the number of operations is always the same in FCN method, revealing that the FCN method is more robust to the dataset and more general to the application.

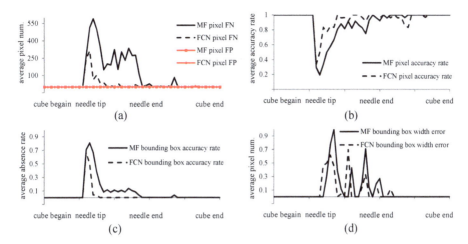

Figure 6. The accuracy performance of FCN and MF based methods. (**a**) Comparison of false negative (FN) and false positive (FP) for the average number needle pixels using two methods; (**b**) Comparison of average accuracy rate of needle pixels using two methods; (**c**) Comparison of average absence rate of bounding box using two methods; (**d**) Comparison of average pixels error for width of bounding box using two methods.

Table 1. MF method inference time.

Stage	Mean (ms)	Variance (ms)
Loading	0.71	0.10
Filtering	7.87	2.22
Detection	17.02	20.55
Total	25.6	22.6

Table 2. FCN method inference time.

Stage	Mean (ms)	Variance (ms)
Loading	0.72	0.11
CNN	97.46	4.67
Fusion	6.63	0.38
Total	121.83	4.91

4. Discussion

This study modified and improved two methods based on the previous work to tackle the challenges for needle segmentation in OCT images: the MF method and the FCN method. The MF based method mainly relies on the needle shadow feature which is hand-crafted by researchers. The FCN method learns the features by itself which is more efficient and less time-consuming as it avoids the cumbersome hand designing phase. The evaluation of two methods is performed with 60 OCT cubes using 7680 B-scan images captured on ex-vivo pig eyes. The experimental results show that FCN method has a better segmentation performance in terms of four evaluation metrics than the MF method. Specifically, the overall average FN needle pixel number is 187.2, and 34.1; the average needle pixel accuracy rate is 90.0%, and 94.7%; the average bounding box accuracy rate is 92.5%, and 97.6% and average pixels error for bounding box is 0.09, and 0.07; for the MF method and FCN method, respectively. Although this study is not targeting on obtaining the highest performance by tunning the FCN network, Our results elucidate that deep learning method indeed generates a powerful model on

Appl. Sci. **2017**, *7*, 748

our small dataset. A future direction is indicated to try different deep learning methods and record more training data. Regarding the runtime, the average inference time of the MF method is four times shorter than FCN method, that is 25.6 ms, and 121.83 ms, respectively. The MF method is a good choice when real-time information is required. The inference time of FCN method can also be improved using parallel programme on more advanced GPU platform. It is possible to get a balance between efficiency and complexity by fusing these two methods together. It is also worth to note that the variance of total inference time for FCN method is lower than MF method for the number of operations in FCN remaining the same in different input images.

Although both methods achieve an average needle pixel accuracy rate above 90.0%, they have problems on segmenting the needle tips, especially for the very beginning part. A potential reason is that the tiny needle segments in B-scan, the image noise as well as the shadows caused by light diffraction principle can be easily mixed with each other. There is a clear improvement from FCN method on the performance for the needle body part, showing almost perfect segmentation results. This also indicates that our future work should focus on improving the needle segmentation performance on the tiny needle tips.

5. Conclusions

We studied the first step of obtaining the needle point cloud for needle pose and position estimation: the needle segmentation in OCT images. We proposed two methods: a conventional needle segmentation method based on morphological features and a fully convolutional neural networks method. These are two typical machine learning methods in image segmentation, manually feature designing based method (MF method) and automatically feature learning based method (FCN method). We analyzed our evaluation results based on the segmentation performance and runtime requirement. Two important insights from our experiments are that the deep learning method generates the discriminative model very well and conventional machine learning method can achieve real-time performance very easily. We believe that our work and insights will be highly helpful and can be referenced as a fundamental work for future needle segmentation research in OCT images. It is also worth to note that the FCN method can segment not only the surgical tools and but also the retinal tissue, which provides additional information to guide the positioning of the surgical tools, such as an alarm warning on the distance between the tool and underlying retinal tissue.

In future, we would like to collect more data and apply the state-of-the-art deep learning model on them. Moreover, we will build up a specific tiny needle tip recognition model in order to overcome the shortness in current performance. A future design can integrate two methods to speed up the processing with better performance. The conventional method could detect the indexes of B-scan images with needle preliminarily in a short time, and then the deep learning is used to segment these B-scan images accurately.

Acknowledgments: Part of the research leading to these results has received funding from the European Union Research and Innovation Programme Horizon 2020 (H2020/2014–2020) under grant agreement No. 720270 (Human Brain Project), and also supported by the German Research Foundation (DFG) and the Technical University of Munich (TUM) in the framework of the Open Access Publishing Program. The authors would like to thank Carl Zeiss Meditec AG. for providing the opportunity to use their ophthalmic imaging devices and Simon Schlegl for part of code contributions.

Author Contributions: M.Z. proposed the idea, designed the experiments and wrote the paper; H.R. performed the experiments; A.E. provided OCT imaging devices and experimental suggestions; G.C. and K.H. provided computation hardware support and suggestions; M.M. and C.P.L. gave the surgical knowledge and experience; A.K. gave the helpful feedback and supported the research. M.A.N. provided experimental materials and designed the experiments.

Conflicts of Interest: The authors declare no conflict of interest.

References

1. World Health Organization (WHO). *Towards Universal Eye Health: A Global Action Plan 2014 to 2019 Report*; WHO: Geneva, Switzerland, 2013.
2. Rizzo, S.; Patelli, F.; Chow, D. *Vitreo-Retinal Surgery: Progress III*; Springer: New York, NY, USA, 2008.
3. Nakano, T.; Sugita, N.; Ueta, T.; Tamaki, Y.; Mitsuishi, M. A parallel robot to assist vitreoretinal surgery. *Int. J. Comput. Assist. Radiol. Surg.* **2009**, *4*, 517–526.
4. Wei, W.; Goldman, R.; Simaan, N.; Fine, H.; Chang, S. Design and theoretical evaluation of micro-surgical manipulators for orbital manipulation and intraocular dexterity. In Proceedings of the 2007 IEEE International Conference on Robotics and Automation, Roma, Italy, 10–14 April 2007; pp. 3389–3395.
5. Bynoe, L.A.; Hutchins, R.K.; Lazarus, H.S.; Friedberg, M.A. Retinal endovascular surgery for central retinal vein occlusion: Initial experience of four surgeons. *Retina* **2005**, *25*, 625–632.
6. Li, Y.; Chen, C.; Huang, X.; Huang, J. Instrument tracking via online learning in retinal microsurgery. In Proceedings of the International Conference on Medical Image Computing and Computer-Assisted Intervention, Boston, MA, USA, 14–18 September 2014; Springer: Cham, Switzerland, 2014; pp. 464–471.
7. Rieke, N.; Tan, D.J.; Alsheakhali, M.; Tombari, F.; di San Filippo, C.A.; Belagiannis, V.; Eslami, A.; Navab, N. Surgical tool tracking and pose estimation in retinal microsurgery. In Proceedings of the International Conference on Medical Image Computing and Computer-Assisted Intervention, Munich, Germany, 5–9 October 2015; Springer: Cham, Switzerland, 2015; pp. 266–273.
8. Sznitman, R.; Ali, K.; Richa, R.; Taylor, R.H.; Hager, G.D.; Fua, P. Data-driven visual tracking in retinal microsurgery. In Proceedings of the International Conference on Medical Image Computing and Computer-Assisted Intervention, Nice, France, 1–5 October 2012; Springer: Berlin, Germany; New York, NY, USA, 2012; pp. 568–575.
9. Kwoh, Y.S.; Hou, J.; Jonckheere, E.A.; Hayati, S. A robot with improved absolute positioning accuracy for CT guided stereotactic brain surgery. *IEEE Trans. Biomed. Eng.* **1988**, *35*, 153–160.
10. McDannold, N.; Clement, G.; Black, P.; Jolesz, F.; Hynynen, K. Transcranial MRI-guided focused ultrasound surgery of brain tumors: Initial findings in three patients. *Neurosurgery* **2010**, *66*, 323.
11. McVeigh, E.R.; Guttman, M.A.; Lederman, R.J.; Li, M.; Kocaturk, O.; Hunt, T.; Kozlov, S.; Horvath, K.A. Real-time interactive MRI-guided cardiac surgery: Aortic valve replacement using a direct apical approach. *Magn. Reson. Med.* **2006**, *56*, 958–964.
12. Vrooijink, G.J.; Abayazid, M.; Misra, S. Real-time three-dimensional flexible needle tracking using two-dimensional ultrasound. In Proceedings of the 2013 IEEE International Conference on Robotics and Automation (ICRA), Karlsruhe, Germany, 6–10 May 2013; pp. 1688–1693.
13. Yasui, K.; Kanazawa, S.; Sano, Y.; Fujiwara, T.; Kagawa, S.; Mimura, H.; Dendo, S.; Mukai, T.; Fujiwara, H.; Iguchi, T.; et al. Thoracic Tumors Treated with CT-guided Radiofrequency Ablation: Initial Experience 1. *Radiology* **2004**, *231*, 850–857.
14. Lam, T.T.; Miller, P.; Howard, S.; Nork, T.M. Validation of a Rabbit Model of Choroidal Neovascularization Induced by a Subretinal Injection of FGF-LPS. *Investig. Ophthalmol. Vis. Sci.* **2014**, *55*, 1204.
15. Huang, D.; Swanson, E.A.; Lin, C.P.; Schuman, J.S.; Stinson, W.G.; Chang, W.; Hee, M.R.; Flotte, T.; Gregory, K.; Puliafito, C.A.; et al. Optical coherence tomography. *Science* **1991**, *254*, 1178.
16. Ehlers, J.P.; Srivastava, S.K.; Feiler, D.; Noonan, A.I.; Rollins, A.M.; Tao, Y.K. Integrative advances for OCT-guided ophthalmic surgery and intraoperative OCT: Microscope integration, surgical instrumentation, and heads-up display surgeon feedback. *PLoS One* **2014**, *9*, e105224.
17. Ehlers, J.P.; Tao, Y.K.; Srivastava, S.K. The Value of Intraoperative OCT Imaging in Vitreoretinal Surgery. *Curr. Opin. Ophthalmol.* **2014**, *25*, 221.
18. Adebar, T.K.; Fletcher, A.E.; Okamura, A.M. 3-D ultrasound-guided robotic needle steering in biological tissue. *IEEE Trans. Biomed. Eng.* **2014**, *61*, 2899–2910.
19. Viehland, C.; Keller, B.; Carrasco-Zevallos, O.M.; Nankivil, D.; Shen, L.; Mangalesh, S.; Viet, D.T.; Kuo, A.N.; Toth, C.A.; Izatt, J.A. Enhanced volumetric visualization for real time 4D intraoperative ophthalmic swept-source OCT. *Biomed. Opt. Express* **2016**, *7*, 1815–1829.
20. Zhang, K.; Kang, J.U. Real-time 4D signal processing and visualization using graphics processing unit on a regular nonlinear-k Fourier-domain OCT system. *Opt. Express* **2010**, *18*, 11772–11784.

21. El-Haddad, M.T.; Ehlers, J.P.; Srivastava, S.K.; Tao, Y.K. Automated real-time instrument tracking for microscope-integrated intraoperative OCT imaging of ophthalmic surgical maneuvers. In Proceedings of SPIE BiOS. International Society for Optics and Photonics, San Francisco, CA, USA, 7 February 2015; p. 930707.

22. Roodaki, H.; Filippatos, K.; Eslami, A.; Navab, N. Introducing augmented reality to optical coherence tomography in ophthalmic microsurgery. In Proceedings of the 2015 IEEE International Symposium on Mixed and Augmented Reality (ISMAR), Fukuoka, Japan, 29 September–3 October 2015; pp. 1–6.

23. Long, J.; Shelhamer, E.; Darrell, T. Fully convolutional networks for semantic segmentation. In Proceedings of the IEEE Conference on Computer Vision and Pattern Recognition, Boston, MA, USA, 7–12 June 2015; pp. 3431–3440.

24. Ahlers, C.; Simader, C.; Geitzenauer, W.; Stock, G.; Stetson, P.; Dastmalchi, S.; Schmidt-Erfurth, U. Automatic segmentation in three-dimensional analysis of fibrovascular pigmentepithelial detachment using high-definition optical coherence tomography. *Br. J. Ophthalmol.* **2008**, *92*, 197–203.

25. Milletari, F.; Navab, N.; Ahmadi, S.A. V-net: Fully convolutional neural networks for volumetric medical image segmentation. In Proceedings of the 2016 Fourth International Conference on 3D Vision (3DV), Stanford, CA, USA, 25–28 October 2016; pp. 565–571.

26. Jia, Y.; Shelhamer, E.; Donahue, J.; Karayev, S.; Long, J.; Girshick, R.; Guadarrama, S.; Darrell, T. Caffe: Convolutional Architecture for Fast Feature Embedding. *arXiv* **2014**, arXiv:1408.5093.

Article

Effects of Temperature Variations during Sintering of Metal Ceramic Tooth Prostheses Investigated Non-Destructively with Optical Coherence Tomography

Cosmin Sinescu [1,†], Adrian Bradu [2,†], Virgil-Florin Duma [3,4,*,†], Florin Topala [1], Meda Negrutiu [1] and Adrian Gh. Podoleanu [2]

1   School of Dental Medicine, Victor Babes University of Medicine and Pharmacy of Timisoara, 2A Eftimie Murgu Place, Timisoara 300070, Romania; minosinescu@yahoo.com (C.S.); florin.topala@gmail.com (F.T.); meda_negrutiu@yahoo.com (M.N.)
2   Applied Optics Group, School of Physical Sciences, University of Kent, Canterbury CT2 7NH, UK; a.bradu@kent.ac.uk (A.B.); a.g.h.podoleanu@kent.ac.uk (A.Gh.P.)
3   3OM Optomechatronics Group, Faculty of Engineering, Aurel Vlaicu University of Arad, 77 Revolutiei Ave., Arad 310130, Romania
4   Doctoral School, Polytechnic University of Timisoara, 1 Mihai Viteazu Ave., Timisoara 300222, Romania
*   Correspondence: duma.virgil@osamember.org; Tel.: +40-751-511-451
†   These authors had equal contributions.

Academic Editor: Michael Pircher
Received: 23 March 2017; Accepted: 24 May 2017; Published: 26 May 2017

Abstract: Calibration loss of ovens used in sintering metal ceramic prostheses leads to stress and cracks in the material of the prostheses fabricated, and ultimately to failure of the dental treatment. Periodic calibration may not be sufficient to prevent such consequences. Evaluation methods based on firing supplemental control samples are subjective, time-consuming, and rely entirely on the technician's skills. The aim of this study was to propose an alternative procedure for such evaluations. Fifty prostheses were sintered in a ceramic oven at a temperature lower, equal to or larger than the temperature prescribed by the manufacturer. A non-destructive imaging method, swept source (SS) optical coherence tomography (OCT) was used to evaluate comparatively the internal structure of prostheses so fabricated. A quantitative assessment procedure is proposed, based on *en-face* OCT images acquired at similar depths inside the samples. Differences in granulation and reflectivity depending on the oven temperature are used to establish rules-of-thumb on judging the correct calibration of the oven. OCT evaluations made on a regular basis allow an easy and objective monitoring of correct settings in the sintering process. This method can serve rapid identification of the need to recalibrate the oven and avoid producing prostheses with defects.

Keywords: metal ceramic dental prostheses; optical coherence tomography (OCT); non-destructive evaluation; sintering technology; material defects

---

1. Introduction

Dental ceramic materials are essential in producing prostheses to replace missing teeth or damaged dental structures. Ceramics are inorganic, non-metallic materials manufactured by heating raw minerals at high temperatures [1]. The issue is that materials like ceramics and glasses are brittle: they display a high compressive strength but a low tensile strength; therefore, they may be fractured under a low strain of only 0.1 to 0.2% [2,3]. As restorative materials, dental ceramics present disadvantages due mostly to their inability to withstand the functional (especially mastication) forces that are present in the oral cavity [3].

While all dental ceramics display low fracture toughness when compared with other dental materials (such as metals), metal ceramic dental prostheses combine both the exceptional esthetic properties of ceramics and the good mechanical properties of metals. Their technology is mature, and advanced materials are used in such constructs [2–4]. However, fractures of the metal ceramic prostheses may occur occasionally [4,5]. Chippings of prosthetic constructs are hard to fix, especially when they are already cemented in the oral cavity.

One of the major causes of the defects that cause cracks and chippings in prosthetic constructs is related to the loss of calibration of the ceramic oven used to fabricate them. Thus, in time (over two or three years [6], and also from our own experience), the ceramic oven may display a different temperature than the real internal temperature. The oven thermal regime is usually adjusted by the dental technician, taking into account the quality of the prostheses produced. Experienced ceramists are guided by the translucency and texture of the ceramic of a fabricated sample using the clearest porcelain powder. However, such a method is performed visually, and therefore relies entirely on the subjective judgment of the operator and it needs a supplemental sample. Quantitative assessments based on Scanning Electron Microscopy are destructive, as they require samples to be cut. Additionally, the very process of cutting the sample introduces artefacts.

The most modern computerized ceramic ovens run a self-diagnostic process to check the quality of the electronics, muffle, and vacuum. Some of the VITA ovens (VITA North America, USA) auto-calibrate themselves every time they are turned on [7]. In general, when a difference in the porcelain is noticed, manufacturers of dental ovens recommend double-checking their calibration by using a tab of clear or window-type porcelain to determine the firing temperature. The ceramists are also advised by the manufacturers to perform the calibration at least once every six months [6]. The problem is that a possible loss of calibration of the oven could happen totally unexpectedly, leading to an improper thermal regime that results in thermal stress inside the ceramic structure. This determines failure of the dental prosthetic treatment [8]. As ovens may process tenths or even hundreds of prostheses each day, such an unexpected and undesired loss of calibration may impact numerous patients negatively. The problem may lay undetected for a certain period of time.

The aim of this paper is to evaluate the utility of an established biomedical imaging technique, optical coherence tomography (OCT), in establishing the correct settings used by ovens. OCT is based on low coherence interferometry [9]. Interferometry confers the method with better depth penetration than conventional microscopy while its low coherence leads to micrometer axial resolution [10,11]. In comparison to other investigation methods for teeth and dental constructs, such as radiography and Cone Beam Computer Tomography, using optical waves of low power, OCT is totally non-invasive. Additionally, OCT exhibits better resolution capability than these methods, essential in the investigation of material granulation [12]. For the present study, the swept source (SS) OCT method is chosen for its superior acquisition speed and sensitivity [13,14]. In order to produce quick images with *en-face* orientation (similar to microscopy), the Master Slave (MS) OCT method for processing the electrical signal is employed [15].

Non-invasiveness and high resolution have promoted OCT towards a wide range of applications, such as biomedical diagnosis (in ophthalmology [9–11], dermatology [16], dentistry [17–20], and endoscopy [21]) and non-biomedical testing (of glass, plastic, and semiconductors [22,23], as well as of profiles of metallic surfaces [24]). The advantages, as well as the limitations/complementarities of OCT in relationship to other investigation methods have been reported in respect to imaging teeth [25–27] or different types of dental constructs [28–33]. Relevant to non- destructive testing, targeted here, the impact of the OCT light source, of its central wavelength, as well as of specular reflection, have already been considered [23].

In this study, we demonstrate that MS/SS-OCT can be used to identify, more frequently than at standardized periods, possible undesired variations of the oven temperature. The application of the MS/SS-OCT is simple, fast and non-invasive, based on the imaging of ceramic constructs only, which are simply produced each day. The aim is to achieve this operation independently to other calibrating procedure recommended by the manufacturer.

With such a method, the technician may immediately infer that a temperature calibration is mandatory.

2. Materials and Methods

For this study, Duceram Kiss ceramics (DeguDent GmbH (a Dentsply Sirona Company), Hanau-Wolfgang, Germany) have been selected for evaluation. This choice has been based on the fact that the composition of each Duceram Kiss ceramic is adapted to meet the needs of its particular fabrication process and framework, with specific parameters to be observed, such as its coefficient of thermal expansion (CTE) [3]. Stress cracks and fractures are thus avoided and strength parameters such as bonding, adhesion and retention are assured. This ceramic material has the following characteristics: CTE Dentine 13.0 μm/m·K; dental ceramics, type 1, classes 2–8 according to DIN EN ISO 6872; metal-ceramic bond characterization, flexural strength and chemical solubility according to DIN EN ISO 9693. The optimization of the number of pigments is also a major part of the solution to the problem of metamerism, which refers to the display of a certain color shade of the dental prostheses depending on the source of light utilized, for example typical indoor lighting or outdoor daylight. In classical high-fusing veneering ceramics, the opalescent effect vanishes after a number of firing cycles [34]. By contrast, in Duceram Kiss, this effect will visibly persist. These superior characteristics simplify our study, as the temperature variations in the sintering oven are likely to be the only source of error/liability of these materials in the manufacturing process of the prostheses. On the other hand, a double check of the technician's work and his supervision during the procedure phases—to assure complying with the manufacturer's requirements—provide a minimization of human errors.

A number of 50 Duceram Kiss metal ceramic crowns were used in the study, split into five groups, with 10 each. They were sintered at different temperatures and then investigated by using the MS/SS-OCT system.

Fifty similar metal copings corresponding to the first maxillary incisor were obtained (Figure 1). The choice of these specific prostheses for the study is justified by how often they are needed by patients in dental practice. From these samples, five groups were considered: **Group N**, for which the ceramic layers were sintered according to the manufacturer's indications (i.e., at 930 °C); **Group L30**, for which the ceramic layers were sintered at 30 °C below the recommended temperature (i.e., at 900 °C); **Group L100**, for which the ceramic layers were sintered at 100 °C under the recommended temperature (i.e., at 830 °C); **Group H30**, with ceramic layers sintered at 30 °C above the recommended temperature (i.e., at 960 °C); **Group H50**, with ceramic layers sintered at 50 °C above the recommended temperature (i.e., at 980 °C). The samples were stored in special boxes, properly marked and prepared for imaging.

Each of these temperature variations were meant to simulate a certain degree of loss of calibration in the ceramic oven in order to evaluate the possibility of detecting it by using MS/SS-OCT.

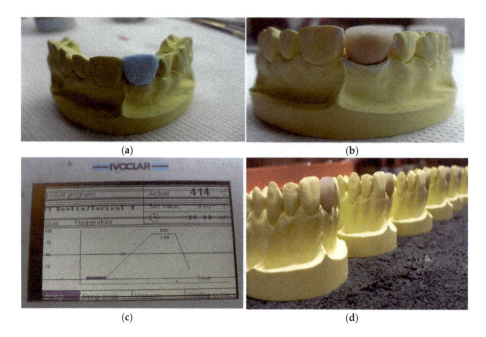

Figure 1. Aspects from the preparation of the samples: (**a**) adding the ceramic material on the copings, before sintering in blue color and larger than the (**b**) sample after its sintering in the oven (which also displays a change in color); (**c**) temperature chart displayed by the dental ceramic oven; (**d**) several samples used in this study.

Figure 2 presents a schematic diagram of the SS-OCT imaging system used in the study. As an optical source, an SS (Axsun Technologies, Billerica, MA, USA), central wavelength 1060 nm, sweeping range 106 nm (quoted at 10 dB), and 100 kHz line rate is used. The use of an optical source emitting at around 1060 nm cannot offer a depth penetration as good as at 1300 nm; however, it offers a better axial resolution, which, in our case, was 10 µm (resolution measured in air by using a mirror as sample and measuring the Full Width Half Maximum (FWHM) of the A-scan peak). Using the procedure of measuring sensitivity described in [35], a sensitivity superior to 95 dB was obtained. This determines the minimum reflectivity measurable within any image.

The interferometer configuration employs two single mode directional couplers, DC_1 and DC_2. DC_1 has a ratio of 20:80 and DC_2 is a balanced splitter, 50:50. DC_2 feeds a balance detection receiver from (Thorlabs, Newton, NJ, USA, model PDB460C), using two photo-detectors, PD_1 and PD_2 and a differential amplifier DA. Twenty percent from the SS power is launched towards the object arm via lens L_1 (focal length 15 mm), which collimates the beam towards a pair of galvoscanners XYSH, (Cambridge Technology, Bedford, MA, USA, model 6115) followed by a scanning lens, L_2, (25 mm focal length), which focuses the light on the sample. The optical fibers used, as well as the lenses L_1 and L_2, determine a lateral resolution of ~10 µm. The power to the object O is 2.2 mW. At the other output of DC1, 80% from the SS power is directed along the reference path via two flat mirrors, M_1 and M_2, both placed on a translation stage, TS to adjust the optical path difference (*OPD*) in the interferometer. Collimating lenses L_3 and L_4 are similar to L_1. The signal from the differential amplifier, DA, is sent to one of the two inputs of a dual input digitizer (Alazartech, Quebec, QC, Canada, model ATS9350, 500 MB/s). The acquired channeled spectra are manipulated via a program implemented in LabVIEW 2013 (National Instruments Corporation, Austin, TX, USA), 64 bit, deployed on a PC equipped with an Intel Xeon processing unit (Santa Clara, CA, USA), model E5646 (clock speed

2.4 GHz, 6 cores). The same program is also used to drive the two galvoscanners via a data acquisition board (model PCI 6110, National Instruments Corporation, Austin, TX, USA). Each B-scan image was built of 500 A-scans; hence, with 100 kHz sweeping speed, 10 ms were required to acquire data for a B-scan (the fast galvoscanner was driven with a triangular wave-form [35]). As a consequence, 500×500 channeled spectra to build an entire 3D volume and also deliver the *en-face* images were acquired in 5 s. Once the full dataset was acquired, *en-face* images were produced using the MS protocol by comparing the signal delivered by the DA with the signal delivered by the DA when a mirror was used as a sample, for an optical path difference in the interferometer adjusted to $z_{\text{en-face}}$, i.e., to 0.75 mm.

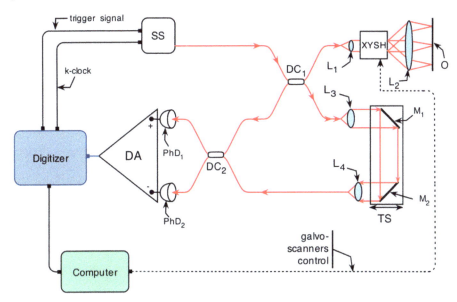

Figure 2. Schematic diagram of the Master-Slave (MS)/Swept Source (SS)-Optical Coherence Tomography (OCT) system. SS: swept source; DC_1, 20/80 single mode directional coupler; DC_2: 50/50 single mode directional coupler; XYSH: two-dimensional XY lateral scanning head; L_1 to L_4: achromatic lenses; O: object to be imaged; PD_1, PD_2: photo-detectors; DA: differential amplifier; TS: translation stage.

All samples were subject to imaging after they have completely cooled down. For their investigations, C-scan/*en-face* OCT images were obtained from a similar depth in all of the samples. The control of the depth was made possible by analyzing the B-scan OCT image of each sample. This depth was chosen at approximately $z_{\text{en-face}} = 0.375$ mm (distance measured in air) from the position corresponding to an optical path difference (*OPD*) of 0 mm considered at the top of each sample in the B-scan image. In all cases, the vestibular surface of the sample was imaged. The system was switched from either real time *en-face* (C-scan) imaging regime to cross section (B-scan) imaging regime, based on its Master Slave procedure.

The lateral size of the *en-face* images was calibrated (size of the image versus voltage applied to the galvoscanners XY) using a high resolution test target. The scale bars in Figures 3–5 were thus $x = y = 3.25$ mm on each of the two directions considered for *en-face* images. Along the z-axis (in-depth, perpendicular on the vestibular surface), the maximum axial range offered by the instrument employed for the study, measured in air is $z = 3.75$ mm. By considering an estimative $n = 1.5$ refractive index for the inspected material, Duceram Kiss ceramics, for the central wavelength of the SS used, the axial range along the vertical in Figure 3 becomes $z/n = 2.5$ mm measured inside the material.

The outer surface of all the samples was positioned at a similar depth ($z_{surface}$) measured from the plane corresponding to a zero *OPD*; then, all *en-face* images were acquired from a similar depth position ($z_{en-face}$). This depth of ~0.375 mm was selected in order to provide as good as possible a reflectivity signal to characterize the granulation in the *en-face* OCT image acquired from inside the material. In this way, as the reference and sample optical power levels were kept constant throughout all measurements, we ensured that changes in the brightness of the images are only due to changes in the optical properties of the samples. Moreover, by placing the samples as explained before, at a similar distance from L_2, we ensure similar lateral resolutions in all of the *en-face* images.

An immediate difficulty when the sample presents a curved input surface, for collecting an *en-face* image from a fixed depth, as determined by the *OPD* of the coherence gate applied, is that the *en-face* slice does not cut a plane surface, but a curved surface from inside the sample. Therefore, the interpretation of what is being seen has to consider the curvature of the sample top surface. Therefore, maintaining the exact depth in all samples of the *en-face* slices collected is somewhat difficult.

Figure 3 depicts the depth selection where the *en-face* images are acquired from. They are not rendered from the volumetric dataset, as performed by conventional OCT technology based on A-scans. They are generated in real time, directly, using a comparison method of optical spectra at the OCT interferometer output. We thus demonstrate that a single OCT *en-face* image, at a suitably selected depth, can provide all of the necessary information for the evaluation of correct values chosen for the temperature settings of the oven.

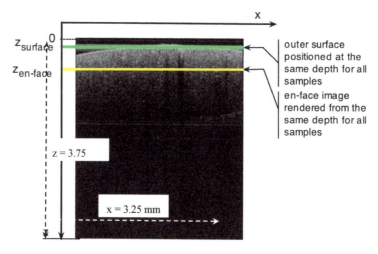

Figure 3. Illustration of the guiding method to choose the depth from where the *en-face* images are selected. The depth is changed by changing the spectrum stored (mask), as described in our previous reports using the MS technology. For all samples, their outer surface at $z_{surface}$ was adjusted to correspond to *OPD* = 0, to insure that the *en-face* images are acquired from a similar depth of $z_{en-face}$ in all cases.

In order to better emphasize the effect of the temperature change, once the *en-face* images are produced, the 500 horizontal lines on the *y*-axis in each image were averaged and the result was displayed as a function of the reflectivity with regard to the position on the *x*-axis. To quantify the effect of the change in temperature on the granulation in the *en-face* images, a simple method was used: (i) each *en-face* image was converted into a binary image and processed (the binary image replaces all pixels in the original gray-scaled image with the value 1 (white) if the gray-scale value is above a threshold (127 has been used on all our images) and replaces all other pixels with the value 0 (black); (ii) To quantify the granulation, the features identified in the binary image are counted and their size is estimated. This fulfills the aim of this study, different from that of imaging, oriented towards providing a simple, fast, and practical assessment of deviations of ovens from their best settings. As a remark, one may consider in such an investigation a bidirectional imaging modality in order to investigate the sample from both sides [36]. However, for the proposed simplicity of the method developed, we have made the investigation strictly for the vestibular surface of the sample, as similar results are expected to be obtained on all the sides of the sintered prostheses; in addition, the depth of the investigation would not be enhanced by a bidirectional imaging.

3. Results

The investigations revealed specific aspects for each of the groups considered. The acquired OCT data was evaluated in the *en-face* imaging plane as well as in the B-scan imaging plane, which was important in order to control the depth of the investigations, as explained before.

The following aspects were determined for the groups of samples studied:

Group N (ceramic layers of the samples sintered at the temperatures indicated by the manufacturers, i.e., 930 °C): a normal distribution of the reflectivity could be noticed (Figure 4(c1)). This can also be seen in the MATLAB (Natick, MA, USA) image processed after obtaining a mean value of the OCT investigation (Figure 4(c2)), obtained from the reflectivity values on the y-axis for each x-position, as explained in the final part of the previous section. B-scan images of the samples from this group have not been relevant with regard to the influence of the temperature variations (Figure 4(c3)); neither are A-scans, i.e., reflectivity profiles on the z-axis. In fact, our study has demonstrated that *en-face* OCT images, and only them, are capable of performing the proposed assessment.

Group L30 (samples sintered at 30 °C below the recommended temperature, i.e., at 900 °C): a decrease of the reflectivity could be seen; an alternation between reflective and less reflective areas (of band-like shapes) is noticed (Figure 4(b1)). These findings could be easily processed and highlighted on images that were processed using MATLAB (Figure 4(b2)).

Group L100 (samples sintered at 100 °C below the recommended temperature, i.e., at 830 °C): the *en-face* OCT image of the investigation revealed a significant decrease of the reflectivity (Figure 4(a1)). The alternation between the reflective and less reflective bands is still visible in the image. However, the non-reflective area spreads much more in comparison with the more reflective area (Figure 4(a2)). A further degradation, in comparison with Group L30, can thus be concluded.

Group H30 (samples sintered 30 °C above the recommended temperature, i.e., at 960 °C): an overall increase of the reflectivity was noticed by comparing the mean value in the normal sample, Group N (Figure 5(a2)) with the mean value corresponding to the Group H30 (Figure 5(b2)). The alternation between the reflective and non-reflective areas disappears and a normal granular dispersion is identified (Figure 5(b2)).

Group H50 (samples sintered 50 °C above the recommended temperature, i.e., at 980 °C): fractures in the ceramic material can be identified in both the *en-face* (Figure 5(c1)) and B-scan (Figure 5(c3)) images. The general aspect of the *en-face* images also shows a significant gradient with an increasing slope in reflectivity (Figure 5(c2)). An important aspect should be pointed out by inspecting the images for Group H30: an approximately 40% increase in the reflectivity (concluded from the images processed in MATLAB, by comparing Group N and Group H30, i.e., Figure 5(a2,b2) have pointed out to a temperature increase of 30 °C inside the oven. Continuing to increase the temperature even more could result in thermal stress that leads to air accumulation inside the ceramic layer or even to fractures at those levels (see the defect D, Figure 5(c1)).

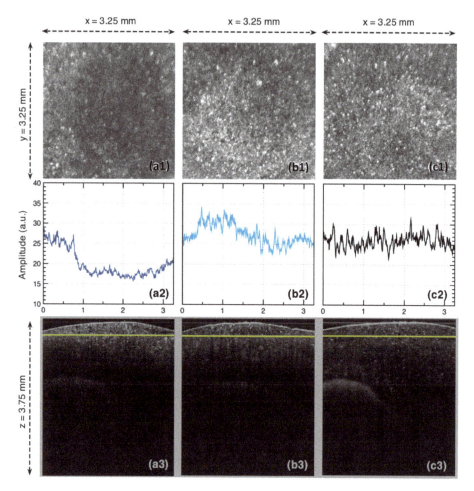

Figure 4. OCT images for the samples obtained at lower than normal temperature in comparison with samples at normal temperature: column (**a**): Group L100; column (**b**): Group L30; column (**c**): Group N. Top row: *en-face* OCT images of samples of Group L100 (**a1**), Group L30 (**b1**), and Group N (**c1**). Middle row, (**a2–c2**): reflectivity average of the 500 horizontal lines making the images shown in the top row. Bottom row, (**a3–c3**): B-scans OCT images corresponding to the three groups. The yellow lines indicate the depth position from where the *en-face* images were rendered (measured from the zero OPD).

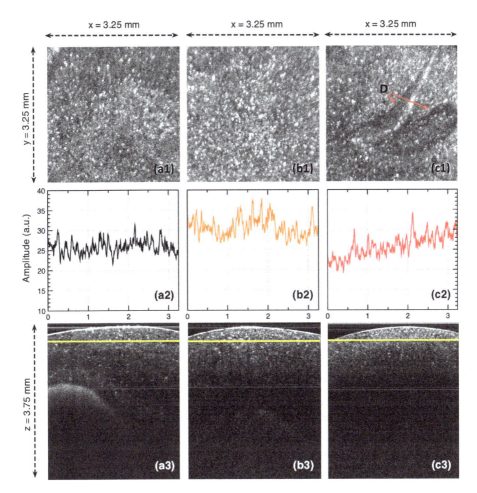

Figure 5. OCT images for the samples obtained at higher than normal temperature in comparison with samples at normal temperature: column (**a**): Group N; column (**b**): Group H30; column (**c**): Group H50. Top row: *en-face* OCT images of samples of Group N (**a1**), Group H30 (**b1**) and Group H50 (**c1**). Middle row, (**a2**–**c2**): reflectivity average of the 500 horizontal lines in the images shown in the top row. Bottom row, (**a3**–**c3**): B-scans OCT images corresponding to the three groups. The yellow lines indicate the position where the *en-face* images were rendered from (measured from the zero OPD).

4. Discussion

Observations correlate well with the typical features characterizing the sintering process. Such processes are commonly divided into three stages [37]. In an initial first stage, a network of particles starts to form, marked by neck formation. The individual powder particles are still distinguishable. In this stage, the relative density of the material increases by roughly 3%.

Most of the densification, as well as the most significant micro-structural changes, occur in what is termed as the intermediate, second stage, where voids form continuously pore channels along three-grain junctions. The pore channels are formed when the individual powder grains become well-bonded during the initial stage. After that, the pore channels tend to shrink due to the diffusion of the atomic vacancies away from them; however, the grain growth occurring at the same time leads to a reduction in the total number of pores and tends to increase their average diameter and length as a consequence of coalescence.

The process continues until the network of pores undergoes the so-called Rayleigh breakup [37,38], first into progressively fewer interconnected elongated pores on the grain edges and finally into isolated pores at the grain corners; at this point, the final, third stage of sintering begins. When the density of a material exceeds 90–94% of the final (theoretical) density, the isolated pores are formed and the densification is slowed down. These can be seen in Figure 4(a1,b1) for Groups L100 and L30, respectively, as the grain density is less than in Figure 4(c1), where the material has been brought to its final, prescribed temperature, for Group N. In the former figures, the process was interrupted, while, in the latter, it has been completed, and a uniform distribution of grains was achieved; this is demonstrated by a more uniform reflectivity in Figure 4(c2) in comparison to the non-uniform reflectivity in Figure 4(a2,b2).

In this third stage of sintering, the isolated pores are eliminated by the transport of mass from the grain boundary to the pore. The densification is pinned by pores and a considerable amount of energy is necessary for further densification. In this stage, the grain growth dominates the sintering process. Therefore, as the temperature is increased above the value prescribed by the manufacturer, as for Groups H30 and H50, the growth of the grains induces mechanic stress inside the material. The worst results are cracks, as seen in Figure 5(c1)—obtained for the highest temperature used for sintering (Group H50). Even for temperatures in the oven that are not so dramatically different from the normal temperature prescribed, a further increase of the grains in size results in a noticeable increase in the reflectivity in the *en-face* image in Figure 5(b2) for Group H30.

In terms of granular dispersion, the *en-face* images were processed further in MATLAB. The result is Figure 6, which is obtained after removing the noise, so after the image is converted into a binary one; in contrast, the graphs in Figures 4 and 5 have been obtained using raw data. The results confirm the reflectivity increase, which is a consequence of the significant increase in the granulation from Group L100 (sintered at 830 °C) towards Group N (sintered at 930 °C, the normal temperature) and Group H50 (sintered at 960 °C). The number of peaks in each plot (Figure 6), as well as their amplitude increase indicates an increase in both the number of the grains in each *en-face* image and in their size, respectively.

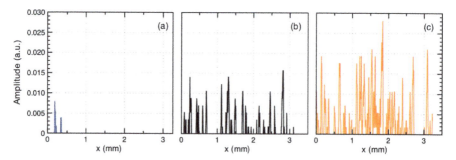

Figure 6. Average of 500 horizontal lines of the *en-face* images, corresponding to sintering temperatures of (**a**) 830 °C (Group L100); (**b**) 930 °C (Group N) and (**c**) 960 °C (Group H50), i.e., the normal and the extreme temperature considered. The number of peaks in each plot indicates the number of "bright particles", while their amplitude indicates their size.

Rules-of-thumb can be extracted from these results, in order to allow for the monitoring of the temperature inside the ceramic oven by OCT imaging of the prostheses sintered:

(i)   A relative uniform reflectivity characterizes the recommended temperature of the oven (Group N, Figure 4(c2)).

(ii)   The normal temperature regime is also correlated with the good uniformity of the grains, which have been fully developed in this case (Figures 4(c1) and 6b).

(iii) A non-uniform reflectivity without a clear pattern (i.e., with ups and downs) characterizes an oven temperature lower than normal (Figure 4(a2,b2)).

(iv)   The lower the oven temperature, the lower the values that the reflectivity reaches, according to Figure 4(a2) in comparison with Figure 4(b2); a similar trend can be seen as well in Figure 6a in comparison with Figure 6b.

(v)   The aspects in the two points above can also be seen directly from the *en-face* image (Figure 4(a1,b1)), although such an evaluation is less certain, because higher temperatures of the oven may also produce such non-uniformity. Such non-uniformities of the grain distribution (and thus of the reflectivity) for lower temperatures, proportional to the deviation of temperature from normal is correlated with the incomplete development of the grains, as discussed above for the sintering process.

(vi)   A global increase in the reflectivity characterizes samples sintered at oven temperatures higher than normal, as shown in Figures 5(b2) and 6c.

(vii) Too high temperatures can also be concluded by comparing an *en-face* image as the one in Figure 5(b1) with an *en-face* image as the one shown in Figure 5(a1). This aspect is correlated with the excessive growth of the grains in the sintering process, as it can be seen from the difference in peak amplitudes in Figure 6c in comparison with Figure 6b.

(viii) A deviation of the oven temperature to even higher values with regard to the prescribed one (as for Group H50) can also be characterized by a positive and constant slope of the reflectivity (Figure 5(c2)).

(ix)   The previous aspect is also the easier and most direct to see on an *en-face* image (Figure 5(c1)), as it produces significant defects (D) that lead to cracks in short periods of time.

(x)   One should not wait for such totally undesired effects to occur. Instead, a re-calibration should be performed immediately as the reflectivity extracted from the *en-face* image increases. Such an increase, of approximately 40%, already corresponds to a 30 °C higher temperature in the oven with regard to normal, as it was concluded in Figure 5(b2) from the *en-face* image in Figure 5(b1).

It should be appreciated that oven temperatures higher than normal are more dangerous for the outcome of the prostheses than oven temperatures that are lower than the normal value. This can be clearly seen from the fact that samples sintered at 50 °C over the normal oven temperature exhibited defects (D, Figure 5(c1)), while samples sintered at oven temperatures with even 100 °C below normal displayed no such defects (Figure 4(a1)).

However, this lack of defects at oven temperatures lower than normal must not be taken lightly or less serious than the loss of calibrations towards oven temperatures that are higher than normal. In the former case, ceramics are not burned properly, the sintering process is incomplete, and therefore the mechanical (and also the esthetic) properties of the prostheses would also be affected, as is well known.

A general overview of the concluded properties of the five groups of samples is made in Table 1.

Table 1. Characteristics of the sintered prostheses, as concluded from the Optical Coherence Tomography (OCT) investigations.

Group *Characteristics*	Group L100	Group L30	Group N	Group H30	Group H50
Grain distribution— from the *en-face* OCT images (Figure 4(a1,b1,c1) as well as Figure 5(a1,b1,c1)	Insufficient number of grains— Figure 4(a1)	Lower-than-normal number of grains— Figure 4(b1)	Normal (as prescribed by the manufacturer)— Figures 4(c1) and 5(a1)	Higher-than-normal number of grains— Figure 5(b1)	The stresses in the material have produced cracks— see defect D, Figure 5(a1)
Reflectivity profile-as obtained from raw data (Figure 4(a2,b2,c2) as well as Figure 5(a2,b2,c2)	Level lower than normal— Figure 4(a2)	Level close to normal, but with supplemental irregularities— Figure 5 (b2)	Normal level (can be considered as a reference for further monitoring)— Figures 4(c2) and 5(a2)	Level higher than normal, with supplemental irregularities— Figure 5(b2)	With a gradient tendency— Figure 5(c2)
Defects	No visible defects (Figure 4(a1,b1)), but the material has low mechanical (and esthetic) properties		No defects— Figures 4(c1) and 5(a1)	Possible cracks (high stress in the material)	Cracks are present—defect D, Figure 5(a1)
Number and size of the grains— in each *en-face* image (Figure 6)	Insufficient number of (too) small grains— Figure 6a	-	Average/normal— Figure 6b	-	Large number of (too) big grains—Figure 6c

5. Conclusions

This study demonstrates the capability of MS/SS-OCT to achieve a simple, fast, and non-invasive monitoring of the temperature inside the ceramic ovens used to obtain metal ceramic dental prostheses. On an every-day basis, the dental technician can regularly evaluate by OCT the ceramic layer at a certain depth inside the material, here ~0.375 mm used. The entire evaluation is based on an inspection of C-scan/*en-face* images only. Similar interpretation as suggested here could be based on using B-scans as well; however, the *en-face* mode is the most illustrative of patterns. The B-scan images have only used the purpose to secure *en-face* display from similar depths in all samples. Using the reflectivity profiles deduced from the *en-face* images, it was concluded, for example, that a variation of more than 40% in the reflectivity of the material in comparison to the level of samples sintered at the prescribed oven temperature should trigger an immediate temperature re-calibration of the oven. An even further increase of the temperature can result in cracks in the material, as demonstrated in our study.

The study may be further refined by evaluating local variance or higher order moments in the reflectivity graphs. However, in the present study, the aim was to provide the practitioners with the simplest possible method to enable them to decide if calibration is needed.

Possible rules-of-thumb were discussed that may allow for monitoring of the sintering process. The procedure suggested, if followed, can prevent thermal stress (for higher than normal temperatures in the oven) or incomplete sintering (for lower than normal temperatures in the oven). Otherwise, incorrect temperature used leads to materials defects and then to fractures inside the ceramic layers of the prostheses manufactured and inserted in the oral cavity.

Acknowledgments: This work was supported by the Romanian National Authority for Scientific Research (CNDI–UEFISCDI project PN-III-P2-2.1-PED-2016-1937 (http://3om-group-optomechatronics.ro/)), including for covering the costs to publish in open access. Adrian Bradu and Adrian Gh. Podoleanu acknowledge the support of the European Research Council (Grant 249889 (http://erc.europa.eu)) and EPSRC (REBOT, Grant EP/N019229/1). Adrian Gh. Podoleanu is also supported by the NIHR Biomedical Research Centre at Moorfields Eye Hospital NHS Foundation Trust and the UCL Institute of Ophthalmology, by the Royal Society Wolfson Research Merit Award, Technical University of Denmark Shape OCT programme, the ERC AMEFOCT Proof of concept 680879 and the UBAPHODESA Marie Curie European Industrial Doctorate, 607627.

Author Contributions: C.S. and V.-F.D. devised and designed the experiments; A.B. and A.Gh.P. devised the OCT system and performed the OCT investigations; C.S., A.B., and V.-F.D. analyzed the data; F.T. and M.N. contributed materials and samples; C.S., A.B., V.-F.D., and A.Gh.P. wrote the paper. C.S., A.B., and V.-F.D. contributed equally to the paper.

Conflicts of Interest: The authors declare no conflict of interest.

References

1. Rosenblum, M.A.; Schulman, A. A review of all ceramic restorations. *J. Am. Dent. Assoc.* **1997**, *128*, 297–307. [CrossRef] [PubMed]
2. Rizkalla, A.S.; Jones, D.W. Mechanical properties of commercial high strength ceramic core materials. *Dent. Mater.* **2004**, *20*, 207–212. [CrossRef]
3. Rizkalla, A.S.; Jones, D.W. Indentation fracture toughness and dynamic elastic moduli for commercial feldspathic dental porcelain materials. *Dent. Mater.* **2004**, *20*, 198–206. [CrossRef]
4. Arango, S.S.; Vargas, A.P.; Escobar, J.S.; Monteiro, F.J.; Restrepo, L.F. Ceramics for dental restorations—An Introduction. *DYNA* **2010**, *77*, 2636.
5. Craciunescu, E.; Sinescu, C.; Negrutiu, M.L.; Pop, D.M.; Lauer, H.-C.; Rominu, M.; Hutiu, Gh.; Bunoiu, M.; Duma, V.-F.; Antoniac, I. Shear Bond Strength Tests of Zirconia Veneering Ceramics after Chipping Repair. *J. Adhes. Sci. Technol.* **2016**, *30*, 666–676. [CrossRef]
6. Patrick, B. Porcelain and Pressing Furnaces. Inside Dental Technology. *AEGIS Commun.* **2011**, *2*, 3.
7. VITA. Available online: http://vitanorthamerica.com/products/equipment/vacumat-6000-m/ (accessed on 23 March 2017).
8. Sinescu, C.; Topala, F.I.; Negrutiu, M.L.; Duma, V.-F.; Podoleanu, A.Gh. Temperature variation in metal ceramic technology analyzed using time domain optical coherence tomography. *Proc. SPIE* **2014**, *8925*, 89250T.
9. Huang, D.; Swanson, E.A.; Lin, C.P.; Schuman, J.S.; Stinson, W.G.; Chang, W.; Hee, M.R.; Flotte, T.; Gregory, K.; Puliafito, C.A.; et al. Optical coherence tomography. *Science* **1991**, *254*, 1178–1181. [CrossRef] [PubMed]
10. Drexler, W.; Fujimoto, J.G. *Optical Coherence Tomography*; Springer: Berlin/Heidelberg, Germany, 2008.
11. Podoleanu, A.Gh. Optical coherence tomography. *J. Microsc.* **2012**, *247*, 209–219. [CrossRef] [PubMed]
12. Wijesinghe, R.E.; Cho, N.H.; Park, K.; Jeon, M.; Kim, J. Bio-Photonic detection and quantitative evaluation method for the progression of dental caries using optical frequency-domain imaging method. *Sensors* **2016**, *16*, 2076. [CrossRef] [PubMed]
13. Wojtkowski, M. High-speed optical coherence tomography: Basics and applications. *Appl. Opt.* **2010**, *49*, D30–D61. [CrossRef] [PubMed]
14. Drexler, W.; Liu, M.; Kumar, A.; Kamali, T.; Unterhuber, A.; Leitgeb, R.A. Optical coherence tomography today: Speed, contrast and multimodality. *J. Biomed. Opt.* **2014**, *19*, 071412. [CrossRef] [PubMed]
15. Podoleanu, A.Gh.; Bradu, A. Master–slave interferometry for parallel spectral domain interferometry sensing and versatile 3D optical coherence tomography. *Opt. Express* **2013**, *21*, 19324–19338. [CrossRef] [PubMed]
16. Lee, K.-S.; Zhao, H.; Ibrahim, S.F.; Meemon, N.; Khoudeir, L.; Rolland, J.P. Three-dimensional imaging of normal skin and nonmelanoma skin cancer with cellular resolution using Gabor domain optical coherence microscopy. *J. Biomed. Opt.* **2012**, *17*, 126006. [CrossRef] [PubMed]
17. Colston, B.W., Jr.; Sathyam, U.S.; DaSilva, L.B.; Everett, M.J.; Stroeve, P.; Otis, L.L. Dental OCT. *Opt. Express* **1998**, *3*, 230–238. [CrossRef] [PubMed]
18. Sinescu, C.; Negrutiu, M.L.; Todea, C.; Balabuc, C.; Filip, L.; Rominu, R.; Bradu, A.; Hughes, M.; Podoleanu, A.Gh. Quality assessment of dental treatments using en-face optical coherence tomography. *J. Biomed. Opt.* **2008**, *13*, 054065. [CrossRef] [PubMed]
19. Hariri, I.; Sadr, A.; Shimada, Y.; Tagami, J.; Sumi, Y. Effects of structural orientation of enamel and dentine on light attenuation and local refractive index: An optical coherence tomography study. *J. Dent.* **2012**, *40*, 387 396. [CrossRef] [PubMed]
20. Shemesh, H.; van Soest, G.; Wu, M.-K.; Wesselink, P.R. Diagnosis of vertical root fractures with optical coherence tomography. *J. Endod.* **2008**, *34*, 739–742. [CrossRef] [PubMed]
21. Carignan, C.S.; Yagi, Y. Optical endomicroscopy and the road to real-time, in vivo pathology: Present and future. *Diagn. Pathol.* **2012**, *7*, 98. [CrossRef] [PubMed]
22. Strakowski, M.R.; Plucinski, J.; Jedrzejewska-Szczerska, M.; Hypszer, R.; Maciejewski, M.; Kosmowski, B.B. Polarization sensitive optical coherence tomography for technical materials investigation. *Sens. Actuators A* **2008**, *142*, 104–110. [CrossRef]
23. Su, R.; Kirillin, M.; Ekberg, P.; Roos, R.; Sergeeva, E.; Mattsson, L. Optical coherence tomography for quality assessment of embedded microchannels in alumina ceramic. *Opt. Express* **2012**, *20*, 4603–4618. [CrossRef] [PubMed]

24. Hutiu, Gh.; Duma, V.-F.; Demian, D.; Bradu, A.; Podoleanu, A.Gh. Surface imaging of metallic material fractures using optical coherence tomography. *Appl. Opt.* **2014**, *53*, 5912–5916. [CrossRef] [PubMed]

25. Feldchtein, F.; Gelikonov, V.; Iksanov, R.; Gelikonov, G.; Kuranov, R.; Sergeev, A.; Gladkova, N.; Ourutina, M.; Reitze, D.; Warren, J. In Vivo OCT imaging of hard and soft tissue of the oral cavity. *Opt. Express* **1998**, *3*, 239–250. [CrossRef] [PubMed]

26. Natsume, Y.; Nakashima, S.; Sadr, A.; Shimada, Y.; Tagami, J.; Sumi, Y. Estimation of lesion progress in artificial root caries by swept source optical coherence tomography in comparison to transverse microradiography. *J. Biomed. Opt.* **2011**, *16*, 071408. [CrossRef] [PubMed]

27. Schneider, H.; Park, K.-J.; Häfer, M.; Rüger, C.; Schmalz, G.; Krause, F.; Schmidt, J.; Ziebolz, D.; Haak, R. Dental applications of optical coherence tomography (OCT) in cariology. *Appl. Sci.* **2017**, *7*, 472. [CrossRef]

28. Isfeld, D.M.; Aparicio, C.; Jones, R.S. Assessing near infrared optical properties of ceramic orthodontic brackets using cross-polarization optical coherence tomography. *J. Biomed. Mater. Res. B Appl. Biomater.* **2014**, *102*, 516–523. [CrossRef] [PubMed]

29. Dsouza, R.; Subhash, H.; Neuhaus, K.; Kantamneni, R.; McNamara, P.M.; Hogan, J.; Wilson, C.; Leahy, M. Assessment of curing behavior of light-activated dental composites using intensity correlation based multiple reference optical coherence tomography. *Lasers Surg. Med.* **2016**, *48*, 77–82. [CrossRef] [PubMed]

30. Jones, R.S.; Staninec, M.; Fried, D. Imaging artificial caries under composite sealants and restorations. *J. Biomed. Opt.* **2004**, *9*, 1297–1304. [CrossRef] [PubMed]

31. Duma, V.-F.; Dobre, G.; Demian, D.; Cernat, R.; Sinescu, C.; Topala, F.I.; Negrutiu, M.L.; Hutiu, Gh.; Bradu, A.; Podoleanu, A.G. Handheld scanning probes for optical coherence tomography. *Romanian Rep. Phys.* **2015**, *67*, 1346–1358.

32. Han, S.-H.; Sadr, A.; Tagami, J.; Park, S.-H. Non-destructive evaluation of an internal adaptation of resin composite restoration with swept-source optical coherence tomography and micro-CT. *Dent. Mater.* **2016**, *32*, E1–E7. [CrossRef] [PubMed]

33. Canjau, S.; Todea, C.; Negrutiu, M.L.; Sinescu, C.; Topala, F.I.; Marcauteanu, C.; Manescu, A.; Duma, V.-F.; Bradu, A.; Podoleanu, A.Gh. Optical Coherence Tomography for Non-Invasive ex vivo Investigations in Dental Medicine—A Joint Group Experience (Review). *Mod. Technol. Med.* **2015**, *7*, 97–115. [CrossRef]

34. DeguDent. Available online: http://www.salloumtrade.com/categories.php?act=show&id=62 (accessed on 26 April 2017).

35. Duma, V.-F.; Tankam, P.; Huang, J.; Won, J.J.; Rolland, J.P. Optimization of galvanometer scanning for Optical Coherence Tomography. *Appl. Opt.* **2015**, *54*, 5495–5507. [CrossRef] [PubMed]

36. Ravichandran, N.K.; Wijesinghe, R.E.; Shirazi, M.F.; Park, K.; Jeon, M.; Jung, W.; Kim, J. Depth enhancement in spectral domain optical coherence tomography using bidirectional imaging modality with a single spectrometer. *J. Biomed. Opt.* **2016**, *21*, 076005. [CrossRef] [PubMed]

37. Rosolowaki, J.H.; Greskovich, C. Theory of the dependence of densification on grain growth during intermediate stage sintering. *J. Am. Ceram. Soc.* **1975**, *58*, 177–182. [CrossRef]

38. Perko, S. Dental Ceramics. Ph.D. Thesis, Jozef Stefan International Postgraduate School, Ljubljana, Slovenia, 2012.

Article

Flow Measurement by Lateral Resonant Doppler Optical Coherence Tomography in the Spectral Domain

Julia Walther [1,2,*] and Edmund Koch [2]

[1] Department of Medical Physics and Biomedical Engineering, Faculty of Medicine Carl Gustav Carus, Technische Universität Dresden, 01307 Dresden, Germany
[2] Clinical Sensoring and Monitoring, Anesthesiology and Intensive Care Medicine, Faculty of Medicine Carl Gustav Carus, Technische Universität Dresden, 01307 Dresden, Germany; edmund.koch@tu-dresden.de
* Correspondence: julia.walther@tu-dresden.de; Tel.: +49-351-458-6132

Academic Editor: Michael Pircher
Received: 20 January 2017; Accepted: 4 April 2017; Published: 11 April 2017

Abstract: In spectral domain optical coherence tomography (SD-OCT), any transverse motion component of a detected obliquely moving sample results in a nonlinear relationship between the Doppler phase shift and the axial sample velocity restricting phase-resolved Doppler OCT (PR-DOCT). The size of the deviation from the linear relation depends on the amount of the transverse velocity component, given by the Doppler angle, and the height of the absolute sample velocity. Especially for very small Doppler angles between the horizontal and flow direction, and high flow velocities, the detected Doppler phase shift approaches a limiting value, making an unambiguous measurement of the axial sample velocity by PR-DOCT impossible. To circumvent this limitation, we propose a new method for resonant Doppler flow quantification in spectral domain OCT, where the scanner movement velocity is matched with the transverse velocity component of the sample motion similar to a tracking shot, where the camera is moved with respect to the sample. Consequently, the influence of the transverse velocity component of the tracked moving particles on the Doppler phase shift is negligible and the linear relation between the phase shift and the axial velocity component can be considered for flow velocity calculations. The proposed method is verified using flow phantoms on the basis of 1% Intralipid solution and diluted human blood.

Keywords: optical coherence tomography; phase-resolved Doppler analysis; flow measurement; oblique sample motion; non-invasive imaging technique; random scattering sample

1. Introduction

Optical coherence tomography (OCT) is a high resolution non-invasive optical imaging modality, which is based on low-coherence interferometry for detecting depth-resolved 2D- and 3D-images of highly scattering semi-transparent samples, e.g., biological tissue [1]. The method uses near-infrared short-coherent light, which is separated within a fiber-coupled or free-space interferometer, to compare the backscattering light from different depth positions within the sample with the reflected light of the reference mirror. Particularly in the early development of the method, the intensity of the interference signal was continuously detected by varying the length of the reference arm to obtain the backscattering light as a function of depth (time domain OCT, TD-OCT). In contrast, the interference signal can be detected in dependence on of the spectral components of the broad band light source (frequency domain OCT, FD-OCT) by a spectrometer configuration (spectral domain OCT, SD-OCT) [2–4] or a wavelength-swept source (swept source OCT, SS-OCT) [5,6]. With the emergence of fast line detectors and commercially available stable wavelength-tunable light sources, SD-OCT and SS-OCT became more attractive, since the entire depth information is simultaneously acquired, without any moving reference mirror.

OCT is generally predestined for morphological imaging of biological tissue due to the reconstruction of cross-sectional images (B-scan) and 3D volumetric visualization on the basis of the multiple detected depth scans (so called A-scans in analogy to ultrasound imaging) of the investigated sample. However, functional imaging has become increasingly attractive for the investigation of tissue dynamics and physiology [7–9]. This is because the backscattered light not only provides information on the depth-dependent amplitude and phase, but also on physical quantities such as the polarization and the optical frequency shift, which are progressively used for functional tissue imaging. An important enhancement is Doppler OCT (DOCT) for blood flow measurements [10,11]. Most of the proposed Doppler approaches are based on the phase shift between subsequently detected OCT signals, which is widely referred to as phase-resolved Doppler OCT (PR-DOCT) [12–14]. For the commonly used spectrometer-based OCT systems, PR-DOCT is still limited by a minimum and ambiguous maximum flow velocity, by interference fringe blurring, due to fast axially moving samples and by a nonlinear relation between the Doppler phase shift and the axial sample velocity for the case of an obliquely moving sample. To increase the dynamic velocity range with high flexibility, Grulkowski et al. have presented adaptive scanning protocols for DOCT, while preserving a high imaging speed [15]. In this work, the Doppler frequency shift related to the phase shift [16] is calculated between consecutive B-scans or segments, instead of subsequent A-scans as has conventionally been employed, resulting in the benefit of mapping low and high flow. To overcome the interference fringe washout for high axial displacement during the A-scan detection time, resonant Doppler imaging was pioneered by Bachmann et al. [17]. This approach uses a moving reference mirror for phase matching the OCT signal by an additional axial reference velocity, to obtain flow-based contrast enhancements of the blood vessels in the OCT images [17,18]. With regard to the nonlinear relationship between the phase shift and axial velocity for an oblique sample movement in SD-OCT [19–21], the discrepancy of the assumed linear relation is dramatic in the case of a highly transverse velocity component. Moreover, the measured phase shift always predicts a smaller absolute flow velocity than actually exists. Besides this, the correlation between adjacent A-scans primarily disappears for high transverse velocity components, resulting in a high mean error of the Doppler phase shift [20]. To overcome these limitations for the case of a small Doppler angle between the horizontal direction and the sample velocity, and thereby a highly dominant transverse velocity component, the present approach proposes the matching of the 2D-scanner in the sample arm to the transverse velocity component of the sample motion, comparable to a tracking shot where the camera moves alongside the object to be detected. Consequently, the backscattering signals of the matched oblique sample movement will be highly correlated, whereas those of static sample structures and slowly moving scatterers will be less correlated and damped, depending on the scanner velocity. The method of the so-called lateral resonant Doppler OCT (LR-DOCT) is validated by means of in vitro flow phantom studies with diluted Intralipid and human blood.

2. Phase-Resolved Doppler Model in SD-OCT

2.1. The Relation Between the Doppler Phase Shift and Axial Sample Velocity

The starting point of the phase-resolved Doppler OCT (PR-DOCT) in SD-OCT is the single complex depth scan $\Gamma_j(z)$ resulting from the Fourier transform of the detected spectrally shaped interference signal $I_j(k,z)$. This A-scan with index j comprises the amplitude $A(z)$ and the random phase $\varphi(z)$ information of the backscattered light from different depth positions z of the volume scattering sample, as shown by Equation (1):

$$FT\{I_j(k,z)\} = \Gamma_j(z) = A_j(z)e^{i\varphi_j(z)} \tag{1}$$

Here, the amplitude enables the structural imaging, while the phase of a scattering sample relates to a random fluctuation term. Since an axial sample movement between consecutive A-scans separated by the period $T_{\text{A-scan}}$ causes a Doppler frequency shift, and with this a phase shift $\Delta\varphi(z)$ of the interference spectrum, the phase information can be used for the measurement of the axial sample velocity $v_z(z)$, in consideration of the center wavelength λ_0, the refractive index n of the sample, and the time $T_{\text{A-scan}}$ needed for one A-scan relating to the sum of the integration time T_{Int} of the line detector of the spectrometer and the detector dead time ΔT due to a possible operating shutter control [21].

$$v_z(z) = \frac{\Delta z}{T_{\text{A-scan}}} = \frac{\Delta\varphi(z)\lambda_0}{4\pi n T_{\text{A-scan}}} \text{ with } \Delta z = v_z \cdot T_{\text{A-scan}} \text{ and } T_{\text{A-scan}} = T_{\text{Int}} + \Delta T \tag{2}$$

The bias-free Doppler phase shift $\Delta\varphi(z)$ is calculated by the multiplication of the complex Fourier coefficient of one A-scan $\Gamma_{j+1}(z)$ and the previous conjugate complex measurement $\Gamma_j^*(z)$ [20,22], where the resulting complex data contains the Doppler phase shift $\Delta\varphi(z)$ as the argument:

$$\Gamma_{\text{res}}(z) = \Gamma_{j+1}(z) \cdot \Gamma_j^*(z) = A_{j+1}(z) A_j(z) e^{i[\varphi_{j+1}(z) - \varphi_j(z)]} \tag{3}$$

Since the noise of the Doppler phase shift increases with the amount of transverse displacement in the case of an oblique sample motion [19,20], low signal-to-noise ratio (SNR) [16], and detector dead time [21], the amplitude weighted averaging by Equation (4) is recommended.

$$\overline{\Delta\varphi(z)} = \arg\left\{\frac{1}{K} \sum_{j=1}^{K} \left[\Gamma_{j+1}(z) \cdot \Delta\Gamma_j^*(z)\right]\right\} \tag{4}$$

The discrepancy of the relationship of $\Delta\varphi(z)$ and $v_z(z)$ for any oblique sample motion detected with spectrometer-based OCT systems has been theoretically described in detail in previous studies [19–21]. To summarize the general description of the Doppler phase shift $\Delta\varphi$ independent of the OCT system parameters, dimensionless coordinates are used as follows:

$$\delta z = \frac{2n\Delta z}{\lambda_0}, \ \delta x = \frac{\Delta x}{w_0}, \ t' = \frac{t}{T_{\text{A-scan}}} \tag{5}$$

where w_0 corresponds to the width of the Gaussian sample beam (FWHM) of the OCT system. Since the dependence of $\Delta\varphi$ on δz and δx cannot be calculated in an analytical way [19–21], the numerical result is given by a universal contour plot, where $\Delta\varphi$ is drawn in steps of $\pi/6$ for δz and δx ranging from zero to four. Additionally, $\Delta\varphi$ is limited to its principle value between 0 and 2π, for which reason phase shifts exceeding 2π are simply wrapped.

For the case of no detector dead time $\Delta T/T_{\text{A-scan}} = 0$, the numerically determined $\Delta\varphi$ is shown by the contour plot in Figure 1a [20]. By considering a partial duty cycle [21] with $\Delta T/T_{\text{A-scan}} = 0.05$, as for the system used in this study, $\Delta\varphi$ as a function of δx and δz is changed as presented in Figure 1b. The phase shift $\Delta\varphi$ arisen from an obliquely moving sample with the Doppler angle ϑ, between the x-direction and the sample velocity, corresponds to a linear slope through the origin $\delta x = \delta z = 0$, as exemplarily shown by the central dashed line. The angle ϑ' in Figure 1, defined as the angle against the horizontal, may not equate with the real world angle ϑ of the experimental setup:

$$\tan\vartheta' = \frac{2nw_0}{\lambda_0} \tan\vartheta \tag{6}$$

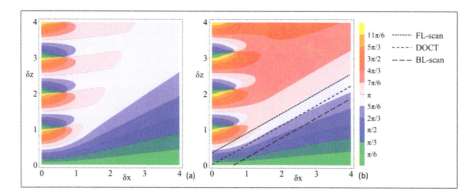

Figure 1. Contour plots of the numerically calculated phase shift $\Delta\varphi$ between consecutive A-scans as a function of the normalized transverse δx and axial δz displacement of the oblique sample motion [20,21] for detector dead times $\Delta T/T_{\text{A-scan}}$ of (**a**) 0.0 and (**b**) 0.05. The central dashed line with ϑ' of 29° corresponds to the expected phase shift of the 1% Intralipid flow experiment presented in Section 4. Additional shifted dashed lines with ϑ' of 29° are given for the LR-scanning explained in Section 2.2.

2.2. The Impact of Lateral Resonant Scanning on PR-DOCT

Since the phase-resolved Doppler measurements in SD-OCT are limited for strong transverse components and high sample velocities, the transverse tracking of the obliquely moving sample is proposed in this research and defined as lateral resonant scanning. With this, the impact of the transverse motion component on the Doppler phase shift can be theoretically reduced. For a better comprehension of the subsequent experimental results, a simple geometric model is defined on the basis of Figure 2 for the Doppler measurement in SD-OCT. As shown, a Gaussian sample beam and an oblique sample motion with an arbitrary Doppler angle ϑ are assumed. In this case, a group of scattering particles is supposed to be seen at the end of the detection intervals, $[T_1, T_1 + T_{\text{Int}}]$, $[T_2, T_2 + T_{\text{Int}}]$, and $[T_3, T_3 + T_{\text{Int}}]$, of three consecutive A-scans, where the respective position of the particles is shown by three blue distances and labeled with I, II, and III. Three cases of Doppler measurement performance are drawn in Figure 2: (a) with a static sample beam, resulting in a conventional M-scan (time-resolved A-scan); (b) with a laterally moving sample beam in the direction of the sample movement, resulting in a so-called forward lateral scan (FL-scan); and (c) with lateral scanning in the opposite direction, so-called backward lateral scan (BL-scan). Case (a) corresponds to the common phase-resolved Doppler measurement without imaging, where the sample beam of the OCT system is static to guarantee the highest phase correlation between adjacent A-scans. Due to the limited Gaussian sample beam, the group of particles seen in the time interval $[T_1, T_1 + T_{\text{Int}}]$ are only partly detectable in the subsequent time interval $[T_2, T_2 + T_{\text{Int}}]$ (shown by the red dotted lines), for which reason the measured axial displacement Δz_d and $\Delta\varphi$ are reduced, especially for small Doppler angles and large transverse displacements [19–21].

Case (b) shows the lateral scanning towards the sample movement, where the sample beam in gray represents the position at $T_1 + T_{\text{Int}}$ and the one drawn in green shows the position at $T_2 + T_{\text{Int}}$. Since the scanner velocity is matched to the transverse velocity of the moving sample, the group of particles detected at time point $T_1 + T_{\text{Int}}$ is tracked over the entire A-scan time $T_{\text{A-scan}}$, until $T_2 + T_{\text{Int}}$. For a temporally invariable sample, the detected movement is purely axial, for which the $\Delta\varphi$-v_z-relation will be linear and the absolute sample velocity can be linearly calculated (cp. Equation (2)). For case (c) of the backward scanning, no phase correlation seems to exist, because the scattering particles detected at $T_1 + T_{\text{Int}}$ within the gray sample beam are not detectable at $T_2 + T_{\text{Int}}$ within the green one. Due to the continuous movement of the scanner and the gradual movement of the scattering particles out of the scanning sample beam within $T_{A\text{-scan}}$, a strongly reduced phase correlation exists, leading to a strongly reduced Δz and a significantly decreased $\Delta \varphi$.

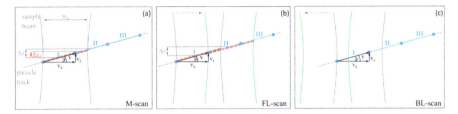

Figure 2. Schematic of the obliquely moving scattering sample relative to (**a**) a static Gaussian sample beam, (**b**) a forward lateral, and (**c**) a backward lateral scanning beam. (**a**) Due to the limited width of the incident Gaussian sample beam and the transverse motion component, the detected axial displacement Δz is reduced to Δz_d. (**b**) If the lateral scanning speed is matched with the transverse velocity of the obliquely moving scatterers, the sample movement is detected as purely axial displacement during $T_{A\text{-scan}}$. (**c**) If the sample beam scanning is opposite to the sample movement, the phase correlation is strongly reduced, for which Δz is measured significantly lower.

Considering the contour plots in Figure 1 describing the Doppler phase shift in dependence on the normalized axial δz and transverse δx displacement, a lateral resonant scanning relative to the transverse component of the oblique sample movement results in a negative offset in the δx-direction of the linear slope, describing the phase shift. Knowing the Doppler angle ϑ from 2D or 3D scans, the absolute sample velocity can be calculated by the linear $\Delta \varphi$-v_z-relation in Equation (2). Finally, the question of whether LR-DOCT is practicable for moving volume scatterers arises. The challenge for flowing emulsions and suspensions with scattering structures for the near-infrared wavelength range is the temporal variability of the particle position within flow channels or blood vessels, which hampers the particle tracking and LR-DOCT. The experimental validation of this question is the subject of the presented research study.

3. Material and Methods

3.1. OCT System Setup and In Vitro Flow Phantom

In the present study, the 1250 nm wavelength band of the dualband spectral domain OCT system of our workgroup is used [23]. This system contains a commercially available supercontinuum laser light source (SuperK Versa Super Continuum Source, Koheras A/S, Birkerød, Denmark), where the wavelength band with a center wavelength λ_0 of 1250 nm has a spectral width $\Delta \lambda$ of 200 nm and an axial resolution of 5.3 µm in air. The system comes with a 3D-scanner, whose Gaussian sample beam has a theoretically calculated width of $w_0 = 9.3$ µm in the focal plane and a Rayleigh length of about 200 µm. The scanning unit for 2D-beam deflection at the sample surface is realized by two galvanometer scanners (cp. Figure 3). The system allows 3D-imaging with an A-scan rate of up to 47 kHz. For high backscattering signals with a strong signal-to-noise ratio (SNR) of the sample, the A-scan rate is set to $f_{A\text{-scan}} = 12$ kHz for the flow measurement. The control of the system, the data acquisition and the processing is realized by means of a personal computer and custom software developed with LabVIEW® (National Instruments, Inc., Austin, TX, USA).

LR-DOCT is experimentally validated by a flow phantom consisting of a flow-controlled infusion pump, a glass capillary with a 312 µm inner diameter, and a tubing system to connect both components. The pumped fluids are a 1% Intralipid emulsion as an established flow phantom model for OCT with well-known flow characteristics, and diluted human blood. Here, the fresh human blood of a young healthy volunteer was extracorporeally stabilized with anticoagulant citrate dextrose solution (ACD-A) and diluted with a solution of 6% dextrose and 0.9% sodium chloride in distilled water with a ratio of

1:2, resulting in a hematocrit of ~15%. Despite the enhanced penetration depth into the blood vessels due to the low absorption of hemoglobin and the reduced scattering in the 1250 nm wavelength band, only the dilution enables the lateral resonant flow measurement over the entire capillary lumen of 312 μm, which is comparable to the inner blood vessel diameter of the in vivo mouse model used in our cooperative research [24–26]. To avoid sedimentation of the solid blood components and to guarantee a homogenous dispersion of erythrocytes within the capillary, a drop distance was installed within the tubing system in front of the capillary entry, in accordance with the schematic drawing in Figure 3. For the 1% Intralipid sample, the flow rate is set to 9.5 mL/h, corresponding to a peak center velocity of about 69 mm/s, and for the diluted human blood, the chosen flow rate is 8 mL/h, resulting in a center flow velocity of about 62 mm/s, which conforms well to the maximum flow velocity of the saphenous artery of an anesthetized mouse during the systole prevalent in our research project [24]. For the analysis of the velocity values, a parabolic flow profile was assumed because the capillary had a length of 80 mm, the inlet path was considered, and the Reynolds number was less than ten for both samples. The Doppler angle ϑ of 1.7° was measured by taking a volume scan with the 3D-scanner.

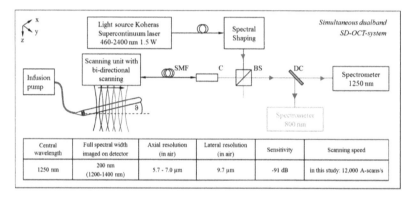

Figure 3. Dualband SD-OCT system, where the long wavelength band centered at 1250 nm with a full spectral width of 200 nm is used with a fiber-coupled customized scanning unit. Abbreviations: SMF: single mode fiber, C: collimator, BS: beam splitter, DC: dichroic mirror, ϑ: Doppler angle. The flow phantom consists of a 312 μm glass capillary connected to an infusion pump by a tubing system and filled with 1% Intralipid and diluted fresh human blood, respectively.

3.2. Bi-Directional Lateral Scanning Protocol and Image Correction for Doppler Analysis

The two galvanometer scanners integrated in the scanning unit are controlled by custom developed software based on LabVIEW® (National Instruments, Inc., Austin, TX, USA), with which arbitrary 2D-scanning patterns along the sample surface are realizable. Conventionally, the orthogonally driven X- and Y-galvanometer scanners are controlled for 2D-beam deflection, to allow 3D-imaging. The corresponding scanning protocol can be found in Figure 4a, where B-scans consisting of an arbitrary number of A-scans are detected by the lateral fast scanning Y-scanner and are spatially separated by the X-scanner. The resulting orthogonally running cross section and longitudinal section of the exemplary glass capillary with flowing 1% Intralipid are presented in Figure 4b.

The modified scan protocol for LR-DOCT is shown in Figure 4d. The first of the three B-scans with the stationary Y-scanner is an M-scan, and within the second B-scan, the X-scanner is driven along the moving direction of the sample (forward lateral scan, FL-scan). The third B-scan implies the reversely driven X-scanner (backward lateral scan, BL-scan). The resulting M- and lateral resonant B-mode images, each consisting of 512 A-scans, detected close to the capillary center with $f_{A\text{-scan}}$ = 12 kHz and a detector dead time $\Delta T / T_{A\text{-scan}}$ of 0.05, are presented in Figure 4e. For the FL-scan, the X-scanner was lateral resonantly driven to the flowing oil particles of the 1% Intralipid emulsion at the capillary center.

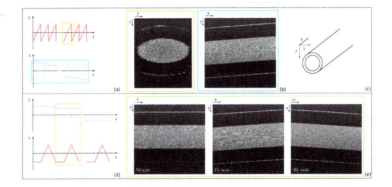

Figure 4. Conventional (**a**) and bi-directional lateral (**d**) scanning protocol of the X- and Y-galvanometer scanners. (**b**) Cross section and longitudinal section of the Intralipid-filled glass capillary generated from the detected 3D-scan. (**c**) Schematic of the imaged glass capillary in respect to the defined coordinate system. (**e**) Resulting M-scan and B-scans (forward and backward lateral scan: FL- and BL-scan) due to the modified scanning protocol for LR-DOCT.

The result of the Doppler analysis by Equation (3) is presented in Figure 5a for the FL-scan of the glass capillary with flowing 1% Intralipid. The calculated Doppler phase shifts are overlaid to the structural information. As the obliquely running direction of the sample hampers the averaging of phase shifts in x-direction, consecutive A-scans are shifted in the z-direction by introducing a k-dependent phase shift. This correction is demonstrated in Figure 5b, using the example of the FL-scan. In analogy to the first shift theorem of the Fourier transform, an additional phase $\varphi_{\text{shifted}}(k,j)$ is inserted into the real-valued interferometric signal $I_j(k,z)$, as shown in Equation (7). According to Equation (8), the fractional shift of each interference spectrum of the lateral resonant scan corresponds to the tangent of the Doppler angle ϑ, where N is the number of points of the Fourier transformation (FT). With this, the originally obliquely running direction of the capillary, and consequently, the Doppler phase shift due to the fast laterally moving X-scanner across the oblique surface of the glass capillary, are eliminated.

$$I_{\text{shifted},j}(k,z) = I_j(k,z)e^{i\varphi_{\text{shifted}}(k,j)} \qquad (7)$$

$$\varphi_{\text{shifted}}(k,j) = \frac{2nkj}{N}\tan\vartheta \qquad (8)$$

Further processing is identical to the conventional OCT and Doppler OCT processing: a Fourier transform is applied to $I_{\text{shifted},j}(k,z)$ in accordance with Equation (1) and the Doppler phase shift $\Delta\varphi$ is calculated by multiplying a complex A-scan with the conjugate complex one (cp. Equation (3)).

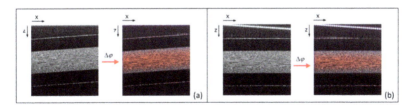

Figure 5. (**a**) Conventional processing of the detected interferometric signals for structural and Doppler flow imaging on the example of the FL-scan of the 1% Intralipid flow within the glass capillary. (**b**) Correction of the obliquely running direction of the glass capillary by insertion of an additional phase term $\varphi_{\text{shifted}}(k,j)$ to the real-valued interference signal and subsequent conventional processing. Note the tilted signal from zero delay at the top.

4. Experimental Results

4.1. LR-DOCT with 1% Intralipid Flow

In order to validate LR-DOCT for moving volume scatterers, Doppler flow experiments are performed by means of a 1% Intralipid emulsion flowing through a glass capillary, as described in Section 3.1. The Doppler angle is $1.7°$ and the maximum flow velocity at the capillary center is estimated to be 69 mm/s. Thus, the transverse displacement δx of the flowing oil particles of the Intralipid emulsion amounts to 0.6 for the case of the detected M-scan, which strongly reduces the phase correlation [20,21,27].

Figure 6 presents the processed flow-relevant parts of the grayscale structural (a) and the Doppler flow images (b) of the detected M-scan, the forward lateral (FL) and the backward lateral (BL) B-scans differing in the speckle pattern, and the resulting Doppler phase shift. The number of analyzed A-scans is chosen to be 512. As expected, highly correlated speckle appears at the capillary border for the M-scan due to the small flow velocity in this area. For the FL-scan, the X-galvanometer scanner velocity is set to 69 mm/s, which corresponds to the transverse velocity component of the Intralipid flow at the capillary center. In this case, the correlation at the border area gets lost, whereas a high correlation occurs at a wide range of the capillary lumen identified by the elongated speckle pattern. In contrast, the speckle seems to be uncorrelated for the case of the BL-scan, where the X-scanner is moved in the opposite direction at 69 mm/s.

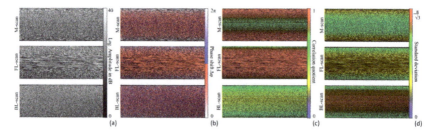

Figure 6. (a) Grayscale structural SD-OCT images of the M-scan, the forward (FL-scan) and backward lateral Doppler scan (BL-scan) of the 1% Intralipid flow through a 312 μm glass capillary. (b) Corresponding phase-resolved Doppler flow images for the measured Doppler angle ϑ of $1.7°$. (c) Colored correlation images representing the correlation quotient CQ in accordance to [20,21]. (d) Standard deviation $\sigma_{\Delta\varphi}$ of the mean Doppler phase shift of 512 single A-scans overlaid to the structural images.

The visual assessment of the Doppler images in Figure 6b suggests a high noise of the Doppler phase shift at the capillary center in the M-scan, at the capillary border in the FL-scan, and over the entire capillary lumen in the BL-scan, which is caused by the reduced correlation of adjacent A-scans. As predicted in Section 2.2, the phase-sensitive image of the BL-scan visually shows the highest noise of the Doppler phase shift. For the quantification of the phase correlation, the correlation quotient CQ, defined in former research [20,21], is calculated as a function of z and is overlaid in color on the structural images, as presented in Figure 6c. CQ is shown in red for the highest correlation between consecutive A-scans. The results of CQ prove the suggestions made by the observed speckle patterns in Figure 6a and the estimated phase shift noise in Figure 6b. Additionally, the standard deviation of the mean Doppler phase shift is given in Figure 6d. As described in previous studies [20,21,27], the distribution of $\Delta\varphi$ cannot be worse than isotropic in the original interval of $[-\pi, \pi]$, resulting in an upper limit of $\pi/\sqrt{3}$, as shown by the red color of the $\sigma_{\Delta\varphi}$ scale bar. By means of the colored structural information for visualizing $\Delta\varphi$, CQ, and $\sigma_{\Delta\varphi}$, the possibility to laterally track the flowing oil particles with a simultaneous decrease of the phase shift noise for small Doppler angles and high flow velocities, becomes apparent.

For quantitative analysis, the averaged phase shift of 512 single measurement points is calculated, as described in Equation (4) and presented in Figure 7a. The measured $\Delta\varphi$ of the M-scan shown by the blue points does not display a quadratic profile, verified by the fitted power law with an exponent of 2.4. The Doppler phase shift at the capillary center is measured to be 1.9 rad, which conforms well to the estimation by means of the contour plot in Figure 1b and the transformed Doppler angle ϑ' of 29°. The result of the forward lateral scan (FL-scan) presented by the red points conforms to a quadratic parabolic profile, with the expected maximum velocity of 69 mm/s at the capillary center. In spite of the increased phase noise at the upper and lower capillary borders, the Doppler phase shift of the tracked Intralipid particles is not limited, and consequently, is truly measured by the linear relation between $\Delta\varphi$ and v_z referred to in Equation (2). The non-resonant imaging at the border of the capillary has no influence on the shape. For the backward lateral scan, the flow profile seems to be a flattened plug profile with a fitted exponent of the power law of 2.6. Despite the strongly increased phase noise $\sigma_{\Delta\varphi}$, an erroneous mean velocity profile is measurable from 512 single complex A-scans.

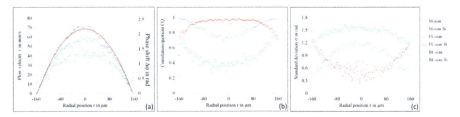

Figure 7. (**a**) Calculated and fitted flow profiles of the flowing 1% Intralipid as a function of the radial position r inside the 312 μm capillary for the M-scan (blue points), the forward (FL-scan, red points) and backward lateral scan (BL-scan, green points). Additionally, the correlation quotient CQ (**b**) and the standard deviation $\sigma_{\Delta\varphi}$ (**c**) are shown versus r for the detected M-scan, the FL-scan and the BL-scan.

The correlation quotient CQ and the standard deviation $\sigma_{\Delta\varphi}$ against the radial position r inside the capillary in Figure 7b,c, show what can be visually expected from Figure 6c,d. For the M-scan (blue points), the backscattering signals of the Intralipid scatterers at the capillary border are highly correlated and show a small standard deviation of $\sigma_{\Delta\varphi} = 0.4$ rad, in comparison to the capillary center with a reduced correlation of CQ = 0.73 and a phase shift noise of $\sigma_{\Delta\varphi} = 1.15$ rad. For the forward lateral scan (red points), the correlation at the capillary center is strongly increased to CQ = 0.97 and the phase noise of the fast moving particles is highly reduced to $\sigma_{\Delta\varphi} = 0.4$ rad compared to the M-scan. The experimentally measured CQ in Figure 7b shows that the maximum correlation of one is not achieved, possibly because of the Brownian motion of the Intralipid. However, it is evident that the velocity of the X-galvanometer scanner is matched very well to the oil particles for the FL-scan, since the behavior of the phase noise $\sigma_{\Delta\varphi}$ of the M-scan (blue points) and the FL-scan (red points) is inverted, as presented in Figure 7c. The result of the BL-scan (green points) shows a strongly damped correlation and a high noise of the Doppler shift $\sigma_{\Delta\varphi}$ over the entire capillary, as expected. The relation of $\sigma_{\Delta\varphi}$ and CQ are empirically determined in [20] and presented as a solid line in Figure 8 besides the very well fitting measurement values. Since the correlation quotient CQ is indirectly proportional to the Doppler noise $\sigma_{\Delta\varphi}$, CQ provides no added value for the analysis of LR-DOCT, for which reason, only $\sigma_{\Delta\varphi}$ is presented for the following in vitro blood flow measurements.

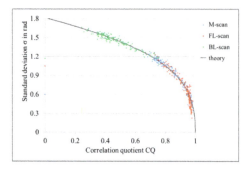

Figure 8. Mean error (standard deviation) $\sigma_{\Delta\varphi}$ is shown theoretically (solid line [20]) and experimentally (measurement points of the M-scan, FL-scan, and BL-scan) as a function of CQ.

4.2. LR-DOCT with In Vitro Blood Flow

In analogy to the Intralipid flow experiment, measurements are performed with the identical capillary model in combination with diluted fresh human blood. The measured Doppler angle ϑ amounts to 1.7°. Since the flow rate is set to 8 mL/h, the measurement parameters concerning the tracked transverse velocities are similar to the Intralipid flow phantom in Section 4.1. Exemplary structural images of the detected M-scans, FL-scans, and BL-scans are found in Figure 9a,d for two data sets, each consisting of 400 A-scans.

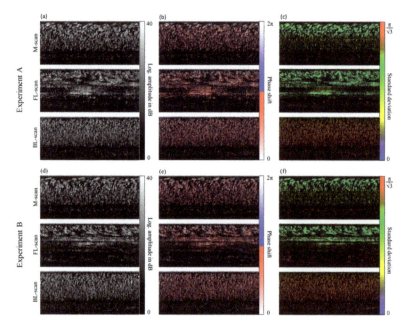

Figure 9. Grayscale structural images (**a,d**) and phase-resolved Doppler images (**b,e**) are shown of the M-scans, FL-scans, and BL-scans each consisting of 400 A-scans for two experiments A and B with identical settings. In addition, the standard deviation of the mean Doppler phase shift of 400 A-scans is overlaid to the structural image for both experiments (**c,f**).

Obviously, the speckle pattern is different to the experiments with Intralipid flow, first due to the larger size of the highly scattering wheel-shaped erythrocytes with a diameter of $\varnothing \approx 7.5$ μm and a thickness of $d \approx 2.0$ μm in comparison to the small spherical oil particles of the Intralipid, with a mean diameter of $\varnothing \approx 100$ nm relative to the size of the sample beam of $w_0 = 9.3$ μm at the focal plane. Secondly, this pattern is due to the inhomogeneous distribution of the erythrocytes of the diluted blood sample within the capillary. Nevertheless, the dilution described in Section 3.1 is necessary for the tracking of erythrocytes at the capillary center, otherwise the erythrocytes are already too dense at $r = 0.5R$. Unfortunately, no backscattering signal is achieved at a larger depth, probably due to the orientation of the erythrocytes in this capillary zone and the resulting angular intensity modulation in the glass capillary cross section [28]. However, the velocity of the galvanometer scanner can be matched with the transverse flow velocity of the erythrocytes at the capillary center with the chosen dilution, as seen in the FL-scan of Figure 9a,d. The FL-scans of the measurements A and B differentiate, because the group of erythrocytes in experiment A is tracked over a shorter distance than the group in B. Therewith, the challenge of LR-DOCT for blood flow measurement becomes apparent, which implies the random catch of erythrocytes at the capillary center and the selected depths within the capillary, respectively. Because of the shear flow-dependent movement and orientation of the erythrocytes, the appearance of erythrocytes at the capillary center in blood samples with reduced hematocrit is small. Generally, for blood samples and especially for diluted blood, sedimentation of the erythrocytes occurs, where the flow velocity is small, as within the syringe and the tubing system connecting the glass capillary with the infusion pump [29]. Consequently, the erythrocytes build aggregates in the form of chain-like stacks [30]. With the falling distance of the connecting tube in front of the capillary entry, sedimentation and agglomeration within the capillary are avoided. The aggregates are dispersed, but erythrocytes are still grouped and randomly orientated at the capillary center in the presented FL-scan. Considering the images with an overlaid Doppler phase shift and standard deviation in Figure 9b,c,e,f, it becomes apparent that the Doppler noise is decreased for the resonantly scanned erythrocytes (FL-scan). On closer examination, the phase shift of the resonantly detected erythrocytes still shows an increased noise in dependence on the amount of erythrocytes flowing above, which is considered in the analysis presented below. The reason for this effect might be the optical inhomogeneity of human blood caused by the higher refractive index inside the erythrocytes. The incident light passing a group of erythrocytes is randomly delayed within the B-scan, which causes random phase fluctuations beneath this area, as in our case at the capillary center. Multiple scattering causing Doppler shadows might worsen the effect.

The mean Doppler phase shift and the corresponding standard deviation of the diluted human blood flow are quantified for both experiments, and the results are presented in Figure 10. As seen, both measurements offer similar characteristics, which show the reproducibility of LR-DOCT. Moreover, the calculated phase shift is slightly higher for the FL-scan than for the M-scan in the depth range of resonantly detected erythrocytes. Additionally, the standard deviation of the mean phase shift is reduced (asterisk in Figure 10b,d) at the radial position r near the center of the capillary. Since the erythrocytes are not completely resonantly detected over the detected 400 A-scans, $\Delta\varphi$ and $\sigma_{\Delta\varphi}$ include the phase information of parts with no significant backscattering signal, which unfortunately increases $\sigma_{\Delta\varphi}$.

For a more precise analysis, parts of the trace with a high signal in the resonant region are selected, as shown in Figure 11. The number of lateral resonantly detected A-scans (originally $N = 400$ Figure 11a) is systematically reduced to $N = 90$ (Figure 11b), $N = 40$ (Figure 11c), and $N = 22$ (Figure 11d), showing an increased correlated backscattering signal at the capillary center with a decreasing number N of selected A-scans. Considering the diagram in Figure 12a, the Doppler phase shift of the resonantly detected erythrocytes at the capillary center differ slightly for $N = 400$ (Figure 11a), $N = 90$ (Figure 11b), and $N = 22$ (Figure 11d). Comparing $\Delta\varphi$ of the M-scan and the FL-scan with $N = 22$, one can identify a difference up to 0.4 rad and 10 mm/s, respectively, for the resonant detection (Figure 12b). The standard deviation of the mean phase shift is maximally reduced for $N = 22$ of the FL-scan and the resonantly scanned depths (cp. Figure 12c,d).

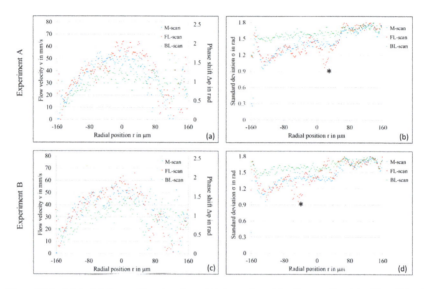

Figure 10. (**a,c**) Calculated mean Doppler phase shift $\Delta\varphi$ and resulting flow profile of the diluted blood flow of experiment A and B are presented as a function of the radial position r for the detected M-scan, FL-scan and BL-scan. (**b,d**) Corresponding standard deviation $\sigma_{\Delta\varphi}$ of the mean phase shift versus the radial position r of the capillary lumen. The asterisk marks the radial position with strongly reduced $\sigma_{\Delta\varphi}$ obtained with the FL-scan near the center of the capillary.

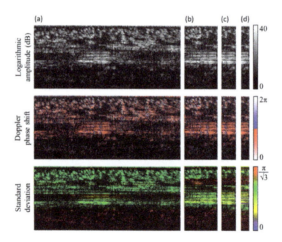

Figure 11. Structural, Doppler flow, and Doppler noise images of the forward lateral scan (FL-scan) of the diluted blood flow through the 312 μm glass capillary are presented for $N = 400$ (**a**), $N = 90$ (**b**), $N = 40$ (**c**), and $N = 22$ (**d**) analyzed A-scans.

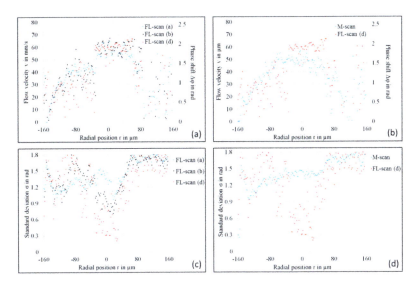

Figure 12. (**a**) Calculated Doppler phase shift of the FL-scans in Figure 11a,b,d are shown as a function of the radial position r. (**b**) Doppler phase shift of the FL-scan in Figure 11d in comparison to the phase shift of the M-scan in Figure 9b. (**c,d**) Corresponding standard deviation of the mean Doppler phase shift presented in (**a,b**).

5. Summary and Conclusions

In the present study, lateral resonant Doppler optical coherence tomography (LR-DOCT) is proposed for velocity measurements of oblique sample motion with very small Doppler angles and high absolute sample velocities in spectral domain OCT (SD-OCT), where the nonlinear relation between the Doppler phase shift $\Delta\varphi$ and the axial velocity component v_z hampers the accurate velocity calculation by the phase-resolved Doppler OCT (PR-DOCT). The described LR-DOCT overcomes the limitation of the nonlinear $\Delta\varphi$-v_z-relationship and allows the calculation of the axial sample velocity v_z by matching the lateral scanner movement velocity with the transverse velocity component v_x of the sample. LR-DOCT is validated by means of two flow phantoms with a 1% Intralipid emulsion and diluted human blood, respectively. The flow velocity at the capillary center is lateral resonantly detected by adapting the scanner velocity to the transverse velocity v_x of the flowing scattering particles. By using an online display of the moving sample and accordingly the capillary flow before data acquisition, the transverse flow velocity v_x can be estimated for the adjustment of the scanner velocity in LR-DOCT. Since the speckle pattern of highly correlated A-scans is clearly identifiable, the adaption of the lateral scanner velocity is simple. Due to the variable scan velocity of the galvanometer scanner, within its specifications, arbitrary flow velocities can be lateral resonantly measured by adapting the scanning protocol. Generally, LR-DOCT works very well for volume scatterers consisting of many small particles (e.g., 1% Intralipid flow) relative to the wavelength λ_0 and the sample beam width w_0 at the focal plane, whereas the method is limited for particles exceeding one or both of these factors (e.g., highly scattering erythrocytes of blood flow). The validation of the presented study has shown that LR-DOCT performs to a limited extent for diluted blood flow, where small groups of obliquely moving erythrocytes are resonantly detectable. This fact probably makes it difficult for in vivo blood flow measurements. Nevertheless, it is conceivable that LR-DOCT could produce precise blood flow measurements of easily accessible blood capillaries running almost perpendicular to the incident sample beam, containing only a few erythrocytes. Even though LR-DOCT has to be evaluated for in vivo conditions in future research, it is still interesting for many other moving samples in ex vivo and in vitro applications, or non-medical and non-biological configurations. Under the discussed conditions, LR-DOCT overcomes the nonlinear relation between the Doppler phase shift $\Delta\varphi$ and axial sample velocity, and with this, the limitation of PR-DOCT in SD-OCT for an oblique sample motion with a strong transverse velocity component. Consequently, LR-DOCT advantageously enables the simple, correct, and unlimited phase-resolved Doppler measurement of sample and flow velocities in SD-OCT.

Acknowledgments: We thank Peter Cimalla for the assistance with the dualband spectral domain OCT system and Christian Schnabel for the voluntary blood donation.

Author Contributions: All authors contributed to this work. Authors Julia Walther and Edmund Koch jointly developed the idea of LR-DOCT, performed the in vitro experiments, processed the acquired data, and analyzed the calculated Doppler information for the validation of LR-DOCT; Julia Walther wrote the paper.

Conflicts of Interest: The authors declare no conflict of interest.

References

1.	Huang, D.; Swanson, E.A.; Lin, C.P.; Schuman, J.S.; Stinson, W.G.; Chang, W.; Hee, M.R.; Flotte, T.; Gregory, K.; Puliafito, C.A.; et al. Optical Coherence Tomography. *Science* **1991**, *254*, 1178–1181. [CrossRef] [PubMed]
2.	Leitgeb, R.; Hitzenberger, C.; Fercher, A. Performance of Fourier domain vs. time domain optical coherence tomography. *Opt. Express* **2003**, *11*, 889–894. [CrossRef] [PubMed]
3.	Choma, M.; Sarunic, M.; Yang, C.; Izatt, J. Sensitivity advantage of swept source and Fourier domain optical coherence tomography. *Opt. Express* **2003**, *11*, 2183–2189. [CrossRef] [PubMed]
4.	De Boer, J.F.; Cense, B.; Park, B.H.; Pierce, M.C.; Tearney, G.J.; Bouma, B.E. Improved signal-to-noise ratio in spectral-domain compared with time-domain optical coherence tomography. *Opt. Lett.* **2003**, *28*, 2067–2069. [CrossRef] [PubMed]
5.	Huber, R.; Wojtkowski, M.; Fujimoto, J.G. Fourier domain mode locking (FDML): A new laser operating regime and applications for optical coherence tomography. *Opt. Express* **2006**, *14*, 3225–3237. [CrossRef] [PubMed]
6.	Kirsten, L.; Domaschke, T.; Schneider, C.; Walther, J.; Meissner, S.; Hampel, R.; Koch, E. Visualization of dynamic boiling processes using high-speed optical coherence tomography. *Exp. Fluids* **2015**, *56*. [CrossRef]
7.	Hitzenberger, C.K.; Gützinger, E.; Sticker, M.; Pircher, M.; Fercher, A.F. Measurement and imaging of birefringence and optic axis orientation by phase resolved polarization sensitive optical coherence tomography. *Opt. Express* **2001**, *9*, 780–790. [CrossRef] [PubMed]
8.	Pircher, M.; Hitzenberger, C.K.; Schmidt-Erfurth, U. Polarization sensitive optical coherence tomography in the human eye. *Prog. Retin. Eye Res.* **2011**, *30*, 431–451. [CrossRef] [PubMed]
9.	Faber, D.J.; Mik, E.G.; Aalders, M.C.G.; van Leeuwen, T.G. Toward assessments of blood oxygen saturation by spectroscopic optical coherence tomography. *Opt. Lett.* **2005**, *30*, 1015–1017. [CrossRef] [PubMed]
10.	Srinivasan, V.J.; Sakadžić, S.; Gorczynska, I.; Ruvinskaya, S.; Wu, W.C.; Fujimoto, J.G.; Boas, D.A. Quantitative cerebral blood flow with optical coherence tomography. *Opt. Express* **2010**, *18*, 2477–2494. [CrossRef] [PubMed]
11.	Leitgeb, R.A.; Werkmeister, R.M.; Blatter, C.; Schmetterer, L. Doppler optical coherence tomography. *Prog. Retin. Eye Res.* **2014**, *41*, 26–43. [CrossRef] [PubMed]
12.	Zhao, Y.H.; Chen, Z.P.; Saxer, C.; Xiang, S.H.; de Boer, J.F.; Nelson, J.S. Phase-resolved optical coherence tomography and optical Doppler tomography for imaging blood flow in human skin with fast scanning speed and high velocity sensitivity. *Opt. Lett.* **2000**, *25*, 114–116. [CrossRef] [PubMed]
13.	Leitgeb, R.; Schmetterer, L.; Drexler, W.; Fercher, A.; Zawadzki, R.; Bajraszewski, T. Real-time assessment of retinal blood flow with ultrafast acquisition by color Doppler Fourier domain optical coherence tomography. *Opt. Express* **2003**, *11*, 3116–3121. [CrossRef] [PubMed]
14.	Vakoc, B.; Yun, S.; de Boer, J.; Tearney, G.; Bouma, B. Phase-resolved optical frequency domain imaging. *Opt. Express* **2005**, *13*, 5483–5493. [CrossRef] [PubMed]
15.	Grulkowski, I.; Gorczynska, I.; Szkulmowski, M.; Szlag, D.; Szkulmowska, A.; Leitgeb, R.; Kowalczyk, A.; Wojtkowski, M. Scanning protocols dedicated to smart velocity ranging in spectral OCT. *Opt. Express* **2009**, *17*, 23736–23754. [CrossRef] [PubMed]

16. Walther, J.; Koch, E. Relation of joint spectral and time domain optical coherence tomography (jSTdOCT) and phase-resolved Doppler OCT. *Opt. Express* **2014**, *22*, 23129–23146. [CrossRef] [PubMed]
17. Bachmann, A.H.; Villiger, M.L.; Blatter, C.; Lasser, T.; Leitgeb, R.A. Resonant Doppler flow imaging and optical vivisection of retinal blood vessels. *Opt. Express* **2007**, *15*, 408–422. [CrossRef] [PubMed]
18. Koch, E.; Hammer, D.; Wang, S.; Cuevas, M.; Walther, J. Resonant Doppler imaging with common path OCT. *Proc. SPIE* **2009**, *7372*, 737220.
19. Koch, E.; Walther, J.; Cuevas, M. Limits of Fourier domain Doppler-OCT at high velocities. *Sens. Actuators A* **2009**, *156*, 8–13. [CrossRef]
20. Walther, J.; Koch, E. Transverse motion as a source of noise and reduced correlation of the Doppler phase shift in spectral domain OCT. *Opt. Express* **2009**, *17*, 19698–19713. [CrossRef] [PubMed]
21. Walther, J.; Koch, E. Impact of a detector dead time in phase resolved Doppler analysis using spectral domain optical coherence tomography. *J. Opt. Soc. Am. A* **2017**, *34*, 241–251. [CrossRef] [PubMed]
22. Wehbe, H.; Ruggeri, M.; Jiao, S.; Gregori, G.; Puliafito, C.A.; Zhao, W. Automatic retinal blood flow calculation using spectral domain optical coherence tomography. *Opt. Express* **2007**, *15*, 15193–15206. [CrossRef] [PubMed]
23. Cimalla, P.; Walther, J.; Mehner, M.; Cuevas, M.; Koch, E. Simultaneous dual-band optical coherence tomography in the spectral domain for high resolution in vivo imaging. *Opt. Express* **2009**, *17*, 19486–19500. [CrossRef] [PubMed]
24. Walther, J.; Mueller, G.; Meissner, S.; Cimalla, P.; Homann, H.; Morawietz, H.; Koch, E. Time-resolved blood flow measurement in the in vivo mouse model by optical frequency domain imaging. *Proc. SPIE* **2009**, *7372*, 73720J.
25. Walther, J.; Mueller, G.; Morawietz, H.; Koch, E. Analysis of in vitro and in vivo bidirectional flow velocities by phase-resolved Doppler Fourier-domain OCT. *Sens. Actuators A* **2009**, *156*, 14–21. [CrossRef]
26. Langbein, H.; Brunssen, C.; Hoffmann, A.; Cimalla, P.; Brux, M.; Bornstein, S.R.; Deussen, A.; Koch, E.; Morawietz, H. NADPH oxidase 4 protects against development of endothelial dysfunction and atherosclerosis in LDL receptor deficient mice. *Eur. Heart J.* **2016**, *37*, 1753–1761. [CrossRef] [PubMed]
27. Vakoc, B.J.; Tearney, G.J.; Bouma, B.E. Statistical properties of phase-decorrelation in phase-resolved Doppler optical coherence tomography. *IEEE Trans. Med. Imaging* **2009**, *28*, 814–821. [CrossRef] [PubMed]
28. Cimalla, P.; Walther, J.; Mittasch, M.; Koch, E. Shear flow-induced optical inhomogeneity of blood assessed in vivo and in vitro by spectral domain optical coherence tomography in the 1.3 μm wavelength range. *J. Biomed. Opt.* **2011**, *16*, 116020. [CrossRef] [PubMed]
29. Douglas-Hamilton, D.H.; Smith, N.G.; Kuster, C.E.; Vermeiden, J.P.W.; Althouse, G.C. Particle distribution in low-volume capillary-loaded chambers. *J. Androl.* **2005**, *26*, 107–114. [PubMed]
30. Friebel, M.; Helfmann, J.; Müller, G.; Meinke, M. Influence of shear rate on the optical properties of human blood in the spectral range 250 to 1100 nm. *J. Biomed. Opt.* **2007**, *12*, 054005. [CrossRef] [PubMed]